Symmetry in Physics

CENTRE DE RECHERCHES MATHÉMATIQUES

Organizers

Pavel Winternitz (chairman) (CRM & DMS, Université de Montréal)
John Harnad (CRM, Concordia University)
C.S. (Harry) Lam (Physics, McGill University)
Jiri Patera (CRM & DMS, Université de Montréal)

Topics covered at the workshop include

Finite and infinite dimensional Lie algebras
Symmetries in nuclear and particle physics
Generating function techniques
Quasi-crystals
Subgroup classifications and applications
Symmetries of differential and difference equations
Classical and quantum integrable systems
Special function theory
Lie algebra contractions
Other related topics

Workshop on Symmetry in Physics
in memory of

Robert T. Sharp

12-14 September, 2002

Confirmed participants
J.L. Birman (City College, CUNY), C. Burgess (McGill University), C. Cummins (Concordia University), S. DasGupta (McGill University), H. de Guise (Lakehead University), J.-P. Gazeau (Université Paris VII), M. Grisaru (McGill University), M. Grundland (CRM & UQTR), F. Gunger (Technical University, Istanbul), J. Harnad (CRM & Concordia University), J. Hurtubise (CRM & McGill University), V. Hussin (CRM & Université de Montréal), M. Havlicek (Czech Technical University), N. Kamran (McGill University), R.C. King (University of Southampton), P. Kramer (University of Tuebingen), C.S. (Harry) Lam (McGill University), F.W. Lemire (University of Windsor), D. Levi (University of Rome III), P. Mathieu (Université Laval), M. de Montigny (University of Alberta), R. Moody (University of Alberta), M. Moshinsky (Universidad Nacional Autonoma Mexico), M. del Olmo (University of Valladoli), Z. Papadopoulos (University of Tuebingen), J. Patera (CRM & Université de Montréal), E. Pelantova (Czech Technical University), G. Pogosyan (Universidad Nacional Autonoma Mexico), P. Ramond (University of Florida), A. Rodriguez (Universidad Complutense de Madrid), C. Roth (McGill University), D.J. Rowe (University of Toronto), G. Shaw (University of California at Irvine), E. Shahbazian (Lockheed Martin, Montreal), M.B. Sheftel (F. Gursey Institute, Istanbul), Y. Smirnov (Universidad Nacional Autonoma Mexico), M. Thoma (Syracuse, USA), Z. Thomova (Stamford, USA), J. Tolar (Czech Technical University), J. Van der Jeugt (University of Gent), L. Vinet (McGill University), H.C. Von Baeyer (W. and M. University), P. Wallace (Vancouver), M. Walton (University of Lethbridge), P. Winternitz (CRM & Université de Montréal)

Information & Registration
http://www.CRM.UMontreal.CA/Sharp

Graphic Design: Seyabzo Inc.

Université de Montréal

Volume 34

CRM PROCEEDINGS & LECTURE NOTES

Centre de Recherches Mathématiques
Université de Montréal

Symmetry in Physics

In Memory of Robert T. Sharp

P. Winternitz
J. Harnad
C. S. Lam
J. Patera
Editors

The Centre de Recherches Mathématiques (CRM) of the Université de Montréal was created in 1968 to promote research in pure and applied mathematics and related disciplines. Among its activities are special theme years, summer schools, workshops, postdoctoral programs, and publishing. The CRM is supported by the Université de Montréal, the Province of Québec (FCAR), and the Natural Sciences and Engineering Research Council of Canada. It is affiliated with the Institut des Sciences Mathématiques (ISM) of Montréal, whose constituent members are Concordia University, McGill University, the Université de Montréal, the Université du Québec à Montréal, and the Ecole Polytechnique. The CRM may be reached on the Web at www.crm.umontreal.ca.

American Mathematical Society
Providence, Rhode Island USA

The production of this volume was supported in part by the Fonds pour la Formation de Chercheurs et l'Aide à la Recherche (Fonds FCAR) and the Natural Sciences and Engineering Research Council of Canada (NSERC).

2000 *Mathematics Subject Classification.* Primary 22–06, 81–06, 81Rxx.

Library of Congress Cataloging-in-Publication Data
Workshop on Symmetries in Physics (2002 : Université de Montréal)
　Symmetry in physics : in memory of Robert T. Sharp / P. Winternitz... [et al.], editors.
　　p. cm. — (CRM proceedings & lecture notes, ISSN 1065-8580 ; v. 34)
　Includes bibliographical references.
　ISBN 0-8218-3409-6 (softcover : acid-free paper)
　1. Symmetry (Physics)—Congresses.　I. Sharp, Robert T.　II. Winternitz, Pavel.　III. Title. IV. Series.

QC174.17.S9W68　2004
539.7′25—dc22
　　　　　　　　　　　　　　　　　　　　　　　　　　　　　　　　　　　　　2003063928

Copying and reprinting. Material in this book may be reproduced by any means for educational and scientific purposes without fee or permission with the exception of reproduction by services that collect fees for delivery of documents and provided that the customary acknowledgment of the source is given. This consent does not extend to other kinds of copying for general distribution, for advertising or promotional purposes, or for resale. Requests for permission for commercial use of material should be addressed to the Acquisitions Department, American Mathematical Society, 201 Charles Street, Providence, Rhode Island 02904-2294, USA. Requests can also be made by e-mail to **reprint-permission@ams.org**.

　Excluded from these provisions is material in articles for which the author holds copyright. In such cases, requests for permission to use or reprint should be addressed directly to the author(s). (Copyright ownership is indicated in the notice in the lower right-hand corner of the first page of each article.)

　　　　　© 2004 by the American Mathematical Society. All rights reserved.
　　　　　　　　The American Mathematical Society retains all rights
　　　　　　　　except those granted to the United States Government.
　　　　　　　　　　　Printed in the United States of America.

　　　∞ The paper used in this book is acid-free and falls within the guidelines
　　　　　　　　established to ensure permanence and durability.
　　　　　This volume was submitted to the American Mathematical Society
　　　　　in camera ready form by the Centre de Recherches Mathématiques.
　　　　　　　Visit the AMS home page at **http://www.ams.org/**

　　　　　　　　　10 9 8 7 6 5 4 3 2 1　　09 08 07 06 05 04

Contents

Preface *Jiří Patera and Pavel Winternitz*	vii
Bob Sharp: The Man and the Scientist *Philip R. Wallace*	ix
Bob Sharp: Teacher, Colleague, Friend *J. Harnad*	xiii
Publications of Robert T. Sharp	xvii
Participants	xxiii
Symmetries and Currents of Massless Neutrino Fields, Electromagnetic and Graviton Fields *Stephen C. Anco and Juha Pohjanpelto*	1
Naturalness and Quintessence *C. P. Burgess*	13
Congruence Subgroups of $PSL(2,\mathbb{Z})$ *C. J. Cummins and S. Pauli*	23
Asymptotic $SU(2)$ and $SU(3)$ Wigner Functions from the Weight Diagram *Hubert de Guise, David J. Rowe, and Barry C. Sanders*	31
Physical Applications of a Five-Dimensional Metric Formulation of Galilean Invariance *M. de Montigny, F. C. Khanna, and A. E. Santana*	43
Variations on Dedekind's Eta *Terry Gannon*	55
Examples of Berezin–Toeplitz Quantization: Finite Sets and Unit Interval *J.-P. Gazeau, T. Garidi, E. Huguet, M. Lachièze-Rey, and J. Renaud*	67
A Modified Weierstrass Representation for CMC-Surfaces in Multi-Dimensional Euclidean Spaces *A. M. Grundland and W. J. Zakrzewski*	77
Boson Realizations of Semi-Simple Lie Algebras *Čestmír Burdík and Miloslav Havlíček*	87

Stretched Littlewood–Richardson and Kostka Coefficients 99
    R. C. King, C. Tollu, and F. Toumazet

Group Actions on Compact Hyperbolic Manifolds and Closed Geodesics 113
    Peter Kramer

Is There an Ultimate Symmetry in Physics? 125
    C. S. Lam

Formal Characters and Resolution of Infinite-Dimensional Simple
A_r-Modules of Finite Degree 131
    Daniel Britten and Frank Lemire

Fusion Rules and the Patera–Sharp Generating-Function Method 141
    L. Bégin, C. Cummins, P. Mathieu, and M. A. Walton

Transient Effects in Wigner Distribution Phase Space of a
Scattering Problem 153
    Marcos Moshinsky

R. T. Sharp and Generating Functions in Group Theory 159
    Jiří Patera

Quasi-Exact Solvability in Nonlinear Optics 165
    G. Álvarez, F. Finkel, A. González-López, and M. A. Rodríguez

Coherent States, Induced Representations, Geometric Quantization, and Their
Vector Coherent State Extensions 177
    D. J. Rowe

Symmetry Math Video Game Used to Train Profound Spatial-Temporal
Reasoning Abilities Equivalent to Dynamical Knot Theory 189
    Mark Bodner and Gordon L. Shaw

Polytope Sums and Lie Characters 203
    Mark A. Walton

Subalgebras of Lie Algebras. Example of $\mathrm{sl}(3,\mathbb{R})$ 215
    Pavel Winternitz

Preface

The Workshop on Symmetries in Physics organized at the Centre de recherches mathématiques, Université de Montréal, September 12–14, 2002, and these Proceedings are dedicated to our colleague and friend Robert T. Sharp. He passed away on October 1, 2001, after a prolonged illness. He had been associated with the Centre de recherches mathématiques, while a professor in the Physics Department at McGill University, virtually since the CRM was created in 1969. Professor Robert T. Sharp—Bob to his friends, colleagues and students—started the tradition of mathematical physics, in particular the use of algebraic and group theoretical methods, in Montréal. We started to collaborate with him in 1970 (J. P.) and 1972 (P. W.), and formed the first team in mathematical physics in Québec.

A special issue of the *Canadian Journal of Physics* was dedicated to Bob Sharp in 1994 on the occasion of his 70th birthday. The idea of putting together an issue of the *Canadian Journal of Physics* celebrating an anniversary of Bob Sharp came to us rather naturally. We felt that Bob's qualities as a teacher, scientist, and friend, although well appreciated by those who came to know him personally in Canada and abroad, deserved to be recognized more generally and made into a more visible example for those who might otherwise be prevented from seeing them by Bob's modesty and unassuming manner.

Having missed the World War II European theatre by weeks, the young, freshly trained Canadian fighter pilot Robert T. Sharp immediately volunteered for the war in the Pacific. Even that war ended before he could reach it, leaving him stranded with his Corsair squadron in England. Returning to Canada, he would consider nothing other than a similarly gallant and thrilling profession: becoming a physicist.

After a short spell on the faculty of the Physics Department of the University of Alberta in Edmonton, the former fighter pilot became physics professor at McGill University in 1953, where he had graduated under Phil Wallace's direction just two years before. At McGill, he continued for several years to be part of the Montreal Auxiliary Squadron and flew Vampire jet fighter planes.

During the last forty years, generations of physicists went through his classes. We know several who gave Bob Sharp the highest marks a teacher may desire: those colleagues who decided in his undergraduate class to become physicists.

For young students of physics the most memorable moment in their career is the first scientific publication, which separates student years from those of an active scientist. Their fellowships, future jobs, and, most importantly, their self-esteem largely depend on that publication. If ever the granting agencies in Canada required lists of publications of students one had supervised, that of Bob Sharp would be one of the most distinguished in Canada. Having always given them specific, well-defined, and difficult problems, and having provided an outstanding example by his personal enthusiasm, intellectual integrity, and often very substantial help in the course of the work, Bob has formed several dozen young Canadian intellectuals.

For anyone who knew Bob well, it was clear that his research was always a large part of his life. Always motivated by physics, his interests often led him to rather mathematical problems, with symmetry properties playing a prominent role.

Bob Sharp started his scientific career very successfully as a nuclear and elementary-particle physicist. For instance, an early short paper of his, published in 1953 jointly with G. K. Horton, is still very much alive. It has received 146 citations since 1982, 20 of them in 2002. Early in his career Bob became interested in the applications of group theory in physics and became one of the pioneers of this type of research in Canada. He developed a simple method for calculating Clebsch–Gordan coefficients and applied it, together with his collaborators and students, to the group $SU(3)$ shortly after this group was proposed by M. Gell-Mann and Y. Ne'eman as the group underlying the classification of hadrons. One of the problems faced by physicists when applying Lie group theory is the appropriate choice of the representation basis. Often, the physically interesting basis is not sufficiently specified. Some quantum numbers are missing, the so-called missing label problem. Bob solved this problem in a highly original way for several important groups: $SU(3)$, in some cases $SU(n)$ for any n, $O(4)$, and $O(5)$.

Since the early 1970s Bob's interests and contributions diversified. He started working on applications of noncompact groups and contributed to the development of methods for classifying their subgroups. Some of his later work was on Kac–Moody algebras and their representations and on graded contractions of Lie algebras and Kac–Moody algebras.

Probably his most important and lasting contribution is his series of articles on generating functions in group representation theory. This is a topic that started in the last century with the work of Cayley and Sylvester, followed by many others, the best known being Poincaré and Molien. For the first half of this century it remained almost dormant until essentially Bob, his students, and collaborators revived it. The majority of the generating functions known today were calculated in his and his co-workers' papers. He knew more about the methods of deriving all sorts of generating functions than anyone in the world and extended their applications to many new group-theoretical problems.

Many of the participants at this workshop were Bob Sharp's personal friends. Indeed, his kindness, fairness and simple human decency, together with his outgoing personality, gained him many friends. He was very much a family man, and many of us remember the warm atmosphere, created by Bob, his wife Kay, and their four children, Susan, Joan, Ted and Doug, in their home. We were very happy that his children could participate in this workshop, or at least in the workshop banquet.

The scope of these Proceedings reflects the influence that Bob Sharp had on physics and also the esteem in which he is held by his colleagues. The unifying theme of this volume is the consistent application of group theory to problems arising in physics and the further development of group theoretical tools for applications in physics.

Jiří Patera and Pavel Winternitz
Montréal

Bob Sharp: The Man and the Scientist

When Pavel Winternitz asked me to speak on this occasion, my first thought was to recall that I had already performed this function, at the banquet recognizing his retirement from McGill. Any of you who may have been at that event might find this story repetitious. If that is so, please bear with me. But Bob did not change in the intervening years.

I should like to remind you of events of the distant past, for Bob and I first met in 1947, when I left the Canadian nuclear energy project to become a professor at McGill. Bob, on the other hand, faced a transition from the air force to academia. But he had prepared himself well; as soon as the war ended he had taken correspondence courses in mathematics and physics in preparation for graduate studies at McGill.

I may also remind you that, up to the war, there had been very little theoretical physics in Canada,—a situation which reflected the scorn for theoreticians of the great Ernest Rutherford. Most Canadian physics departments did not recognize theorists, relegating them to mathematics departments where they were not fully accepted as real mathematicians. The "Applied Mathematics" department at Toronto, led by J. L. Synge and Leopold Infeld, straddled the fence and nurtured in the following years a slow but constant flow of theorists, mostly isolated from their faculty colleagues in experimental physics.

What has this to do with Bob? Simply that, to the best of my knowledge, Bob was the first addition to the modest body of postwar theoretical physicists, the first product of a renaissance which would grow exponentially in the following decades

Bob was also my first graduate student.

While I had done my Ph.D. in relativity with Leopold Infeld, my wartime occupation with nuclear reactors had led me down a quite different path. Graphite was an important component of these reactors, and it came in macroscopic blocks. It was essential to understand its properties. But solid state physics was in its infancy, and we had in the project no experts in the field. So I was sent off on a quick course with Neville Mott at Bristol. It was in this context that I found a Ph.D. problem for Bob.

It didn't take long for Bob's brilliance to become evident to me. I found that complicated calculations were no obstacle to him, nor were the underlying physical insights.

I will skip to Bob's defense of his thesis, before a very distinguished committee. These events usually take about an hour, but in Bob's case it brought a seemingly endless barrage of questions. When it was over, he asked me what was wrong that elicited such a grilling. What I had to explain to him was that they were so impressed by the ease and insight of his responses that they did not want to stop enjoying it.

Once he had his degree in hand, I discovered something that I had not discerned—that there was a real and important difference in our outlook and interests. Bob told me that he did not want to continue working in solid state physics; that this was, for him, too specialized and not interesting enough mathematically. In short, I understood that he had loftier goals. It is Eugene Wigner who raised the question: Why is it that mathematics is so naturally adapted to our way of understanding the world? And I believe that this is the kind of question that intrigued Bob. At this point I touch lightly on the subject of this conference: Is not symmetry at the core of our understanding of the physical world? Emmy Noether's linking of dynamic variables to symmetries was an early example. Obviously I, as someone whose major field of interest has been in the physics of crystalline solids, caught up in the same net, and when I added strong magnetic fields, it was strengthened even more.

And as we are still struggling to reach an unequivocal understanding of quantum theory itself; the role of symmetry still haunts us So when I speak of the divergence of the paths of Bob and me, it may be summarized in the fact that I had a fascination with the details of things, and Bob was concerned with the grand fundamental principles. This divide was, I believe, constructive because we complemented each other, linking the language of mathematics to the message.

The theoretical group started with Ted Morris, Bob and me, supplemented a bit later by Dave Jackson. By the sixties, when a perceptive Dean (Ken Hare) decided that it was time to move us to the physics department where we belonged. This move was both logical and in tune with the times, but was accepted with some misgivings by the existing physics department. By this time our group had expanded to cover a wide range of physics. A new post-war generation of theorists had been assembled, too numerous to recite here, and Bob was in the core of it.

There was one respect in which Bob stood out. In the late sixties and the early seventies there was a global student revolution, which led to a more inclusive role for students, one consequence of which was the introduction of student evaluations of all their courses and their teachers. You will remember the scheme of rating on a scale of 1 to 5. 1 for excellence, down to 5 for dismal failure. I remind you of this because Bob usually rated $1 + x$ with at $x < 0.5$. While many of us struggled with scores of 2+ or 3+, Bob was easily the most popular of us with the students. This was not because his courses were easy—in a lecture he could fill the blackboard several times over with rather daunting and/or complicated formulae. The secret was that everything was precise and logical. Julian Schwinger was like that too; before lecturing he had smoothed out all unnecessary complications, but with Bob it was not calculated but a manifestation of the clarity of his mind.

It is the word logical which is the key to the phenomenon—and if there is one thing that stands out in describing Bob's special characteristics it is that he had a LOGICAL mind, not just applied to physics or even science alone but to all of life.

Since it is difficult to argue against logic, Bob was not an argumentative person, whatever the subject. But this did not conflict with firm convictions, in any context. On principles we could agree, but the difference between Bob's clear logic and my "yes, buts" was always present. Harry Lam recounted to me that Bob had rebuked him for buying lottery tickets. Harry did not disagree with the contention that the odds were against him, but yet made a reasonable response—that the price of the tickets was utterly insignificant, and could make no impression on his life. But if he won, he could retire and live the rest of his life as a free man. This I would categorize as a "yes, but," out of the scope of simple logic.

Sometimes Bob's logic took an unexpected turn. He served in the Air force in the war—the mark of a good patriot. Yet when the Canadian parliament was debating the issue of a uniquely Canadian flag, with a fervor that shook the whole country and aroused bitter parliamentary debate, Bob made a radical proposal: "Wouldn't it be nice to be the only country in the world that did not HAVE a flag." Thinking about it afterwards, it struck me that Bob's logic had been stronger than his emotions—that, like myself, he could find no rational case for worshiping Mr. Landry's "piece of red cloth" or his blue *fleur de lys*. One's debt to society needed a basis more substantial than this corruptible and (at that time) contentious symbol.

Another of Bob's characteristics was an utter immunity to fear. Fear was simply not in his nature. That was one of the things, I suppose, which made him enjoy the perils of flying and his experience as an air force pilot during the war. He carried his love of flying into peacetime, apparently enjoying most the most daring of its challenges.

On one occasion, he asked me if I would like to fly with him at weekend. He "rented" small planes and kept his flying skills alive. On this particular occasion, when we went to the airport office to arrange the flight, he was told that there was a warning of an approaching storm. Bob's reaction was to tell me to hurry so that we could take off before it was forbidden by the airport authorities. For Bob, flying through a storm would make the flight more exciting.

But the storm did not catch up with us so that, once in the air, Bob asked me if I would like to do a loop-the-loop. I had not envisioned this, but said to Bob "no, I think I would rather not." But Bob's goal was some excitement, so he tried again, asking if I would like a dive, pulling out at the last moment. Once again, I disappointed him; my enthusiasm for that maneuver was even less than for the loop. So he resigned himself to a boring afternoon, probably muttering beneath his breath, "Why did I ask this guy to come with me?"

But some time later I redeemed myself. Bob asked a student if he would like to go flying. The student acquiesced. But when it came to the acrobatics, the student, recognizing that Bob was his superior, felt compelled to agree to his aerial acrobatics. Bob did the loop, he did the dive, and the student threw up. When the flight was over, he took off and left Bob to deal with the mess. From thence on, I suspect, Bob was more cautious in choosing his passengers,

What Bob did not lack, however, was a sense of humor. At one point our group had bridge evenings, in which one rotated partners and adversaries and tallied up accumulated scores at the end. Bob had some prizes for the winners, but he added an extra one at the end for the player with the lowest score. His prize was a light bulb. This, he announced, was to bring some light into the darkness. It turned out that I was the recipient. The thought that went through my head was that I played bridge somewhat as Bob flew an airplane, but without any of that risk or skill that made the activity exciting.

In these tales, I have left out one of Bob's most distinctive features. It was almost impossible not to like Bob, and he seemed always to see only the best in others. There was an intrinsic friendliness and goodwill in his social and professional relationships; he rarely criticized others, and if he did it was always without rancor or ill will. Academic politics is often bitter and contentious, but he could rise above it. It was this that led me, on the occasion of his retirement, to describe him as an academic Mr. Chips, if any of you can remember that ancient story and movie about the teacher whom everyone revered. But Bob added an extra dimension which led me to add an additional feature, leaving the picture as that of Mr. Chips in aviator's goggles.

He was a good man, a modest man with a brilliant mind and a good friend to all those who were made better human beings by knowing him. He left his imprint on all of us.

Philip R. Wallace
Victoria

Bob Sharp: Teacher, Colleague, Friend

Perhaps a better title for this would have been "Waiting for Bob Sharp," since my own experience involved several delays before finally succeeding in meeting him. I first learned about him—by reputation—while still an undergraduate physics student at McGill in the 1960's. Somehow, I seem to have twice missed the opportunity of meeting someone rumored to be one of the "really good" physics professors at McGill—and who was specially interested in group theoretical methods in physics. This was the subject that I discovered with great delight in my final year of the honors physics program (the year was 1967) and was thrilled to learn about from Wigner's book. This was a revelation, and I remember being amazed by the beauty and "magical" quality of the results that it led to. However that year, exceptionally, the course was not taught by Bob Sharp, since he was away on sabbatical leave at the University of Illinois.

That was the first missed opportunity. The next year, I happened to follow the same track, going to the University of Illinois, with the intention of doing my doctorate there. I heard of the visiting professor from McGill who had been there the previous year, and had taught everybody a great deal—and who was always spoken of with a certain degree of awe and respect, but who had gone back by then to Montréal.

In fact it was only several years later that I finally got to know Bob Sharp, and learn that the respect and warmth with which he was always spoken of had indeed a very valid basis, which included much more than his special talents as teacher or specialist in group theoretical methods in physics. This was after I had already completed my doctorate, not in Illinois, but as it turned out, through the help and kind encouragement of Professor Phil Wallace (and a generous Shell scholarship) at Oxford, under the supervision first of Roger Phillips, and then J. C. Taylor. That was in the days of transition between the declining "analyticity-bootstrap" approach to high energy theory, and the beginning of the great success of unified gauge theory. My thesis work unfortunately was still on the old tack—Regge and Lorentz poles—however, from a group theoretical viewpoint.

It was only after having spent a couple of years as an I.P.P. postdoc at Carleton in Ottawa, and then, in the rather depressed job market in theoretical high energy physics that existed at that time, finding myself back in Montréal with just a temporary CEGEP professor's job during a particularly hard year—1975–76—that I got to know Bob Sharp. This was the year of the public service strikes in Quebec that eventually helped to bring about the downfall of the Bourassa government. After attending the necessary quota of rather rowdy union meetings, trying to win the good-will of students for the CEGEP teachers' cause, but finding myself unable to return to the classroom, I decided to while away the time, not on the picket lines, but at the CRM. Here, to my delight, I found there was a group of genuine "mathematical physicists"—Pavel Winternitz, Jiří Patera and also Bob Sharp (who was often there working with them)—specially interested in symmetries and the "group theoretic approach." For me, this was rather a difficult period; it seemed as though any hope of a future career in research was fading in the nearly non-existent job market of the time. But, being at the CRM, even unofficially, was a help, since there seemed to be a "local fluctuation" there that was contrary to the general trend, with genuinely interesting work apparently going on.

Then a small gesture was made—by Bob Sharp—that psychologically meant a great deal to me. He came to me one day and said "since you are working here anyway, I think we should somehow make this official," and he proceeded to arrange to have a certain amount paid to me from his grant for the rest of the year, as a visiting research associate. Later, he also arranged for me to have an office at McGill. Bob didn't know me very well at the time—perhaps he had just attended one or two seminars I gave, but this positive gesture of encouragement, made very quietly, with no previous discussion—just seemingly a spontaneous decision on his part, made an enormous difference to me. Afterward, Pavel Winternitz and Jiří Patera, and later Steve Shnider, all contributed to creating a longer term research associate position for me at the CRM, which gave me the courage to give up the CEGEP job and focus fully on research for a number of years to follow. But the single instant of Bob's gesture was absolutely critical, and probably made the difference between my trying to continue, despite all, to pursue a career in research, or giving up. I suspect there are others who have been helped by him in various ways, perhaps also just when it was truly needed, and that it was often in such a quiet, almost incidental way that he intervened when he did, just doing what seemed to him as natural. I have always thought of him at least as much in terms of his spontaneous, simple, human good-will, and understanding, as his love of scientific research and the human milieu in which it is done. However I did have occasion, in the following years, through his lectures, and discussions, to learn of his very deep understanding of group theoretical methods and his own special approaches to the solution of problems in physics by group theoretical means.

Later, I also got to know him well as a friend with a very warm, good nature, and as a great canoeing excursion companion. This is something that he, and Doug, and Pavel Winternitz with his sons and friends had been doing for years before I was initiated into the experience. I won't talk about my own unusual predilections for ending up in the water, together with all the contents and supplies of whichever canoe had the misfortune to be carrying me. But I do remember one particular occasion at La Vérendrye Park when, on arriving a little late to our campsite, Bob took a misstep on disembarking and ended up completely soaked. What was striking was that what to others might have been experienced as at least a mildly unpleasant mishap did not seem to diminish his good humor in the slightest. He continued to be as good-natured, good spirited and sociable, while trying to dry himself off, together with his drenched things, as at any other time—maybe even more so. And if we look at his characteristic smile on the pictures posted around us—which is the way I always visualize him—it was not diminished in the least after this dunking (which, if I remember correctly, actually occurred twice in succession). His pleasure in the environment and his good nature simply did not allow any diminishing of the pleasure of the moment.

He often used to keep us entertained by stories, at the campfire or elsewhere—and maybe it was from these, or perhaps on some other occasion that I formed an impression that there was one special time in his life in which he had been particularly happy. This was when he was a fighter pilot trainer during the war. One of the images on the posters shows him at this time, and it makes me think of one of the stories that he told of his experiences during that time. Probably others who are here, especially his family, will know of this story in more detail—or others from that time—but it made a particular impression on me because of the way he told it. It seems that at some point while on a flying mission, his plane stalled, and he started into a dive. He described the moment, saying he thought he had better get the engine going again—which after a time, he did succeed in doing—but without any apparent recollection of distress—as though he had just been curious about whether he would be able to do it in time or not. He did not describe it as a harrowing, frightening, or distressing experience—but rather as a challenging moment whose outcome was not at all clear, and he described it with the same good-humored smile and tone as one sees in the photograph of him as a pilot. Perhaps at the time, he didn't have quite such a smile for the whole time before he got out of the stall, but somehow, I think that it must have returned to him very quickly afterward, and the recollection of the incident just did not cause him any subsequent distress. He was certainly very happy in that period of his life—but I think that he was also a man who found reason for happiness in all the phases of his life.

That story came to mind many years later, when I came to visit him in the hospital after a multiple bypass operation, during which he was experiencing quite serious complications. He was in a pretty poor state, but I remembered that there had been many, many times in which health problems had loomed threateningly, yet he had managed somehow to get through them, in as good cheer as always, and bringing as little attention to the seriousness of the situation as he could. I couldn't help thinking "will you be able to get through this one too, and back into flight again, as you were able so many times in the past?" And it really seemed, for a short while, that perhaps he could do it. I briefly saw him, just once more, at the CRM, some time after he had been released from hospital. He seemed again in good cheer—I even remember discussing with him possible places to go that summer—in particular, a conference in Crete that seemed particularly to appeal to him. But unfortunately, it was not to be.

When thinking of Bob, however, the images that come to mind are inevitably the smiling ones, like the one on the poster where he is standing before a blackboard—which really recalls the animated spirit that he brought to his lectures, or a mental one of him at a campfire telling or hearing stories, from the recent or more distant past, and enjoying it in great spirits, or one of him at home, at Rosemère, with his family and the dogs, or—perhaps, the one which precedes what most of us, even his oldest friends here, could have been present to share with him—namely, the one in pilot's uniform. He must have valued and enjoyed life very much—he certainly gave a great deal to all those around him, and I'm sure he received much love in return, both from his family, and from all those, colleagues, friends, students who knew him, learned from him, shared with him and benefited so greatly from the experience of having known him.

J. Harnad
Montréal

Publications of Robert T. Sharp

1. G. K. Horton and R. T. Sharp, *Approximate wave functions for unbound relativistic particles in a Coulomb field*, Phys. Rev. **89** (1953), 885–886.
2. G. K. Horton and R. T. Sharp, *Variational approach to the many-electron problem*, Phys. Rev. **90** (1953), 317.
3. R. T. Sharp, *Direct method of identifying statistical quantities with their thermodynamic analogs*, Amer. J. Phys. **23** (1955), 69–70.
4. J. D. Jackson and R. T. Sharp, *Orthogonality properties of Furry–Sommerfeld–Maue wave function and their use in perturbation theory*, Phys. Rev. **98** (1955), 1128.
5. G. Bach and R. T. Sharp, *Time reversal effects on muon decay*, Canad. J. Phys. **35** (1957), 1199–1203.
6. R. T. Sharp, *Two-nucleon potential from Chew–Low theory*, Nuovo Cimento **9** (1958), 23–36.
7. R. T. Sharp, *Hypernuclei*, Phys. in Canada **15** (1959), 11–17.
8. R. T. Sharp, *Self and external meson fields*, Canad J. Phys. **37** (1959), 515–520.
9. R. C. Smith and R. T. Sharp, *Three-nucleon forces according to Chew–Low theory*, Nuclear Forces and the Few-Nucleon Problem (London, 1959) (T. C. Griffith, ed.) Pergamon, Oxford, 1960, pp. 245–248.
10. R. T. Sharp, *Simple derivation of the Clebsch–Gordan coefficients*, Amer. J. Phys. **28** (1960), 116–118.
11. R. T. Sharp and R. C. Smith, *Central three-body forces in heavy nuclei*, Canad. J. Phys. **38** (1960), 1154–1167.
12. C. Roth and R. T. Sharp, *Distortion of the triton core in the hypernucleus* $_\Lambda H^4$, Canad. J. Phys. **40** (1962), 1457–1460.
13. R. T. Sharp and H. C. von Baeyer, *Polynomial bases and isoscalar factors for* SU(3), J. Mathematical Phys. **7** (1966), 1105–1122.
14. C. Chew and R. T. Sharp, *On the degeneracy problem in* SU(3), Canad. J. Phys. **44** (1966), 2789–2796.
15. R. T. Sharp, *Stretched X-coefficients*, Nuclear Phys. A **95** (1967), 222–228.
16. R. T. Sharp, *Generalized Regge identity for Wigner coefficients*, Nuovo Cimento **47** (1967), 860–868.
17. C. Chew and R. T. Sharp, SU(3) *isoscalar factors*, Nuclear Phys. B **2** (1967), 697–712.
18. W. J. Pardee, D. G. Ravenhall and R. T. Sharp, *Connection between* SU(3) *and* O(4), Phys. Rev. **164** (1967), 1950–1955.
19. S. C. Pieper and R. T. Sharp, O(5) *polynomial bases*, J. Mathematical Phys. **9** (1968), 663–667.
20. R. T. Sharp, SU(3) *crossing matrix*, Phys. Rev. **168** (1968), 1753–1755.
21. R. T. Sharp, *Notes on* O(4) *representations*, J. Mathematical Phys. **47** (1968), 359–365.
22. S. C. Pieper, R. T. Sharp, and H. C. von Baeyer, *Polynomial bases and Wigner coefficients for* SU(3) $\supset R_3$, Nuclear Phys. A **127** (1969), 513.
23. C. S. Lam and R. T. Sharp, *Internal-labeling problem*, J. Mathematical Phys. **10** (1969), 2033–2038.
24. G. Jakimow and R. T. Sharp, SU(4) *polynomial bases*, Canad. J. Phys. **47** (1969), 2137–2142.
25. K. Ahmed and R. T. Sharp, O(5) *bases for seniority model*, J. Mathematical Phys. **11** (1970), 1112–1117.

26. R. T. Sharp and H. C. von Baeyer, *Clebsch–Gordan coefficients for* $SU(3) \supset R_3$ *in different bases*, Nuclear Phys. A **140** (1970), 118–128.
27. R. T. Sharp, *Internal labelling: the classical groups*, Proc. Cambridge Philos. Soc. **68** (1970), 571–578.
28. D. Lee and R. T. Sharp, $SU(4)$ *van der Waerden invariant*, Rev. Mexicana Fís. **20** (1971), 203–215.
29. R. T. Sharp, $SU(n-2) \times SU(2) \times U(1)$ *bases for* $SU(n)$, J. Mathematical Phys. **13** (1972), 183–186.
30. K. Ahmed and R. T. Sharp, *Wigner supermultiplet bases*, Ann. Physics **71** (1972), 421–437.
31. J. Patera and R. T. Sharp, *Angular-momentum basis for a general* j^n *configuration*, Nuovo Cimento A (11) **12** (1972), 365–376.
32. C. S. Lam and R. T. Sharp, *Multiparticle momentum correlations and elastic angular distribution*, Phys. Rev. D (3) **8** (1973), 278–286.
33. E. G. Kalnins, J. Patera, R. T. Sharp, and P. Winternitz, *Two-variable Galilei Group expansion of nonrelativistic scattering amplitudes*, Phys. Rev. D (3) **8** (1973), 2552–2572.
34. E. G. Kalnins, J. Patera, R. T. Sharp, and P. Winternitz, *Potential scattering and Galilei invariant expansions of scattering amplitudes*, Phys. Rev. D (3) **8** (1973), 3527–3538.
35. J. Patera, R. T. Sharp, and P. Winternitz, *Nagel–Moshinsky operators for discrete unitary representations of* $U(p, q)$, Rev. Mexicana Fís. **23** (1974), 81–98.
36. M. Hongoh, R. T. Sharp, and D. E. Tilley, *Explicit* $O(5)$ *Wigner coefficients*, J. Mathematical Phys. **15** (1974), 782–788.
37. M. Moshinsky, J. Patera, R. T. Sharp, and P. Winternitz, *Isotopic spin conservation and charge distribution in multipion production*, Phys. Rev. D (3) **10** (1974), 1587–1594.
38. E. G. Kalnins, J. Patera, R. T. Sharp, and P. Winternitz, *Elementary particle reactions and the Lorentz and Galilei groups*, Group Theory and its Applications. Vol. III (E. M. Loebl, ed.), Academic Press, New York, 1975, pp. 369–464.
39. R. T. Sharp, *Internal-labeling operators*, J. Mathematical Phys. **16** (1975), 2050–2053.
40. W. McKay, J. Patera, and R. T. Sharp, *Eigenstates and eigenvalues of labelling operators for* $O(3)$ *bases of* $U(3)$ *Representations*, Comput. Phys. Comm. **10** (1975), 1–10.
41. M. Moshinsky, J. Patera, R. T. Sharp, and P. Winternitz, *Everything you always wanted to know about* $SU(3) \supset O(3)$, Ann. Physics **95** (1975), 139–169.
42. E. Chacón, M. Moshinsky, and R. T. Sharp, $U(5) \supset O(5) \supset O(3)$ *and the exact solution for the problem of nuclear quadrupole vibrations*, Group Theoretical Methods in Physics (Nijmegen, 1975), Lecture Notes in Phys., vol. 50, Springer, Berlin, 1976, p. 337.
43. J. Patera, R. T. Sharp, P. Winternitz, and H. Zassenhaus, *Casimir operators of subalgebras of the Poincaré Lie algebra and of real Lie algebras of low dimension*, Group Theoretical Methods in Physics (Nijmegen, 1975), Lecture Notes in Phys., vol. 50, Springer, Berlin, 1976, pp. 500–515.
44. E. Chacón, M. Moshinsky, and R. T. Sharp, $U(5) \supset O(5) \supset O(3)$ *and the exact solution of the nuclear quadrupole vibrarion problem*, J. Mathematical Phys. **17** (1976), 668–676.
45. J. Patera, R. T. Sharp, P. Winternitz, and H. Zassenhaus, *Subgroups of the Poincaré group and their invariants*, J. Mathematical Phys. **17** (1976), 977–985.
46. E. Chacón, M. Moshinsky, and R. T. Sharp, *Applications of traceless boson operators in the nuclear collective model*, Proceedings of the International Symposium on Mathematical Physics (Mexico, 1976), CONACYT, Mexico, 1976, 65.
47. J. Patera, R. T. Sharp, and P. Winternitz, *Applications of group representation indices of higher order*, Proceedings of the International Symposium on Mathematical Physics (Mexico, 1976), CONACYT, Mexico, 1976, pp. 213–223.
48. J. Patera, R. T. Sharp, P. Winternitz, and H. Zassenhaus, *Casimir operators of real Lie algebras*, Proceedings of the International Symposium on Mathematical Physics (Mexico, 1976), CONACYT, Mexico, 1976, pp. 225–233.
49. J. Patera, R. T. Sharp, P. Winternitz, and H. Zassenhaus, *Invariants of real low dimension Lie algebras*, J. Mathematical Phys. **17** (1976), 986–994.
50. J. Patera, R. T. Sharp, P. Winternitz, and H. Zassenhaus, *Subgroups of the similitude group of three-dimensional Minkowski space*, Canad. J. Phys. **54** (1976), 950–961.
51. A. Peccia and R. T. Sharp, *Number of independent missing label operators*, J. Mathematical Phys. **17** (1976), 1313–1314.

52. W. McKay, J. Patera, and R. T. Sharp, *Branching rules and Clebsch–Gordan series for* E_8, J. Mathematical Phys. **17** (1976), 1371–1375.
53. C. P. Boyer, R. T. Sharp, and P. Winternitz, *Symmetry breaking interactions for the time dependent Schrödinger equation*, J. Mathematical Phys. **17** (1976), 1439–1451.
54. J. Patera, R. T. Sharp, and P. Winternitz, *Higher indices of group representations*, J. Mathematical Phys. **17** (1976) 1972–1979.
55. J. Patera, R. T. Sharp, P. Winternitz, and H. Zassenhaus, *Continuous subgroups of the fundamental groups of physics. III. The de Sitter groups*, J. Mathematical Phys. **18** (1977), 2259–2288.
56. R. Gaskell, A. Peccia, and R. T. Sharp, *Generating functions for polynomial irreducible tensors*, J. Mathematical Phys. **19** (1978), 727–733.
57. J. Patera, R. T. Sharp, and P. Winternitz, *Polynomial irreducible tensors for point groups*, J. Math. Phys. **19** (1978), 2362–2376.
58. J. Patera and R. T. Sharp, *Generating function techniques pertinent to spectroscopy and crystal physics*, Recent Advances in Group Theory and Their Application to Spectroscopy (Antigonish, 1978) (J. Donini, ed.), NATO Adv. Study Inst. Ser. B: Physics, vol. 43, Plenum, New York, 1979, pp. 219–248.
59. P. E. Desmier and R. T. Sharp, *Polynomial tensors for double point groups*, J. Math. Phys. **20** (1979), 74–82.
60. J. Patera and R. T. Sharp, *Generating functions for characters of group representations and their application*, Group Theoretical Methods in Physics, Lecture Notes in Phys., vol. 94, Springer, Berlin, 1979, pp. 175–183.
61. J. Patera and R. T. Sharp, *Generating functions for plethysms of continuous and finite groups*, J. Phys. A **13** (1980), 397–416.
62. M. Couture and R. T. Sharp, *Reduction of enveloping algebras of low-rank groups*, J. Phys. A **13** (1980), 1925–1945.
63. J. Patera, R. T. Sharp, and R. Slansky, *On a new relation between semisimple Lie algebras*, J. Math. Phys. **21** (1980), 2335–2341.
64. M. Couture and R. T. Sharp, *Structure of enveloping algebras of low-rank groups*, Group Theoretical Methods in Physics, Lecture Notes in Phys., vol. 135, Springer, Berlin, 1980, pp. 473–477.
65. D. Phaneuf and R. T. Sharp, *Polynomial space group tensors*, Group Theoretical Methods in Physics, Lecture Notes in Phys., vol. 135, Springer, Berlin, 1980, pp. 517–519.
66. J. Patera and R. T. Sharp, *Generating functions for* SU(2) *plethysms with fixed exchange symmetry*, J. Math. Phys. **22** (1981), 261–266.
67. J. Patera and R. T. Sharp, *On the triangle anomaly number of* SU(n) *Representations*, J. Math. Phys. **22** (1981), 2352–2356.
68. R. Gaskell, G. Rosensteel and R. T. Sharp, Sp(6) *states in an* SU(3) × U(1) *basis*, J. Math. Phys. **22** (1981), 2732–2735.
69. R. Gaskell and R. T. Sharp, *Generating functions for* G_2 *characters and subgroup branching rules*, J. Math. Phys. **22** (1981), 2736–2739.
70. J. McKay, J. Patera, and R. T. Sharp, *Second and fourth indices of plethysms*, J. Math. Phys. **22** (1981), 2770–2774.
71. J. Patera and R. T. Sharp, SU(3) × SU(2) × U(1) *content of all* SU(5) *representations*, Phys. Rev. D (3) **25** (1982), 1141–1142.
72. R. C. King, J. Patera and R. T. Sharp, *On finite and continuous little group representations of semisimple Lie groups*, J. Phys. A **15** (1982), 1143–1158.
73. P. E. Desmier, J. Patera and R. T. Sharp, *Analaytic* SU(3) *states in a finite subgroup basis*, J. Math. Phys. **23** (1982), 1393–1398.
74. J. Patera and R. T. Sharp, *Signatures of all finite representations of* SU(p,q), $p + q \leq 4$, Kinam Rev. Fĩs. **4** (1982), 93–98.
75. J. Bystricky, R. Gaskell, J. Patera, and R. T. Sharp, *Generalized* SU(2) *spherical harmonics*, J. Math. Phys. **23** (1982), 1560–1565.
76. P. E. Desmier, J. Patera and R. T. Sharp, *Finite subgroup bases for compact Lie groups*, Phys. A **114** (1982), 336–340.
77. R. W. Gaskell and R. T. Sharp, G_2 *van der Waerden invariant*, J. Math. Phys. **23** (1982), 2016–2018.

78. M. V. Jarić, L. Michel, and R. T. Sharp, *Invariant formulation for the zeros of covariant vector fields*, Group Theoretical Methods in Physics (Istanbul, 1982), Lecture Notes in Phys., vol. 180, Springer, Berlin, 1983, pp. 317–318.
79. T. H. Seligman and R. T. Sharp, *Internal labels of degenerate representations*, J. Math. Phys. **24** (1983), 769–771.
80. R. V. Moody, J. Patera, and R. T. Sharp, *Character generators for elements of finite order in simple Lie groups A_1, A_2, A_3, B_2 and G_2*, J. Math. Phys. **24** (1983), 2387–2396.
81. M. V. Jarić, L. Michel, and R. T. Sharp, *Zeros of covariant vector fields for the point groups: invariant formulation*, J. Phys. **45** (1984), 1–27.
82. M. Couture, Y. Giroux, and R. T. Sharp, *Degenerate enveloping algebras of $SU(3)$, $SO(5)$, G_2 and $SU(4)$*, J. Phys. A **17** (1984), 715–725.
83. J. Patera and R. T. Sharp, *Signatures of finite $SU(p,q)$ representations*, J. Math. Phys., **25** (1984), 2128–2132.
84. R. W. Gaskell and R. T. Sharp, *Fixed symmetry and fixed class generating functions*, J. Math. Phys. **25** (1984), 2144–2148.
85. J. W. B. Hugues, J. Van der Jeugt, and R. T. Sharp, *application of generating function techniques to Lie superalgebras*, J. Math. Phys. **26** (1985), 901–912.
86. D. Phaneuf and R. T. Sharp, *Polynomial tensors for the space groups $\wp m$ and $p4m$*, J. Math. Phys. **26** (1985), 1534–1539.
87. M. Moshinsky, M. Nicolescu and R. T. Sharp, *Collectivity and geometry. IV. $Sp(6) \supset Sp(2) \times O(3)$ basis states for open shells*, J. Math. Phys. **26** (1985), 2995–2998.
88. Y. Giroux and R. T. Sharp, *A folk theorem revisited: degenerate representations*, J. Math. Phys. **28** (1987), 1671–1672.
89. J. Van der Jeugt, B. Morel, J. Patera, and R. T. Sharp, *Indices of representations of simple superalgebras*, J. Math. Phys. **28** (1987), 1672–1682.
90. L. Michel, J. Patera and R. T. Sharp, *The Demazure–Tits subgroup of a simple Lie group*, J. Math. Phys. **29** (1988), 777–796.
91. M. Couture and R. T. Sharp, *Irreducible embeddings and polynomials tensors*, J. Phys. A **22** (1989), 1525–1542.
92. J. Patera and R. T. Sharp, *Branching rules for representations of simple Lie algebras through Weyl group orbit reduction*, J. Phys. A **22** (1989), 2329–2340.
93. J. Van der Jeugt, J. Patera, and R. T. Sharp, *New gradings of $sl(3,\mathbb{C})$ representations*, J. Math. Phys. **30** (1989), 2763–2769.
94. M. Couture and R. T. Sharp, *A reply to Scutaru's letter on the generalized exponents of $sl(3,\mathbb{C})$*, J. Phys. A **23** (1990), 1827–1828.
95. M. Couture, C. J. Cummins and R. T. Sharp, *Generating functions, elementary multiplets and Young tableaux*, J. Phys. A **23** (1990), 1929–1957.
96. X. Leng, J. Patera, and R. T. Sharp, *Subjoining of affine Kac–Moody algebras*, J. Phys. A **23** (1990), 3397–3407.
97. H. de Guise and R. T. Sharp, *Polynomials states for $SU(3)$ and $SO(5)$ in a Demazure–Tits basis*, J. Phys. A **24** (1991), 557–568.
98. M. Couture, J. Patera, R. T. Sharp, and P. Winternitz, *Graded contractions of $sl(3,\mathbb{C})$*, J. Math. Phys. **32** (1991), 2310–2318.
99. F. Gingras, J. Patera, and R. T. Sharp, *Orbit-orbit branching rules between simple low-rank algebras and equal-rank subalgebras*, J. Math. Phys. **33** (1992), 1618–1626.
100. F. Bégin and R. T. Sharp, *Weyl orbits and their expansions in irreducible representations for affine Kac–Moody algebras*, J. Math. Phys. **33** (1992), 2343–2356.
101. Č. Burdik, C. J. Cummins, R. W. Gaskell, and R. T. Sharp, *Complete branching rules generating function for $SO(7) \supset SU(2)^3$ and polynomial basis states*, J. Phys. A **25** (1992), 4835–4846.
102. L. Farell, C. S. Lam, and R. T. Sharp, *G_2 generator matrix elements for degenerate representations in an $SU(3)$ basis*, J. Phys. A **27** (1994), 2761–2771.
103. B. Champagne, M. Kjiri, J. Patera, and R. T. Sharp, *Description of reflection generated polytopes using decorated Coxeter diagrams*, Canad. J. Phys. **73** (1995), 566–584.
104. N. Hambli and R. T. Sharp, *Generator matrix elements for $G_2 \supset SU(3)$. II. Generic representations*, J. Phys. A **28** (1995), 2581–2588.
105. N. Hambli, J. Michelson, and R. T. Sharp, *Character states and generator matrix elements for $Sp(4) \supset SU(2) \times U(1)$*, J. Math. Phys. **37** (1996), 3022–3031.

106. R. T. Sharp and M. Thoma, *Orbit-orbit branching rules for families of classical Lie algebra-subalgebra pairs*, J. Math. Phys. **37** (1996), 4750–4757.
107. R. T. Sharp and M. Thoma, *Orbit-orbit branching rules between classical simple Lie algebras and maximal reductive subalgebras*, J. Math. Phys. **37** (1996), 6570–6581.
108. C. S. Lam, J. Patera and R. T. Sharp, *Generating functions for the Coxeter group* H_4, J. Phys. A **29** (1996), 7705–7719.
109. R. T. Sharp and M. Thoma, *Orbit-orbit branching rules for algebra-subalgebra pairs* $C_{m+n} \supset C_m \oplus C_n$. Proceedings of the IV Wigner Symposium (Guadalajara, 1995), World Sci. Publishing, River Edge, NJ, 1996, 265–269.
110. N. Hambli and R. T. Sharp, *On character generators*, Advances in Mathematical Sciences: CRM's 25 years (Montréal, 1994), CRM Proc. Lecture Notes, vol. 11, Amer. Soc., Providence, RI, 1997, 415–419.
111. R. T. Sharp and M. Thoma, *Branching rules for* $SO(n) \supset SO(n-2) \times U(1)$. Proceedings of the 5th Wigner Symposium (Vienna, 1997), World Sci. Publishing, River Edge, NJ, 1998, 82–84.
112. R. T. Sharp and M. Thoma, *Weyl orbit-orbit branching rules for affine Lie algebras: an example* $C_{m+n}^{(1)} \downarrow C_m^{(1)} \oplus C_n^{(1)}$. Proceedings of the 5th Wigner Symposium (Vienna, 1997), World Sci. Publishing, River Edge, NJ, 1998, 85–87.
113. H. de Guise, J. Patera and R. T. Sharp, $SL(3,\mathbb{C})$ *generator matrix elements in a Pauli subgroup basis*, J. Math. Phys. **41** (2000), 4860–4880.

Participants

Nibaldo Alvarez
  Université de Montréal
  Montréal, Canada

Stephen Anco
  Brock University
  St. Catharines, Canada

Angel Ballesteros
  Universidad de Burgos
  Burgos, Spain

Luc Bégin
  Université de Moncton
  Moncton, New Brunswick

Marco Bertola
  Concordia University
  Montréal, Canada

Stéphane Bourque
  Université Laval
  Québec, Canada

Cliff Burgess
  McGill University
  Montréal, Canada

Jiří Bystricky
  CEA/Saclay
  Gif-sur-Yvette, France

Alain Caillé
  Université de Montréal
  Montréal, Canada

Mikhail B. Cheftel
  Feza Gürsey Institute
  Istanbul, Turkey

Andreas P. Contogouris
  McGill University
  Montréal, Canada

Michel Couture
  Canadian Nuclear Safety Commission/
  Commission canadienne de sûreté nucléaire
  Ottawa, Canada

Chris J. Cummins
  Concordia University
  Montréal, Canada

Subal Dasgupta
  McGill University
  Montréal, Canada

Hubert de Guise
  Lakehead University
  Thunder Bay, Canada

Marc de Montigny
  University of Alberta
  Edmonton, Canada

Tamara Diaz Chang
  Université de Montréal
  Montréal, Canada

Hassan Firouzhaji
  McGill University
  Montréal, Canada

Jean-François Fortin
  Université Laval
  Québec, Canada

Terry Gannon
  University of Alberta
  Edmonton, Canada

Jean-Pierre Gazeau
  Université Paris 7 — Denis-Diderot
  Paris, France

David Gomez-Ullate
  Université de Montréal
  Montréal, Canada

Cezary Gonera
  Uniwersytet Łódzki
  Łódź, Poland

Marcus Grisaru
  McGill University
  Montréal, Canada

Michel Grundland
  Université du Québec à Trois-Rivières
  Trois-Rivières, Canada

John Harnad
  Concordia University
  Montréal, Canada

Miloslav Havlíček
  Czech Technical University in Prague
  Prague, Czech Republic

Francisco Jose Herranz
  Universidad de Burgos
  Burgos, Spain

Véronique Hussin
  Université de Montréal
  Montréal, Canada

Patrick Jacob
  Université Laval
  Québec, Canada

PARTICIPANTS

Niky Kamran
McGill University
Montréal, Canada

Ronald King
University of Southampton
Southampton, UK

Siaka Kone
Université de Cocody
Abidjan, Côte d'Ivoire

Peter Kramer
Eberhard-Karls-Universität
Tübingen, Germany

C. S. Lam
McGill University
Montréal, Canada

Paul H. C. Lee
National Central University
Taiwan, Republic of China

Frank Lemire
University of Windsor
Windsor, Canada

Jean LeTourneux
Université de Montréal
Montréal, Canada

Yves Vincent Marcoz
Université de Montréal
Montréal, Canada

Pierre Mathieu
Université Laval
Québec, Canada

John McKay
Concordia University
Montréal, Canada

Ruxandra Moraru
McGill University
Montréal, Canada

Marcos Moshinsky
Universidad Autónoma de México
México, Mexico

Zorka Papadopolos
Eberhard-Karls-Universität
Tübingen, Germany

Manu Paranjape
Université de Montréal
Montréal, Canada

Popat Patel
McGill University
Montréal, Canada

Jiří Patera
Université de Montréal
Montréal, Canada

Alexei V. Penskoi
Université de Montréal
Montréal, Canada

Manuel F. Ranada
Universidad de Zaragoza
Zaragoza, Spain

Miguel A. Rodriguez
Universidad Complutense
Madrid, Spain

Charles Roth
McGill University
Montréal, Canada

David J. Rowe
University of Toronto
Toronto, Canada

Norbert Schlomiuk
Université de Montréal
Montréal, Canada

Elisa Shahbazian
Lockheed Martin Canada
Montréal, Canada

Douglas Sharp
Palo Alto, USA

Joan Sharp
Simon Fraser University
Burnaby, Canada

Roger Sharp
Baltimore, USA

Susan Sharp
Baltimore, USA

Ted Sharp
Toronto, Canada

Gordon Shaw
M.I.N.D. Institute
Costa Mesa, USA

Miloslav Svec
Dawson College
Montréal, Canada

Kanehisa Takasaki
Kyoto University
Kyoto, Japan

Piergiulio Tempesta
Université de Montréal
Montréal, Canada

Martin Thoma
Friedrich-Alexander-Universität
Nürnberg, Germany

Zora Thomova
SUNY Institute of Technology
Utica, USA

Sébastien Tremblay
Université de Montréal
Montréal, Canada

Luc Vinet
McGill University
Montréal, Canada

Philip Wallace
Victoria, Canada

Mark Walton
University of Lethbridge
Lethbridge, Canada

Pavel Winternitz
Université de Montréal
Montréal, Canada

Symmetries and Currents of Massless Neutrino Fields, Electromagnetic and Graviton Fields

Stephen C. Anco and *Juha Pohjanpelto*

ABSTRACT. A recent complete, explicit classification of all locally constructed symmetries and currents for free spinorial massless spin s fields on Minkowski space is summarized and extended to give a classification of all covariant symmetry operators and conserved tensors. The results, for physically interesting cases, are also presented in tensorial form for electromagnetic and graviton fields ($s = 1, 2$) and in Dirac 4-spinor form for neutrino fields ($s = \frac{1}{2}$).

1. Introduction

One of the earliest applications of group theory in the foundations of both classical and quantum field theory was to the study of the fundamental linear spinorial equations for free relativistic fields on Minkowski space [14, 16]. These field equations arise in a natural group theoretical manner by providing unitary irreducible representations modulo a sign of the Poincaré group—the isometry group of Minkowski space—realized on spinorial fields on space-time. As shown by Wigner and Bargmann [5], the representations are characterized in terms of mass $m \geq 0$ and spin $s = 0, \frac{1}{2}, 1, \frac{3}{2}, 2, \ldots$ of the field, which are given by eigenvalues of the Casimir operators of the Lie algebra of the Poincaré group. In particular, the square of the translation operator yields m^2 while the square of the (Pauli–Lubanskí) spin operator yields $s(s+1)m^2$ for massive fields. The spin for massless fields has a special characterization given by the magnitude of the helicity $\pm s = 0, \pm\frac{1}{2}, \pm 1, \pm\frac{3}{2}, \pm 2, \ldots$ which arises from an equality between the translation operator and spin operator holding for irreducible representations when $m = 0$.

The most important cases of physical interest are the spinorial fields with zero mass $m = 0$ and nonzero spin $s = \frac{1}{2}, 1, 2$, respectively describing neutrino fields, electromagnetic fields and graviton fields (i.e., linearized gravitation). Gravitino fields, described by $m = 0$ and $s = \frac{3}{2}$, are of theoretical interest in supersymmetric field theory. Due to their linear nature, all these fields have a rich structure of conserved currents and symmetries, which have interesting physical applications: currents provide conserved quantities associated with the propagation of the fields

---

2000 *Mathematics Subject Classification.* Primary: 81R20, 70S10; Secondary: 81R25.
This is the final form of the paper.

©2004 American Mathematical Society

on space-time, while symmetries lead to invariant solutions and are connected with separation of variables for the field equations.

In recent work [3,4] by means of spinorial methods, we have obtained a complete, explicit classification of all locally constructed spinorial symmetries and currents for massless fields of every spin $s \geq \frac{1}{2}$, extending some earlier results [2,12] obtained for the electromagnetic case $s = 1$. As this classification uses the spinorial formulation of the field equations, the symmetries and currents are derived in a gauge invariant and coordinate invariant spinor form. For physical applications, however, a tensorial form for integer spin fields and a Dirac 4-spinor form for half-integer spin fields is the most appropriate formulation.

In this paper we present the symmetries and currents in tensorial form for electromagnetic and graviton fields and in Dirac 4-spinor form for neutrino and gravitino fields. In addition, we extend our previous results to give a complete classification of all Poincaré covariant conserved tensors and symmetry operators for massless spinorial fields of every spin $s \geq \frac{1}{2}$. Throughout we use the index notation and conventions of [14].

2. Spin s symmetries and currents

On Minkowski space $M = (\mathbb{R}^4, \eta_{ab})$, recall that the Pauli spin matrices (and identity matrix) $\sigma_a{}^{AA'}$ provide an isomorphism between the tangent space of M and the space of real spinorial vectors over spinor space $(\mathbb{C}^2, \epsilon_{AB})$, where η_{ab} is the Minkowski metric and ϵ_{AB} is the spin metric, related by $\eta_{ab} = \sigma_a{}^{AA'} \sigma_b{}^{BB'} \epsilon_{AB} \epsilon_{A'B'}$. Hereafter we will omit $\sigma_a{}^{AA'}$ wherever convenient and simply write $a = AA'$ to identify vector and tensor fields with vectorial and tensorial spinor fields on M.

Massless spin s fields are described by symmetric spinor fields $\phi_{A_1 \cdots A_{2s}}(x)$ on M satisfying the field equation

$$(2.1) \quad \Delta_{A'A_2 \cdots A_{2s}} \equiv \partial_{A'}^{A_1} \phi_{A_1 \cdots A_{2s}}(x) = 0,$$

where $\partial_{A'}^A$ denotes the spinorial coordinate derivative operator associated with standard Minkowski coordinates $x^{CC'}$. The vector space of C_0^∞ solutions of (2.1) defines an irreducible representation of the double cover $\mathrm{ISL}(2, \mathbb{C})$ of the Poincaré group of M, with the group action generated by Lie derivatives with respect to Killing vectors ξ^c on M,

$$(2.2) \quad \mathcal{L}_\xi \phi_{A_1 \cdots A_{2s}}(x) = \xi^{CC'} \partial_{CC'} \phi_{A_1 \cdots A_{2s}}(x) + s \partial_{C'(A_1} \xi^{C'C} \phi_{A_2 \cdots A_{2s})C}(x),$$

where $\mathcal{L}_\xi \eta_{ab} = 0$. The Poincaré Lie algebra generated by $\mathcal{L}_\xi$ comprises translations $\xi^a \mathcal{G}_a$ and rotations/boosts $\partial^{[a} \xi^{b]} \mathcal{G}_{ab}$ defined by $(1/i)\mathcal{L}_\xi$ via the corresponding Killing vectors ($\xi^a = \mathrm{const}$, $\partial^{[a} \xi^{b]} = \mathrm{const}$, respectively). The Pauli–Lubanski spin operator is defined by $\mathcal{S}_a = \epsilon_a{}^{bcd} \mathcal{G}_b \mathcal{G}_{cd}$. On C_0^∞ solutions of the field equation, these are self-adjoint operators that satisfy $\mathcal{G}_a \mathcal{G}^a = \mathcal{S}_a \mathcal{S}^a = \mathcal{S}_a \mathcal{G}^a = 0$ and $\mathcal{S}_a = -s \mathcal{G}_a$, from which the helicity of $\phi_{A_1 \cdots A_{2s}}(x)$ is defined to be $-s$. A similar discussion applies to the complex conjugate massless spin s field $\bar{\phi}_{A'_1 \cdots A'_{2s}}(x)$ satisfying $\bar{\Delta}_{AA'_2 \cdots A'_{2s}} = \partial_A^{A'_1} \bar{\phi}_{A'_1 \cdots A'_{2s}}(x) = 0$, with helicity $+s$ as defined by the equality $\mathcal{S}_a = s \mathcal{G}_a$ holding on C_0^∞ solutions of this field equation.

The field equation (2.1) possesses an important local solvability property by which, for each $q \geq 1$, the values of $\phi_{A_1 \cdots A_{2s}}(x_o)$ and all of its symmetrized derivatives $\partial_{(C_1'}^{C_1'} \cdots \partial_{C_p'}^{C_p'} \phi_{A_1 \cdots A_{2s})}(x_o)$ for $p \leq q$ at any given point $x_o^{AA'}$ in M are freely specifiable data on solutions, as explained by Penrose [**14**] using the notion of "exact set of fields". Thus, it is convenient to work with the associated coordinate space,

$$(2.3) \qquad J_\Delta^q(\phi) \equiv \left\{ \left(x^{CC'}, \phi_{A_1\cdots A_{2s}}, \phi_{(A_1\cdots A_{2s},C_1')}^{C_1'}, \ldots, \phi_{(A_1\cdots A_{2s},C_1'\cdots C_q')}^{C_1'\cdots C_q'} \right) \right\},$$

$0 \leq q \leq \infty$, known as the solution jet space of the field equation (2.1), where a point in $J_\Delta^q(\phi)$ corresponds to the values of the field and all symmetrized derivatives of the field up to order q at a point in M. This is a subspace of the full jet space $J^q(\phi) \supset J_\Delta^q(\phi)$ whose coordinates are defined by $x^{CC'}$, $\phi_{A_1\cdots A_{2s}}$, $\phi_{A_1\cdots A_{2s},C_1'\cdots C_p'}^{C_1'\cdots C_p'}$, $1 \leq p \leq q$. In the sequel we will employ a multi-index notation and write

$$(2.4) \qquad \phi_{\mathbf{A}_{2s},\mathbf{C}_p'}^{\mathbf{C}_p'} = \phi_{A_1\cdots A_{2s},C_1'\cdots C_p'}^{C_1'\cdots C_p'}, \qquad \phi_{\mathbf{C}_{2s+p}}^{\mathbf{C}_p'} = \phi_{(\mathbf{C}_{2s},\mathbf{C}_{p,2s}')}, \quad p \geq 0,$$

with multi-indices defined to be completely symmetric in their constituent indices: $\mathbf{B}_p = (B_1 \cdots B_p)$, $\mathbf{B}_{p,q} = (B_{1+q} \cdots B_{p+q})$. We will use the convention that a multi-index with $p = 0$ stands for an empty set containing no index.

We let $D_{CC'}$ denote the total derivative operator with respect to $x^{CC'}$ on $J^\infty(\phi)$ and write $\mathcal{D}_{CC'}$ for its restriction to $J_\Delta^\infty(\phi)$ given by

$$(2.5) \qquad \mathcal{D}_C^{C'} = \partial_C^{C'} + \sum_{q \geq 0} \left(\phi_{\mathbf{A}_{2s+q}C}^{\mathbf{A}_q'C'} \partial_\phi{}_{\mathbf{A}_q'}^{\mathbf{A}_{2s+q}} + \text{c.c.} \right),$$

where $\partial_\phi{}_{\mathbf{A}_p'}^{\mathbf{A}_{2s+p}}$ is the partial derivative operator with respect to $\phi_{\mathbf{A}_{2s+p}}^{\mathbf{A}_p'}$ and c.c. denotes the complex conjugate of the preceding term. We write higher order symmetrized derivatives as $\partial^{(p)}{}_{\mathbf{C}_p'}^{\mathbf{C}_p'} = \partial_{(C_1'}^{C_1'} \cdots \partial_{C_p')}^{C_p'}$ and $\mathcal{D}^{(p)}{}_{\mathbf{C}_p'}^{\mathbf{C}_p'} = \mathcal{D}_{(C_1'}^{C_1'} \cdots \mathcal{D}_{C_p')}^{C_p'}$. Note that we can lift the Lie derivative (2.2) for any Killing vector $\xi^{CC'}$ to define an operator $\mathfrak{L}_\xi$ on $J_\Delta^\infty(\phi)$ given by

$$\mathfrak{L}_\xi \phi_{\mathbf{A}_{2s+p}}^{\mathbf{A}_p'} = -\xi_C^{C'} \phi_{\mathbf{A}_{2s+p}C}^{\mathbf{A}_p'C'} + \left(s + \frac{p}{2} \right) \xi_{(A_{2s+p}}{}^C \phi_{\mathbf{A}_{2s+p-1})C}^{\mathbf{A}_p'} - \frac{p}{2} \bar{\xi}_{C'}^{(A_p'} \phi_{\mathbf{A}_{2s+p}}^{\mathbf{A}_{p-1}')C'},$$

where $\xi_{BC} = \partial_{C'B} \xi_C^{C'}$.

2.1. Conformal Killing vectors and Killing–Yano tensors. The classification of local symmetries and local currents of massless spin s fields given in [**3, 4**] relies on the properties of Killing spinors, which are spinorial generalizations of Killing vectors related to twistors [**14**]. For the results presented here, we need Killing spinors of two types. A real spinor function $\xi_A^{A'}(x)$ satisfying $\partial_{(B}^{(B'} \xi_{A)}^{A')} = 0$ represents a conformal Killing vector ξ^a [**14, 16**], which generates a conformal isometry of Minkowski space. A symmetric spinor function $Y^{A'B'}(x)$ satisfying $\partial_C^{(C'} Y^{A'B')} = 0$ represents a conformal Killing–Yano tensor $Y^{ab} = \epsilon^{AB} Y^{A'B'}$ [**7, 14**] that is self-dual, $*Y^{ab} = iY^{ab}$, where $*$ denotes the Hodge dual operator. These Killing spinors have a direct generalization $\zeta_{\mathbf{A}_k}^{\mathbf{A}_k'}(x)$ and $\Upsilon^{\mathbf{A}_{2k}'}(x)$ to ones of any rank $k \geq 1$. Their explicit form is given by polynomials in $x^{CC'}$ of degree up to $2k$,

$$(2.6) \qquad \zeta_{\mathbf{A}_k}^{\mathbf{A}_k'} = \sum_{q \geq p = 0}^{k} \alpha_{\mathbf{B}_q'(\mathbf{A}_{k-p,p}}^{\mathbf{B}_p'(\mathbf{A}_{k-q,q}'} x^{(q)\mathbf{A}_q')\mathbf{B}_q} x^{(p)}{}_{\mathbf{A}_p)\mathbf{B}_p'} + \text{c.c.},$$

$$\Upsilon^{\mathbf{A}'_{2k}} = \sum_{p=0}^{2k} \beta_{\mathbf{B}_p}^{(\mathbf{A}'_{2k-p},p} x^{(p)\,\mathbf{A}'_p)\mathbf{B}_p}, \tag{2.7}$$

where $x^{(p)\,\mathbf{C}'_p}_{\mathbf{C}_p} = x^{C'_1}_{C_1} \cdots x^{C'_p}_{C_p}$, with the coefficients $\alpha^{\mathbf{B}'_p \mathbf{A}'_{k-q}}_{\mathbf{B}_q \mathbf{A}_{k-p}}$ and $\beta^{\mathbf{A}'_{2k-p}}_{\mathbf{B}_p}$ being arbitrary constant spinors. There are respectively

$$(k+1)^2(k+2)^2(2k+3)/12, \quad (2k+1)(2k+2)(2k+3)/3 \tag{2.8}$$

linearly independent Killing spinors (2.6) and (2.7) over the reals.

An important property of these Killing spinors is that they possess a factorization into sums of symmetrized products of conformal Killing vectors $\xi^{A'}_A$ and conformal Killing–Yano tensors $Y^{A'B'}$:

$$\zeta^{\mathbf{A}'_k}_{\mathbf{A}_k} = \sum_{\xi} \xi^{A'_1}_{(A_1} \cdots \xi^{A'_k}_{A_k)}, \quad \Upsilon^{\mathbf{A}'_{2k}} = \sum_{Y} Y^{(A'_1 A'_2} \cdots Y^{A'_{2k-1} A'_{2k})}. \tag{2.9}$$

This is a consequence of the more general factorization of Killing spinors into sums of symmetrized products of twistors and dual-twistors, holding in Minkowski space.

2.2. Symmetries. From a group theoretical perspective, a local symmetry of the massless spin s field equation (2.1) is a one-parameter (ε) local transformation group [6] on the coordinate space $J^\infty(\phi)$ that preserves the contact ideal (i.e., derivative relations among the coordinates [6,13]) and maps solutions $\phi_{\mathbf{A}_{2s}}(x)$ into solutions. It is well known that the infinitesimal action of any such transformation on $\phi_{\mathbf{A}_{2s}}(x)$ is the same as one in which there is no motion on $x^{CC'}$,

$$x^{CC'} \to x^{CC'}, \quad \phi_{\mathbf{A}_{2s}} \to \phi_{\mathbf{A}_{2s}} + \varepsilon Q_{\mathbf{A}_{2s}}(x,\phi^{[r]}) + O(\varepsilon^2), \tag{2.10}$$

with the prolongation $\phi^{\mathbf{A}'_p}_{\mathbf{A}_{2s+p}} \to \phi^{\mathbf{A}'_p}_{\mathbf{A}_{2s+p}} + \varepsilon \mathcal{D}^{(p)\,\mathbf{A}'_p}_{(\mathbf{A}_p} Q_{\mathbf{A}_{2s},p)}(x,\phi^{[r]}) + O(\varepsilon^2)$ for $p \geq 1$. The spinor function $Q_{\mathbf{A}_{2s}}(x,\phi^{[r]})$ is called the symmetry characteristic of the local transformation group (2.10) and satisfies the determining equation

$$\mathcal{D}^{A_{2s}}_{A'} Q_{\mathbf{A}_{2s}}(x,\phi^{[r]}) = 0. \tag{2.11}$$

Here $\phi^{[r]}$ denotes the set of coordinates $\phi^{\mathbf{A}'_p}_{\mathbf{A}_{2s+p}}, \bar{\phi}^{\mathbf{A}_p}_{\mathbf{A}'_{2s+p}}$, with $0 \leq p \leq r$. The infinitesimal generator of the resulting local transformation group (defined by formal exponentiation [6] of the generator) is given by

$$\mathbf{X}_Q = Q_{\mathbf{A}_{2s}} \partial_\phi{}^{\mathbf{A}_{2s}}, \tag{2.12}$$

which we will call a local spin s symmetry of order r. More geometrically, a local symmetry can be understood [13] to be a tangent vector field on the solution jet space $J^\infty_\Delta(\phi) \subset J^\infty(\phi)$ that preserves the contact ideal associated with the coordinates.

If $Q_{\mathbf{A}_{2s}}$ depends only on $x^{CC'}$, so that $Q_{\mathbf{A}_{2s}}(x)$ is a solution of (2.1), we call $\mathbf{X}_Q$ an elementary spin s symmetry. A spin s symmetry $\mathbf{X}_Q$ is a classical point symmetry [6,13] if it has the form $Q_{\mathbf{A}_{2s}}(x,\phi^{[1]}) = \zeta^{CC'} \phi_{\mathbf{A}_{2s}CC'} + \varrho_{\mathbf{A}_{2s}}$ for some spinor functions $\zeta^{CC'}(x,\phi^{[0]}), \varrho_{\mathbf{A}_{2s}}(x,\phi^{[0]})$. Allowing for complexification, the point symmetries admitted by massless spin s fields consist of the scaling and duality rotation symmetries

$$Q^{\mathrm{S}}_{\mathbf{A}_{2s}}(\phi^{[0]}) = \phi_{\mathbf{A}_{2s}}, \quad Q^{\mathrm{S}}_{\mathbf{A}_{2s}}(i\phi^{[0]}) = i\phi_{\mathbf{A}_{2s}}, \tag{2.13}$$

as well as the space-time symmetries

$$Q^{\text{K}}_{\mathbf{A}_{2s}}(\xi, \phi^{[1]}) = \hat{\mathcal{L}}_\xi \phi_{\mathbf{A}_{2s}}, \quad Q^{\text{K}}_{\mathbf{A}_{2s}}(\xi, i\phi^{[1]}) = i\hat{\mathcal{L}}_\xi \phi_{\mathbf{A}_{2s}}, \tag{2.14}$$

arising from the action of the group of conformal isometries of Minkowski space generated by conformal Killing vectors ξ^c, where the operator

$$\hat{\mathcal{L}}_\xi = \mathcal{L}_\xi + \frac{1-s}{4} \operatorname{div} \xi, \quad \operatorname{div} \xi \equiv \partial_a \xi^a \tag{2.15}$$

is, geometrically, a conformally-weighted Lie derivative [3, 14].

Massless spin s fields, remarkably, also admit non-classical local symmetries involving conformal Killing–Yano tensors, given by

$$Q^{\text{C}}_{\mathbf{A}_{2s}}(Y, \phi^{[2s]}) = \sum_{p=0}^{2s} \frac{4s-p+1}{4s+1} \binom{2s}{p} \partial^{(p)}{}_{\mathbf{B}'_p(\mathbf{A}_p} Y^{\mathbf{B}'_{4s}} \bar{\phi}_{|\mathbf{B}'_{4s-p,p}|\mathbf{A}_{2s-p,p})}, \tag{2.16}$$

where $Y^{\mathbf{B}'_{4s}}$ is any self-dual conformal Killing–Yano tensor of rank $2s$. Symmetries of this type were first found in tensorial form in the electromagnetic case $s=1$ by Fushchich and Nikitin [9, 10, 15]. The generalization (2.16) for all $s \geq \frac{1}{2}$ was derived in [4]. We call (2.16) chiral symmetries of order $2s$ since $Q^{\text{C}}_{\mathbf{A}_{2s}}$ depends on the positive helicity spin s field $\bar{\phi}_{\mathbf{A}'_{2s}}$, in contrast to the dependence of the space-time symmetries $Q^{\text{K}}_{\mathbf{A}_{2s}}$ on the opposite helicity spin s field $\phi_{\mathbf{A}_{2s}}$.

We now state the main classification result for local spin s symmetries. First, note that given any local spin s symmetry $\mathbf{X}_Q$ of order $r \geq 0$, we can obtain higher order symmetries by replacing $\phi^{[r]}$ in $Q_{\mathbf{A}_{2s}}(x, \phi^{[r]})$ with repeated Lie derivatives $(\hat{\mathcal{L}}_\xi)^n \phi^{[r]}$ for any conformal Killing vector ξ^c, since $\hat{\mathcal{L}}_\xi \phi_{\mathbf{A}_{2s}}(x)$ is a solution of the massless spin s field equation whenever $\phi_{\mathbf{A}_{2s}}(x)$ is one. We denote by $Q_{\mathbf{A}_{2s}}(\xi^{(n)}; x, \phi^{[r+n]})$ the resulting symmetry characteristic for $n \geq 0$.

THEOREM 2.1. *Every local symmetry* (2.12) *of the massless spin s field equation* (2.1) *is a sum of an elementary symmetry and a linear symmetry that is given by, to within a scaling and duality rotation, a sum of space-time symmetries* (2.14), *chiral symmetries* (2.16), *and their higher order extensions*

$$\sum_{\substack{n \geq 0 \\ \xi, Y}} Q^{\text{K}}_{\mathbf{A}_{2s}}(\xi^{(n)}; \xi, \phi^{[1+n]}) + iQ^{\text{K}}_{\mathbf{A}_{2s}}(\xi^{(n)}; \xi, \phi^{[1+n]}) + Q^{\text{C}}_{\mathbf{A}_{2s}}(\xi^{(n)}; Y, \phi^{[2s+n]})$$

involving real conformal Killing vectors ξ and self-dual conformal Killing–Yano tensors Y of rank $2s$.

2.3. Currents. A local conserved current of the massless spin s field equation (2.1) is a real vector function on the coordinate space $J^\infty(\phi)$ that it is divergence free on all solutions $\phi_{\mathbf{A}_{2s}}(x)$ of (2.1). Without loss of generality, it is convenient to restrict local currents to be divergence-free vector functions $\Psi_a(x, \phi^{[r]})$ on the solution jet space $J^\infty_\Delta(\phi)$,

$$\mathcal{D}^a \Psi_a(x, \phi^{[r]}) = 0. \tag{2.17}$$

Consider a space-like hyperplane Σ, with a future time-like normal t^a. For any current $\Psi_a(x, \phi^{[r]})$, the associated conserved quantity for C^∞_0 solutions $\phi_{\mathbf{A}_{2s}}(x)$ is then $\int_\Sigma t^a \Psi_a(x, \phi^{[r]}(x)) \, d^3x$ where $t^a \Psi_a(x, \phi^{[r]}(x))$ is the conserved density expression. This quantity is finite and time-independent. Thus, a local current $\Psi_a(x, \phi^{[r]})$ is considered trivial if it agrees with a curl $\Psi_{AA'} = D_A^{B'}\Theta_{A'B'} + \text{c.c.}$ on $J_\Delta^\infty(\phi)$, for some symmetric spinor function $\Theta_{A'B'}(x,\phi^{[r]})$, since the resulting conserved quantity vanishes by Stokes' theorem. Consequently, two local currents are considered equivalent if their difference is a trivial current.

The massless spin s field equation (2.1) does not possess a local Lagrangian formulation in terms of $\phi_{\mathbf{A}_{2s}}$ and its derivatives (and their complex conjugates). As a result, local spin s currents do not arise from local spin s symmetries via Noether's theorem but instead are related to adjoint symmetries of the field equation (2.1) as follows. A spin s adjoint symmetry of order r is a spinor function $P_A^{\mathbf{A}'_{2s-1}}(x,\phi^{[r]})$ that satisfies the adjoint of the symmetry determining equation (2.11)

$$(2.18) \qquad \mathcal{D}^{A(A'_{2s}} P_A^{\mathbf{A}'_{2s-1})}(x,\phi^{[r]}) = 0.$$

Every spin s adjoint symmetry generates a local conserved current through a homotopy integral formula

$$(2.19) \qquad \Psi_{AA'}(P) = \int_0^1 d\lambda\, \bar\phi_{A'\mathbf{A}'_{2s-1}} P_A^{\mathbf{A}'_{2s-1}}(x,\lambda\phi^{[r]}) + \text{c.c.}$$

which is derived from the adjoint relation between equations (2.18) and (2.11). Conversely, as shown in [**3**], every local spin s current (2.17) is equivalent to one given by the integral formula (2.19) for some spin s adjoint symmetry. Note when $P_A^{\mathbf{A}'_{2s-1}}$ depends only on $x^{CC'}$, so that $P_A^{\mathbf{A}'_{2s-1}}(x)$ is a solution of the adjoint spin s field equation, we obtain the elementary, linear currents of the massless spin s field equation (2.1).

Quadratic currents depending on Killing vectors have long been known in the electromagnetic case $s=1$, corresponding to conservation of energy, momentum, angular and boost momentum given via the electromagnetic stress-energy tensor. Moreover, so-called zilch quantities for electromagnetic fields are known to arise in a similar fashion from Lipkin's zilch tensor [**15**]. Analogous local currents and tensors are also known in the graviton case $s=2$ [**8**]. Generalizations of these currents in spinorial form for all $s \geq \frac{1}{2}$ were first obtained in [**3**], given by

$$(2.20) \qquad \Psi_{AA'}^{\text{K}}(\zeta,\phi^{[0]}) = \zeta^{\mathbf{A}_{2s-1}\mathbf{A}'_{2s-1}} \bar\phi_{A'\mathbf{A}'_{2s-1}} \phi_{A\mathbf{A}_{2s-1}},$$

$$(2.21) \qquad \Psi_{AA'}^{\text{Z}}(\xi,\zeta,\phi^{[1]}) = i\zeta^{\mathbf{A}_{2s-1}\mathbf{A}'_{2s-1}} \bar\phi_{A'\mathbf{A}'_{2s-1}} \hat{\mathcal{L}}_\xi \phi_{A\mathbf{A}_{2s-1}} + \text{c.c.}$$

for any real conformal Killing vectors $\xi^{CC'}$ and real conformal Killing tensors $\zeta^{\mathbf{A}_{2s-1}\mathbf{A}'_{2s-1}}$ of rank $2s-1$. We will refer to (2.20) and (2.21) as the space-time currents and zilch currents. These currents possess even parity under duality rotations of the spin s field. Remarkably, the massless spin s field equation also admits odd parity currents, first found in tensorial form in the electromagnetic case $s=1$ by Fushchich and Nikitin [**10**] using non-invariant coordinate methods. In [**3**] the currents were generalized to all $s \geq \frac{1}{2}$ in spinorial form,

$$(2.22) \quad \Psi_{AA'}^{\text{C}}(\xi,Y,\phi^{[1]})$$
$$= \left(Y^{\mathbf{A}'_{2s}\mathbf{B}'_{2s}} \bar\phi_{\mathbf{B}'_{2s}\mathbf{A}'_{2s}A} + \frac{2s+1}{4s+1}\partial_{AA'_{2s}} Y^{\mathbf{A}'_{2s}\mathbf{B}'_{2s}} \bar\phi_{\mathbf{B}'_{2s}}\right) \hat{\mathcal{L}}_\xi \bar\phi_{A'\mathbf{A}'_{2s-1}} + \text{c.c.}$$

for any conformal Killing–Yano tensors $Y^{\mathbf{A}'_{4s}}$ of rank $2s$ and any conformal Killing vectors $\xi^{CC'}$. Since (2.22) is of opposite parity to (2.20) and (2.21), we call (2.22) the chiral currents.

A complete classification of local spin s currents arises from $\Psi_{AA'}(P)$ by a classification of local spin s adjoint symmetries similarly to theorem 2.1. As was the case for local symmetries, given any local spin s current $\Psi_{AA'}(x, \phi^{[r]})$ of order $r \geq 0$, we can replace $\phi^{[r]}$ by repeated Lie derivatives $(\hat{\mathcal{L}}_\xi)^n \phi^{[r]}$ to obtain higher order currents, which we will denote by $\Psi_{AA'}(\xi^{(n)}; x, \phi^{[r+n]})$, $n \geq 0$.

THEOREM 2.2. *Every local current (2.17) of the massless spin s field equation (2.1) is equivalent to a sum of an elementary linear current and a quadratic current given by a sum of space-time currents (2.20), zilch currents (2.21), chiral currents (2.22), and their higher order extensions*

$$\sum_{\substack{n\geq 0 \\ \xi, \zeta, Y}} \Psi_a^K(\xi^{(n)}; \zeta, \phi^{[n]}) + \Psi_a^Z(\xi^{(n)}; \xi, \zeta, \phi^{[1+n]}) + \Psi_a^C(\xi^{(n)}; \xi, Y, \phi^{[1+n]})$$

involving real conformal Killing vectors ξ and Killing tensors ζ of rank $2s-1$, and self-dual conformal Killing–Yano tensors Y of rank $2s$.

2.4. Covariant conserved tensors and symmetry operators.

We now extend the previous classification results to covariant conserved tensors and symmetry operators of the massless spin s field equation (2.1). To begin, recall a spinor function is said to be Poincaré covariant if it transforms equivariantly under the double cover ISL$(2,\mathbb{C})$ of the Poincaré group acting on $\phi^{[r]}$ and hence depends purely on the coordinates $\phi^{[r]}$ and spin metric ϵ_{AB}. On the solution jet space $J_\Delta^\infty(\phi)$, a covariant conserved tensor $T^{A'\mathbf{B}'_q}_{A\mathbf{B}_p}(\phi^{[r]})$ of order r is then a spinor function that is Poincaré covariant and divergence free, $\mathcal{D}^A_{A'} T^{A'\mathbf{B}'_q}_{A\mathbf{B}_p}(\phi^{[r]}) = 0$, and a covariant symmetry operator $X_{\mathbf{A}_{2s}\mathbf{B}_p}^{\mathbf{B}'_q}(\phi^{[r]})\partial_\phi^{\mathbf{A}_{2s}}$ of order r is characterized by a spinor function that is Poincaré covariant and satisfies the symmetry equation $\mathcal{D}^{A_{2s}}_{A'} X_{\mathbf{A}_{2s}\mathbf{B}_p}^{\mathbf{B}'_q}(\phi^{[r]}) = 0$.

By contracting any covariant conserved tensor or symmetry operator with products of an arbitrary constant spinor κ^B and its conjugate $\bar{\kappa}^{B'}$, we obtain a local current or symmetry, respectively. Conversely, if the Killing spinors $\xi^{C'}_C$, $\zeta^{\mathbf{A}'_{2s-1}}_{\mathbf{A}_{2s-1}}$, $Y^{\mathbf{A}'_{4s}}$ in any local current or symmetry are set to equal products of $\kappa^B, \bar{\kappa}^{B'}$ and factored out, then we clearly obtain a covariant conserved tensor or symmetry operator. The classification theorems 2.1 and 2.2 now lead (as shown with the methods of [3, 4]) to the following results.

THEOREM 2.3. *Every covariant spin s symmetry operator is a complex linear combination of space-time and chiral symmetry operators,*

$$X_{\mathbf{A}_{2s}\mathbf{B}_p}^{\mathbf{B}'_p} = \phi_{\mathbf{B}_p\mathbf{A}_{2s}}^{\mathbf{B}'_p}, \quad X_{\mathbf{A}_{2s}\mathbf{B}_p}^{\mathbf{B}'_{4s+p}} = \bar{\phi}_{\mathbf{B}_p\mathbf{A}_{2s}}^{\mathbf{B}'_{4s+p}}, \quad \text{for } p \geq 0,$$

in addition to the elementary operator $X_{\mathbf{A}_{2s}}^{\mathbf{B}_{2s}} = \delta_{\mathbf{A}_{2s}}^{\mathbf{B}_{2s}}$. Every covariant spin s conserved tensor is equivalent to a complex linear combination of the elementary tensor $T^{A'B'}_{A\mathbf{B}_{2s-1}} = \epsilon^{A'B'}\phi_{A\mathbf{B}_{2s-1}}$, and space-time tensors, zilch tensors, and chiral tensors,

$$T^{A'\mathbf{B}'_{2s+2p-1}}_{A\mathbf{B}_{2s+2p-1}} = \bar{\phi}^{A'(\mathbf{B}'_{2s+p-1}}_{(\mathbf{B}_{p,2s+p-1}}\phi^{\mathbf{B}'_{p,2s+p-1})A}_{\mathbf{B}_{2s+p-1})A}, \quad T^{A'\mathbf{B}'_{2s+2p}}_{A\mathbf{B}_{2s+2p}} = i\bar{\phi}^{A'(\mathbf{B}'_{2s+p}}_{(\mathbf{B}_{p+1,2s+p}}\phi^{\mathbf{B}'_{p,2s+p})}_{\mathbf{B}_{2s+p-1})A},$$

$$T^{A'\mathbf{B}'_{4s+2p+1}}_{A\mathbf{B}_{2p+1}} = \bar{\phi}^{(\mathbf{B}'_{4s+p+1}}_{A(\mathbf{B}_p}\bar{\phi}^{\mathbf{B}'_{p,4s+p+1})A'}_{\mathbf{B}_{p+1,p})}, \quad \text{for } p \geq 0,$$

in addition to their complex conjugates.

3. Results for spin $s = \frac{1}{2}, 1, \frac{3}{2}, 2$

Real conformal Killing vectors $\xi^a = \xi^{AA'}$ and self-dual conformal Killing–Yano tensors $Y^{ab} = \epsilon^{AB} Y^{A'B'}$ satisfy the tensorial equations

$$\partial^{(a}\xi^{b)} = \tfrac{1}{4}\eta^{ab}\partial_c\xi^c, \quad \partial^{(a}Y^{b)d} = \tfrac{1}{3}\eta^{ab}\partial_c Y^{cd} + \tfrac{1}{3}\eta^{d(a}\partial_c Y^{b)c} \tag{3.1}$$

whose solutions are quadratic polynomials in the coordinates x^a,

$$\xi^a = \alpha_1{}^a + \alpha_2{}^{ab}x_b + \alpha_3 x^a + \alpha_4{}^c x_c x^a - \tfrac{1}{2}\alpha_4{}^a x^c x_c, \tag{3.2}$$

$$Y^{ab} = \beta_1{}^{ab} + \beta_2{}^{[a}x^{b]+} + \beta_3{}^{c[a}x^{b]+}x_c \tag{3.3}$$

with constant coefficients (respectively, real and complex valued)

$$\alpha_1{}^a, \alpha_2{}^{ab} = \alpha_2{}^{[ab]}, \alpha_3, \alpha_4{}^c, \beta_1{}^{ab} = \beta_1{}^{[ab]+}, \beta_2{}^a, \beta_3{}^{ab} = \beta_3{}^{[ab]-}, \tag{3.4}$$

where we use $+/-$ superscripts to denote self-/antiself-dual projections as defined by $\tfrac{1}{2}(\mathbf{1} \mp \mathrm{i}*)$. There are 15 linearly independent conformal Killing vectors (3.2) and 20 linearly independent self-dual conformal Killing–Yano tensors (3.3) over the reals. Hereafter, we write $\mathcal{L}_\xi^{(w)} = \mathcal{L}_\xi - (w/4)\operatorname{div}\xi$ where $\mathcal{L}_\xi = \hat{\mathcal{L}}_\xi - \tfrac{1}{4}\operatorname{div}\xi$ is the ordinary Lie derivative operator [16], satisfying the conformal Killing equation $\mathcal{L}_\xi \eta_{ab} = \tfrac{1}{2}\eta_{ab}\operatorname{div}\xi$.

3.1. Electromagnetic fields.

In tensorial form a spin $s = 1$ field is represented by the electromagnetic field tensor

$$F_{ab} = \epsilon_{AB}\bar{\phi}_{A'B'} + \text{c.c.}, \tag{3.5}$$

which is real, antisymmetric, and satisfies the Maxwell field equations

$$\partial^a F_{ab}(x) = \partial^a *F_{ab}(x) = 0, \tag{3.6}$$

where $*$ is the Hodge dual, $*F_{ab} = \mathrm{i}\epsilon_{AB}\bar{\phi}_{A'B'} + \text{c.c.}$. It is convenient to decompose F_{ab} into its self-dual and antiself-dual parts

$$F^+{}_{ab} = \tfrac{1}{2}(F_{ab} - \mathrm{i}*F_{ab}) = \epsilon_{AB}\bar{\phi}_{A'B'}, \tag{3.7}$$

$$F^-{}_{ab} = \tfrac{1}{2}(F_{ab} + \mathrm{i}*F_{ab}) = \overline{F^+{}_{ab}} = \epsilon_{A'B'}\phi_{AB}. \tag{3.8}$$

The electromagnetic scaling and duality rotation symmetries are given by $Q^{\mathrm{S}}_{ab} = F_{ab}$, $Q_{ab} = *F_{ab} = *Q^{\mathrm{S}}_{ab}$, while the space-time symmetries depending on real conformal Killing vectors ξ^c have the form

$$Q^{\mathrm{K}}_{ab} = \mathcal{L}_\xi F_{ab}, \quad Q_{ab} = \mathcal{L}_\xi *F_{ab} = *Q^{\mathrm{K}}_{ab}, \tag{3.9}$$

reflecting the invariance [14,16] of (3.6) under conformal scaling of η_{ab}. The chiral symmetries are given by

$$Q^{\mathrm{C}}_{ab} = \sum_Y Y^{+\ de}_{(2)\ c[b}\partial_{a]}\partial^c F^+{}_{de} + \tfrac{8}{5}\partial_{[a|}Y^{+\ de}_{(2)\ c|b]}\partial^c F^+{}_{de} \\ + \tfrac{1}{5}\partial_{[a|}\partial^c Y^{+\ de}_{(2')\ c|b]} F^+{}_{de} + \text{c.c.} \tag{3.10}$$

which depend on self-dual conformal Killing–Yano tensors Y^{ab}, where we have introduced the product tensors

$$Y^{+\ cdef}_{(2)} = Y^{cd}Y^{ef}, \quad Y^{+\ cdef}_{(2')} = Y^{cd}Y^{ef} - 4Y^{c[e}Y^{f]+d} \tag{3.11}$$

associated with terms arising in the factorization (2.9) of rank-two self-dual conformal Killing–Yano tensors in tensorial form.

The space-time currents and zilch currents are given by
$$\Psi_a^{\mathrm{K}} = \xi_b F^+{}_{ac} F^{-bc} + \text{c.c.}, \quad \Psi_a^{\mathrm{Z}} = \sum_\xi i\xi_b F^{-bc} \mathcal{L}_\xi F^+{}_{ac} + \text{c.c.}, \tag{3.12}$$

and the chiral currents have the form
$$\Psi_a^{\mathrm{C}} = \sum_{Y,\xi} (Y_{(2)}^{+\,bcde} \partial_b F^+{}_{de} + \tfrac{1}{5} \partial_b Y_{(2')}^{+\,bcde} F^+{}_{de}) \mathcal{L}_\xi F^+{}_{ae} + \text{c.c.}. \tag{3.13}$$

3.2. Graviton fields. The tensorial form of a spin $s = 2$ field consists of a real trace-free tensor with Riemann symmetries,
$$C_{abcd} = C_{[cd][ab]} = \epsilon_{AB}\epsilon_{CD}\bar\phi_{A'B'C'D'} + \text{c.c.}, \quad C_{adc}{}^d = *C_{adc}{}^d = 0, \tag{3.14}$$
representing the graviton field strength, where $*C_{abcd} = i\epsilon_{AB}\epsilon_{CD}\bar\phi_{A'B'C'D'} + \text{c.c.}$ is the dual tensor. The graviton field equations
$$\partial^a C_{abcd}(x) = \partial^a *C_{abcd}(x) = 0 \tag{3.15}$$
are analogous to Maxwell's equations, but with conformal scaling weight $w = 1$. Decomposition of C_{abcd} into self-dual and antiself-dual parts gives
$$C^+{}_{abcd} = \tfrac{1}{2}(C_{abcd} - i*C_{abcd}) = \epsilon_{AB}\epsilon_{CD}\bar\phi_{A'B'C'D'}, \tag{3.16}$$
$$C^-{}_{abcd} = \tfrac{1}{2}(C_{abcd} + i*C_{abcd}) = \overline{C^+{}_{abcd}} = \epsilon_{A'B'}\epsilon_{C'D'}\phi_{ABCD}. \tag{3.17}$$

The scaling and duality rotation symmetries are given by $Q_{abcd}^{\mathrm{S}} = C_{abcd}$, $Q_{abcd} = *C_{abcd} = *Q_{abcd}^{\mathrm{S}}$, and the space-time symmetries depending on real conformal Killing vectors ξ^c are given by
$$Q_{abcd}^{\mathrm{K}} = \mathcal{L}_\xi^{(1)} C_{abcd}, \quad Q_{abcd} = \mathcal{L}_\xi^{(1)} *C_{abcd} = *Q_{abcd}^{\mathrm{K}}. \tag{3.18}$$
The chiral symmetries have the lengthy form
$$\begin{aligned}Q_{abcd}^{\mathrm{C}} = \sum_Y \Big(& Y_{(4)}^{+\,ghjk}{}_{e[b|f[d}\partial_{c]}\partial_{|a]}\partial^f \partial^e C^+{}_{ghjk} \\
& + \tfrac{32}{9} \partial \mathfrak{S}_{([a|} Y_{(4)}^{+\,ghjk}{}_{e|b]+f[d}\partial_{c])} \partial^f \partial^e C^+{}_{ghjk} \\
& + \tfrac{2}{9} \partial^{(2)} \mathfrak{S}_{([a|}{}^e Y_{(4')}^{+\,ghjk}{}_{e|b]f[d}\partial_{c])} \partial^f C^+{}_{ghjk} \\
& + \tfrac{4}{9} \partial^{(2)} \mathfrak{S}_{([c|[a|} Y_{(4')}^{+\,ghjk}{}_{e|b]+f|d]+)} \partial^e \partial^f C^+{}_{ghjk} \\
& + \tfrac{8}{21} \partial^{(3)} \mathfrak{S}_{([a|[c|}{}^e Y_{(4'')}^{+\,ghjk}{}_{f|d]+e|b])} \partial^f C^+{}_{ghjk} \\
& + \tfrac{1}{21} \partial^{(4)}{}_{[a|[c|}{}^{ef} Y_{(4''')}^{+\,ghjk}{}_{f|d]e|b]} C^+{}_{ghjk} \Big) + \text{c.c.}\end{aligned} \tag{3.19}$$
depending on self-dual conformal Killing–Yano tensors Y^{ab}, where
$$Y_{(4)}^{+\,ghjkedcb} = Y^{gh} Y^{jk} Y^{ed} Y^{cb}, \tag{3.20}$$
$$Y_{(4')}^{+\,ghjkedcb} = Y^{gh} Y^{jk} (Y^{cb} Y^{ed} - 12 Y^{c[e} Y^{d]+b}), \tag{3.21}$$
$$Y_{(4'')}^{+\,ghjkedcb} = ((Y^{cb} Y^{ed} - 4 Y^{c[e} Y^{d]+b}) Y^{gh} - 8 Y^{c[g} Y^{h]+b} Y^{ed}) Y^{jk}, \tag{3.22}$$
$$\begin{aligned}Y_{(4''')}^{+\,ghjkedcb} = & Y^{gh} Y^{ed} Y^{cb} Y^{jk} + \tfrac{32}{3} Y^{e[g} Y^{h]+d} Y^{c[j} Y^{k]+b} \\
& - 8(Y^{e[g} Y^{h]+d} Y^{jk} + Y^{e[j} Y^{k]+d} Y^{gh}) Y^{cb}\end{aligned} \tag{3.23}$$

are product tensors arising from the tensorial form of the factorization (2.9) of rank-four self-dual conformal Killing–Yano tensors. Here

$$\partial^{(2)}{}_{ab} = \partial_a\partial_b - \tfrac{1}{4}\eta_{ab}\partial_c\partial^c, \quad \partial^{(3)}{}_{abc} = \partial_a\partial_b\partial_c - \tfrac{1}{2}\eta_{(ab}\partial_{c)}\partial_d\partial^d, \tag{3.24}$$

$$\partial^{(4)}{}_{abcd} = \partial_a\partial_b\partial_c\partial_d - \tfrac{3}{4}\eta_{(ab}\partial_c\partial_{d)}\partial_e\partial^e + \tfrac{1}{16}\eta_{(ab}\eta_{cd)}(\partial_e\partial^e)^2, \tag{3.25}$$

are the trace-free derivatives, and the index operator $\mathfrak{S}$ is defined by symmetrization over two pairs of skew indices $[ab][cd]$.

The space-time currents and zilch currents are given by

$$\Psi^{K}_a = \sum_\xi \xi^c \xi_e \xi_f C^{+}{}_{abcd} C^{-\,bedf} + \text{c.c.}, \tag{3.26}$$

$$\Psi^{Z}_a = \sum_\xi \mathrm{i}\xi^c \xi_e \xi_f C^{-\,bedf} \mathcal{L}^{(1)}_\xi C^{+}{}_{abcd} + \text{c.c.}, \tag{3.27}$$

and the chiral currents have the form

$$\Psi^{C}_a = \sum_{Y,\xi} \Big(Y^{+\,cdefghbj}_{(4'')} \partial_b C^{+}{}_{cdef} + \tfrac{1}{3}\partial_b Y^{+\,bjghcdef}_{(4''')} C^{+}{}_{cdef} \Big) \mathcal{L}^{(1)}_\xi C^{+}{}_{ajgh} \tag{3.28}$$
$$+ \text{c.c.}.$$

3.3. Neutrino and gravitino fields. We first recall the gamma matrices [14]

$$\gamma_a = \sqrt{2}\begin{pmatrix} 0 & \sigma_{aB'}{}^{C} \\ \sigma_{aB}{}^{C'} & 0 \end{pmatrix}, \quad \gamma_5 = \frac{1}{4!}\epsilon^{abcd}\gamma_a\gamma_b\gamma_c\gamma_d = \begin{pmatrix} \mathrm{i} & 0 \\ 0 & -\mathrm{i} \end{pmatrix}. \tag{3.29}$$

The Dirac 4-spinor form of a spin $s = \tfrac{1}{2}$ neutrino field is represented by a Majorana spinor satisfying the massless Dirac field equation

$$\psi = \begin{pmatrix} \bar\phi_{C'} \\ \phi_C \end{pmatrix}, \quad \gamma^a \partial_a \psi = 0. \tag{3.30}$$

The scaling and duality rotation symmetries are simply $Q^{S} = \psi$, $Q = \gamma_5\psi = \gamma_5 Q^{S}$, while the space-time symmetries are given by

$$Q^{K} = \mathcal{L}^{(-1)}_\xi \psi, \quad Q = \mathcal{L}^{(-1)}_\xi \gamma_5 \psi = \gamma_5 Q^{K} \tag{3.31}$$

which depend on real conformal Killing vectors ξ^a. Note $\mathcal{L}^{(-1)}_\xi = \hat{\mathcal{L}}_\xi$ appears due to the conformal scaling weight $w = -1$ of the Dirac operator $\gamma^a\partial_a$. The chiral symmetries have the simple form

$$Q^{C} = \sum_Y \widetilde{Y}^{ab}\gamma_a\partial_b\psi + \tfrac{1}{3}(\partial_b\widetilde{Y}^{ab} - \partial_b *\widetilde{Y}^{ab}\gamma_5)\gamma_a\psi \tag{3.32}$$

depending on real conformal Killing–Yano tensors $\widetilde{Y}^{ab} = \tfrac{1}{2}(Y^{ab} + \overline{Y}^{ab})$.

The space-time currents reduce here to $\Psi_a = \psi^\dagger \gamma_a \psi$, which physically describes the neutrino particle density current, where † denotes the transpose spinor. The zilch currents are given by

$$\Psi^{Z}_a = (\mathcal{L}^{(-1)}_\xi \psi)^\dagger \gamma_5 \gamma_a \psi, \tag{3.33}$$

while the chiral currents have the form

$$\Psi^{C}_a = \sum_{Y,\xi} (\mathcal{L}^{(-1)}_\xi \psi)^\dagger (\widetilde{Y}^{bc}\gamma_a\gamma_c\partial_b\psi + \tfrac{1}{3}(\partial_b\widetilde{Y}^{bc} - \partial_b *\widetilde{Y}^{bc}\gamma_5)\gamma_a\gamma_c\psi). \tag{3.34}$$

Finally, a spin $s = \frac{3}{2}$ gravitino field is represented by a hybrid antisymmetric tensor/Majorana 4-spinor of the form

$$\psi_{ab} = \begin{pmatrix} \epsilon_{AB}\bar{\phi}_{A'B'C'} \\ \epsilon_{A'B'}\phi_{ABC} \end{pmatrix}, \quad \gamma^a\psi_{ab} = 0, \tag{3.35}$$

with left and right handed parts $\psi^\pm{}_{ab} = \frac{1}{2}(1 \mp i\gamma_5)\psi_{ab} = \frac{1}{2}(1 \mp i*)\psi_{ab}$, related by conjugation, $\psi^\pm{}_{ab} \equiv \overline{\psi^\mp{}_{ab}}$. The gravitino field equation is

$$\gamma^c\partial_c\psi_{ab} = 0, \quad \text{or equivalently,} \quad \partial^a\psi_{ab} = 0, \tag{3.36}$$

the latter being conformally scaling invariant.

The gravitino scaling and duality symmetries as well as the space-time symmetries are analogous to those for neutrino fields, $Q^S_{ab} = \psi_{ab}$, $Q^K_{ab} = \mathcal{L}_\xi\psi_{ab}$, while the space-time and zilch currents are given by

$$\Psi^K_a = \sum_\xi \xi_b\xi^c(\psi^{bd})^\dagger \gamma_c\psi_{ad}, \quad \Psi^Z_a = \sum_\xi \xi_b\xi^c(\psi^{bd})^\dagger \gamma_5\gamma_c\mathcal{L}_\xi\psi_{ad}. \tag{3.37}$$

In contrast, the chiral symmetries and currents have a more complicated form than those in the neutrino case,

$$Q^C_{ab} = \sum_Y \left(Y^{+\ fged}_{(3)}{}_{c[b}\partial_{a]}\partial_d\partial^c\gamma_e\psi^+{}_{fg} + \frac{6}{7}\partial_c Y^{+\ fgedc}_{(3)}{}_{[b}\partial_{a]}\partial_d\gamma_e\psi^+{}_{fg} \right.\tag{3.38}$$
$$+ \frac{12}{7}\partial_{[a}Y^{+\ fgedc}_{(3)}{}_{b]+}\partial_c\partial_d\gamma_e\psi^+{}_{fg} + \frac{1}{7}\partial^{(2)}{}_{c[a}Y^{+\ fgedc}_{(3')}{}_{b]}\partial_d\gamma_e\psi^+{}_{fg}$$
$$\left. + \frac{2}{7}\partial^{(2)}{}_{cd}Y^{+\ fgedc}_{(3')}{}_{[b}\partial_{a]+}\gamma_e\psi^+{}_{fg} + \frac{12}{35}\partial^{(3)}{}_{cd[a}Y^{+\ fgedc}_{(3''')}{}_{b]}\gamma_e\psi^+{}_{fg} \right)$$
$$+ \text{c.c.},$$

$$\Psi^C_a = \sum_{Y,\xi} (Y^{+\ fgdebc}_{(3'')} (\partial_b\psi^+{}_{fg})^\dagger + \frac{4}{7}\partial_b Y^{+\ fgbcde}_{(3')}(\psi^+{}_{fg})^\dagger)\gamma_c\gamma_a\mathcal{L}_\xi\psi^+{}_{de} + \text{c.c.}, \tag{3.39}$$

owing to the presence of the product tensors

$$Y^{+\ fgedcb}_{(3)} = Y^{fg}Y^{ed}Y^{cb}, \tag{3.40}$$

$$Y^{+\ fgedcb}_{(3')} = (Y^{fg}Y^{ed} - 8Y^{f[e}Y^{d]+g})Y^{cb}, \tag{3.41}$$

$$Y^{+\ fgedcb}_{(3'')} = (Y^{fg}Y^{ed} - 4Y^{f[e}Y^{d]+g})Y^{cb} - 4Y^{c[e}Y^{d]+b})Y^{fg}, \tag{3.42}$$

$$Y^{+\ fgedcb}_{(3''')} = Y^{fg}(Y^{ed}Y^{cb} - \frac{4}{3}Y^{c[e}Y^{d]+b}), \tag{3.43}$$

which are associated with the factorization (2.9) of rank-three self-dual conformal Killing–Yano tensors.

4. Concluding remarks

Our results on local currents provide a complete set of conserved quantities on Minkowski space for the propagation of electromagnetic and graviton fields described using tensorial field strengths, as well as massless neutrino and gravitino fields described in Dirac 4-spinor form. In addition, our results on local symmetries hold interest for the study of connections between symmetry operators and separation of variables for these physical field equations. Local symmetries and currents, moreover, are important in the investigation of nonlinear interactions allowed for massless fields [1, 11].

A classification of further symmetries and currents involving the familiar electromagnetic and graviton potentials will be given elsewhere by an application of cohomology results for the solution jet space of the massless spin s field equation.

References

1. S. C. Anco, *Construction of locally-symmetric Lagrangian field theories from variational identities*, Mathematical Aspects of Classical Field Theory (Seattle, WA, 1991), Contemp. Math. vol. 132, Amer. Math. Soc., Providence, RI, 1992, pp. 27-50.
2. S. C. Anco and J. Pohjanpelto, *Classification of local conservation laws of Maxwell's equations*, Acta Appl. Math. **69** (2001), no. 3, 285–327.
3. _____, *Conserved currents of massless fields of spin $s \geq \frac{1}{2}$*, R. Soc. Lond. Proc. Ser. A Math. Phys. Eng. Sci. **459** (2003), no. 2033, 1215–1239.
4. _____, *Generalized symmetries of massless free fields*, 2003 (preprint).
5. V. Bargmann and E. P. Wigner, *Group theoretical discussion of relativistic wave equations*, Proc. Nat. Acad. Sci. U. S. A. **34** (1948). 211–223.
6. G. W. Bluman and S. C. Anco, *Symmetry and integration methods for differential equations*, Appl. Math. Sci., vol. 154, Spinger-Verlag, New York, 2002.
7. W. Dietz and R. Rüdiger, *Space-times admitting Killing–Yano tensors. I*, Proc. Roy. Soc. London Ser. A **375** (1981), no. 1762, 361–378.
8. K. S. Feldman, *The zilch in general relativity*, Nuovo Cimento (10) **37** (1965), 104–109.
9. W. I. Fushchich and A. G. Nikitin, *On the new invariance algebras and superalgebras of relativistic wave equations*, J. Phys. A **20** (1987), no. 3, 537–549.
10. _____, *Symmetries of equations of quantum mechanics*, Allerton Press, New York, 1994.
11. M. Henneaux, *Consistent interactions between gauge fields: the cohomological approach*, Secondary Calculus and Cohomological Physics (Moscow, 1997), Contemp. Math. vol. 219, Amer. Math. Soc., Providence, RI, 1998, pp. 93-109.
12. E. G. Kalnins, R. G. McLenaghan, and G. C. Williams, *Symmetry operators for Maxwell's equations on curved space-time*, Proc. Roy. Soc. London Ser. A **439** (1992), no. 1905, 103–113.
13. P. Olver, *Applications of Lie groups to differential equations*, 2nd ed., Grad. Texts in Math., vol. 107, Springer-Verlag, New York, 1993).
14. R. Penrose and W. Rindler, *Spinors and space-time*, vol. I, Cambridge Monogr. Math. Phys., Cambridge Univ. Press, Cambridge, 1984; vol. II, 1986.
15. J. Pohjanpelto, *Symmetries, conservation laws, and Maxwell's equations*, Advanced Electromagnetism: Foundations, Theory and Applications, (T. W. Barrett and D. M. Grimes, D.M. eds.) World Scientific, Singapore, 1995, pp. 560–589.
16. R. M. Wald, *General relativity*, Univ. of Chicago Press, Chicago, IL, 1984.

DEPARTMENT OF MATHEMATICS, BROCK UNIVERSITY, ST. CATHARINES, ON L2S 3A1, CANADA
E-mail address: sanco@brocku.ca

DEPARTMENT OF MATHEMATICS, OREGON STATE UNIVERSITY, CORVALLIS, OR 97331-4605, USA
E-mail address: juha@math.orst.edu

Naturalness and Quintessence

C. P. Burgess

Dedicated to happy memories of Bob Sharp.

ABSTRACT. Although quintessence models have many attractive cosmological features, they face two major difficulties. First, it has not yet been possible to find one which convincingly realizes the goal of explaining present-day cosmic acceleration generically using only attractor solutions. Second, quintessence has proven difficult to obtain within realistic microscopic theories, largely due to two major obstructions. Both of these difficulties are summarized in this article, together with a recent proposal for circumventing the second of them within a brane-world context. It is shown that this proposal leads to a broader class of dynamics for the quintessence field, in which its couplings slowly run (or: "walk") over cosmological time scales. The walking of the quintessence couplings opens up new possibilities for solving the first problem: that of obtaining acceptable transitions between attractor solutions.

1. Recollections about Bob Sharp

I first arrived as a new assistant professor at McGill in January 1987, and moved into an office two doors down from Bob Sharp's on the 3rd floor of the Rutherford Physics Building. Bob was very much the friendly neighbor, whose door was always open for students and for stupid questions from rookie colleagues. Because his office was in the corner, he would pass mine to go to and from class, and he would often pop his head into my office—typically completely covered in a layer of chalk dust—to share an anecdote or joke on his way.

There were many lessons which I learned from Bob's example about how to successfully perform a professor's job. One of the key ones, in retrospect, was not to sweat the small stuff. No matter how late I thought I was with an exam or a grant application, I found I was never *really* late because the administration was still waiting for Bob's. I soon learned that this was *not* because he was a poor teacher or poorly funded, but rather that he thought that such things would take as much time as you gave them, and this would be time spent away from more

---

2000 *Mathematics Subject Classification.* 83.

This research was supported by the Ambrose Monell Foundation, and by grants from N.S.E.R.C. (Canada) and F.C.A.R. (Québec).

This is the final form of the paper.

important things like physics or family. So all the better if you are efficient enough to perform them quickly just before deadline.

Bob's professional niche in our department was mathematical physics in general, and the exploitation of symmetries in particular. Although my contribution to this volume concerns cosmology I came into this subject from string theory and brane physics, where the exploitation of unusual symmetries like duality have been enormously fruitful. I like to think that he would have appreciated and enjoyed these modern developments.

2. Quintessence cosmology

Modern cosmology is now entering a somewhat paradoxical period. On one hand the Hot Big Bang paradigm is now being tested in a redundant way by increasingly precise measurements of the microwave background together with more and more detailed surveys of the properties of extremely distant objects. The success of the Hot Big Bang predictions shows that our overall picture of post-nucleosynthesis cosmology is basically correct.

On the other hand, the same tests provide several independent lines of evidence which show that the Universe is currently dominated by no less than *two* distinct types of unknown forms of matter: the Universal energy density is roughly 25% "Dark Matter" [14, 20, 26] and 70% "Dark Energy" [9, 21, 24]. Incredibly, at most a few percent of the energy density can consist of ordinary baryonic matter, and so the dark matter and dark energy *cannot* simply consist of mundanely dark versions of ordinary protons and neutrons. This discovery has truly Copernican implications for our picture of the Universe as a whole.

It is absolutely breathtaking that so little is known about these two most abundant forms of matter. What is known is usually expressed in terms of their equations of state, through the ratio of pressure to energy density, $w = p/\rho$. The formation of galaxies and galaxy clusters by gravitational attraction appears to require the dark matter to be "cold," with w close to zero. Similarly, the current acceleration which the universal expansion is undergoing indicates $w \lesssim -0.3$ for the dark energy.

Perhaps the simplest explanation for the dark energy is that it is simply the energy density of the quantum vacuum, since this satisfies $w = -1$ and is generically nonzero in realistic quantum field theories. The problem with this explanation is that the predicted vacuum energy is typically at least 10^{56} times larger than what is observed. The Cosmological Constant Problem [28] is the recognition that at present no way is known to naturally obtain a vacuum energy anywhere near the required size within a realistic microscopic theory.

Quintessence models [12, 13, 25, 32] take a different track to explain the dark energy. In these models the dark energy is attributed to the dynamics of a scalar field, ϕ, which is currently evolving in a cosmologically interesting way. Since $w = (K-V)/(K+V)$, where $K = \frac{1}{2}\dot\phi^2$ and V are the scalar's kinetic and potential energies, the condition that w be negative requires the scalar's evolution at present must be slow in the sense that $K \lesssim V$. (The vacuum energy is a special case of this kind of evolution, where $K = 0$.)

These models do not solve the Cosmological Constant Problem, since they do not provide natural reasons for V to be currently so small, but they can (potentially) explain why an evolving scalar field could naturally have an energy density which is now so similar to that of other forms of matter, like the dark matter. They can do so, firstly because their equations of motion often admit "tracking" solutions, within which the scalar energy density closely follows (or tracks) the dominant energy density of the Universe as it evolves. Furthermore, the late-time evolution of the scalar field is often drawn to these "tracking" solutions for wide choices of initial conditions, because these solutions are also "attractors" for the scalar equations of motion.

Unfortunately, since these tracker solutions typically require the scalar itself not to be the dominant energy density, in order to become the dark energy the scalar must eventually leave the tracker solution. Although one might hope that this could also be naturally achieved—such as being perhaps due to a transient behaviour due to the crossover from radiation to matter domination—so far it has proven difficult to make a completely convincing cosmology along these lines.

The purpose of this article is to describe a new category of quintessence model, which could be called "Walking Quintessence" [2,3], that may offer new ways to accomplish this crossover. They may do so because within these models the couplings of the scalar field slowly run (or walk) as the Universe evolve, and this walking may help facilitate the crossover between tracking solutions. What is remarkable is that these models were developed in an attempt to address a completely different set of (very serious) problems which arise once one tries to obtain quintessence from a realistic model of microscopic physics.

Since they play such an important part in the motivation of the models, the bulk of the article is devoted to these serious microphysical problems. The problems themselves are first summarized in the next section, followed by a description how they are addressed within the attractive brane-world picture which has emerged as a potential low-energy consequence of string theory. The results of a sample cosmology built on this model are then briefly presented, intended as first step towards a more systematic exploration of the cosmologies which are suggested by this class of models.

3. Naturalness issues

From a microscopic perspective any viable quintessence model must have two very remarkable properties, which turn out to be quite difficult to arrange in realistic theories of microscopic physics. As is explained below, they must have:

- **Extremely Light Scalars**, whose mass must be at most $m_Q = 10^{-32}$ eV, and;
- **No New Long-Range Forces**, to the extent that these would ruin the agreement between General Relativity and observations on Earth, within the Solar System and beyond [1, 29, 30].

3.1. Light scalars. The first requirement (very light scalars) is a very generic property of explanations of the dark energy in terms of a rolling scalar field. Its necessity may be seen from the scalar field equation of motion within a cosmological context:

$$(3.1) \qquad \ddot{\phi} + 3H\dot{\phi} + \frac{\partial V}{\partial \phi} = 0, \quad \text{and} \quad H^2 = \frac{\rho}{3M_{\rm p}^2},$$

where $M_{\rm p} = 10^{18}$ GeV is the rationalized Planck mass, and ρ is the total energy density (which is at present dominated by the dark energy).

To quantify how small this requires the scalar mass to be, it is instructive to consider the very broad class of models within which

$$V(\phi) = \mu^4 U(\phi/f). \tag{3.2}$$

Here μ and f are arbitrary mass scales, whose size may be determined if the dimensionless function $U(x)$ and its derivatives are at present $O(1)$. In this case the present value of the scalar potential and its derivatives are $V \sim \mu^4$, $dV/d\phi \sim \mu^4/f$, etc., and the square of the scalar mass is of order $d^2V/d\phi^2 \sim \mu^4/f^2$. This turns out to be of order $(10^{-33}\,\text{eV})^2$ given the values which are required for μ and f.

The value $\mu \sim 10^{-3}\,\text{eV}$ is inferred by recognizing that these estimates also apply to the total scalar field energy, $\rho = K + V \sim V$, since present observations require the scalar field to be rolling with K less than but of the same order of magnitude as V. In this way we see that μ controls the present dark energy density, $\rho \sim \mu^4 \sim (10^{-3}\,\text{eV})^4$, and so also $H \sim \mu^2/M_\text{p}$ and $\dot\phi \sim \sqrt{K} \sim \sqrt{V} \sim \mu^2$.

The value $f = M_\text{p}$ is determined from the scalar field equation, (3.1), together with the above slow-roll conditions, since this implies the $\ddot\phi$ term should be much smaller than the other two. Except for the case of a pure cosmological constant (for which only the last term is important), we therefore have $H\dot\phi \sim \partial V/\partial\phi$ and so $\mu^4/M_\text{p} \sim \mu^4/f$, from which we learn $f \sim M_\text{p}$ and hence $m_Q \sim \mu^2/f \sim 10^{-33}\,\text{eV}$.

Such an extraordinarily small scalar mass is extremely difficult to achieve in a realistic microscopic theory. There are two separate aspects to this difficulty.

HIERARCHY PROBLEM 1. How does such a small nonzero mass arise as a combination of microscopic parameters?

HIERARCHY PROBLEM 2. Given that such a small mass is predicted by the theory of microscopic physics, how does it remain small as one integrates out all the physics between these microscopic scales and the cosmological scales at which it is measured? This is a problem because, for instance, a particle of mass M and coupling $1/f$ shifts the scalar mass by an amount

$$\delta m \sim \frac{M^2}{4\pi f} \tag{3.3}$$

when it is integrated out.

Both of these problems are the direct analogs of two aspects of the famous heirarchy problem as applied to the Higgs field which breaks electroweak gauge symmetry within the Standard Model of particle physics. There the weak scale, $M_\text{w} \sim 100\,\text{GeV}$ is controlled by the scalar Higgs mass, and one asks how this can be so much smaller than, say, the Planck scale, $M_\text{p} \sim 10^{18}\,\text{GeV}$.

The problem for quintessence models, however, is arguably much worse for two reasons. First, the quintessence scalar is many more orders of magnitude lighter than the basic microscopic physics scale M_w than M_w itself is from M_p. Second, the range of scales between M_w and m_Q is well studied by experiment, which presumably makes it harder to hide the other degrees of freedom which are typically invoked to alleviate Hierarchy Problem 2 (such as the superpartners which do the job if supersymmetry is the solution).

Almost none of the extant quintessence models address these issues, with the exception being those based on pseudo-Goldstone bosons [16], which address Hierarchy Problem 2. The models presented in the later sections are unique among those yet proposed in that *both* of these problems are related to the same microscopic quantities which explain why $M_w \ll M_p$. This turns out to have potentially observable implications both for table-top tests of the gravitational force law as well as for high-energy accelerators. These quintessence models are therefore unusual in the richness of their implications for laboratory experiments.

3.2. Long-range forces. The incredibly small quintessence scalar mass also raises a related observational problem, since it implies that the exchange of this scalar must mediate a very long-range force. Furthermore, the strength with which this force couples to particles of energy E is of order $E/f \sim E/M_p$, which makes it comparable to gravity. This is problematic, since many observations now strongly constrain the existence of gravitational-strength forces having a range longer than about 100 microns.

In the models which are described in subsequent sections this problem is evaded because the scalar couplings turn out to evolve over cosmological time scales. In the examples given it happens that these couplings evolve to become extremely small during the present epoch, which is when all of the very constraining observations have been made.

Another way to ensure acceptably small couplings is to arrange the scalar to couple to quantities, such as spin, which ordinary matter in bulk does not carry. This kind of coupling can occur in pseudo-Goldstone-boson models [16].

4. Large extra dimensions

New insights into naturalness problems, like the heirarchy between M_w and M_p, have been made based on the recently much-discussed brane world picture, in which all observed elementary particles are confined to a domain-wall-like surface (or "brane") which sits within a higher-dimensional "bulk" spacetime. The simplest choice puts us on a 4-dimensional surface (or 3-brane) within a bulk space which has anywhere from 5 to 11 dimensions. By contrast, gravitational interactions in this picture are not confined to these branes. This kind of picture is very well motivated within string theory.

The fact that gravity and other interactions do not see the same number of dimensions lies at the root of the surprising realization that the fundamental string scale, M_s, can be much smaller than M_p [5, 17–19, 31], and of the new perspective this has allowed for understanding low-energy naturalness problems. In particular, it can imply the relationship

$$(4.1) \qquad \left(\frac{M_s}{M_p}\right)^2 \sim \frac{\alpha^2}{(M_s r)^n},$$

where α is the (open) string coupling and r is the radius of the n extra dimensions. For supersymmetric systems one generically has $M_w/M_p \sim (M_s/M_p)^2$ and so this remarkable formula allows the observed ratio $M_w/M_p \sim 10^{-16}$ (Hierarchy Problem 1, above) to be understood using parameters themselves no smaller than $\alpha \sim 1/(M_s r) \sim 0.01$ if $M_s \sim 10^{11}$ GeV and there are $n = 6$ extra dimensions [10, 11]. (In this picture it is supersymmetry which accounts for Hierarchy Problem 2.)

4.1. Natural quintessence and the brane world.

The brane-world variant which is of most interest for the present purposes, puts the string scale as low as is consistent with experiment, $M_s \sim M_w$ [6–8]. From (4.1), one sees this is permitted (even if α is not small) so long as $n = 2$ and $r \sim 0.1$ mm (or $1/r \sim 10^{-3}$ eV). In this picture $M_w/M_p \sim \alpha/(M_s r)$, so the heirarchy problem is not solved so much as translated into the problem of understanding the origin of the large heirarchy $M_s r/\alpha \sim 10^{16}$.

A remarkable feature of this scenario is that it provides a framework within which scalars can be naturally as light as 10^{-33} eV, and it is instructive to see how this works for specific examples. The success of the models of [2, 3] relies on choosing the extra dimensions to be torii and the quintessence field to be a component of the extra-dimensional metric—the radion, r.

This construction directly solves Hierarchy Problems 1 as follows. First, because the radion field starts life as a component of the six-dimensional metric its kinetic term has the same origin as does the 4D graviton. Dimensionally reducing the 6D Einstein-Hilbert action to four dimensions gives

$$\text{(4.2)} \qquad \frac{L_{\text{kin}}}{\sqrt{-g}} = -\frac{M_p^2}{2} g^{\mu\nu}\left[R_{\mu\nu} + \frac{4}{r^2}\partial_\mu r \partial_\nu r\right],$$

from which it follows that $f \sim M_p$.

For toroidal compactifications, direct dimensional reduction of the higher-dimensional Einstein–Hilbert action gives no radion potential at all (provided that the cosmological constant is chosen to vanish, as usual). This is an artifact of the classical approximation, however, and a potential for r is generated once quantum effects are included, such as through the Casimir effect which predicts (for large r) a potential of the form $V \sim 1/r^4$ [2, 23]. If this potential can be stabilized to have a minimum for some $r = r_0$ (more about this later), then it predicts $\mu \sim 1/r_0$.

More remarkably, this model also addresses Hierarchy Problem 2, since these predictions are protected from being ruined as physics between the weak scale and the quintessence mass is integrated out. Although the stability of the prediction for f against quantum corrections is fairly trivial, it is the stability of the potential which bears closer examination.

There turn out to be two reasons for this success. The Casimir effect predicts $\mu \sim 1/r$ largely because the potential is generated when modes having energies of order $1/r$ are integrated out. Now consider the effect of integrating out modes with energies lower than $1/r$. For scales $M \lesssim 1/r$ the effective theory is four-dimensional, and the naive corrections to $V(r)$ are correct, since no symmetries preclude generating a radion potential. Integrating out a mode of mass M then contributes to $V(r)$ terms of order $\delta\mu^4 \sim M^4/(4\pi)^2$, which is not dangerous since it predicts $\delta\mu \sim M/\sqrt{4\pi} \lesssim 1/r$.

The key is at energies above the scale $1/r$, where the effective theory is six-dimensional and so is constrained by additional symmetries like 6D general covariance. For these scales the result of integrating out modes with $M \gg 1/r$ may be expanded in powers of the 6D curvature R_{mnpq}. If the extra dimensions were to have spherical geometry then $R_{mnpq} \propto 1/r^2$, and these curvature terms constitute dangerously large contributions to the radion potential. For flat spaces like torii, however, R_{mnpq} is independent of r and these terms are not dangerous. The question for torii becomes whether it is possible to keep the internal dimensions flat despite the existence of large vacuum energies on the various branes on which our observed particles live. It is a special feature of two dimensions that this is so, since Einstein's equations predict locally flat geometries around point sources [15, 27].

We see that in this picture the quintessence scales μ and f are predicted to be respectively given by $1/r$ and M_p because the quintessence field has its microscopic origin as part of the higher-dimensional geometry. The success of the predicted dark energy density, $V \sim 1/r^4$, and quintessence mass, $m_Q \sim 1/(M_\mathrm{p} r^2)$, then follow from the choice $1/r \sim 10^{-3}$ eV which is required in any case to solve the electroweak heirarchy, since $M_\mathrm{w}/M_\mathrm{p} \sim 1/(M_\mathrm{w} r)$.

5. Radius stabilization and "walking" quintessence

In order to more precisely pin down the cosmology the explicit form for the quintessence potential is required, and within the brane world picture being presented this amounts to providing a mechanism for stabilizing the radion at large values. This section relates a concrete proposal for doing so, as made in [2].

Besides providing an interesting cosmology in its own right, there is another reason for exploring this specific proposal in more detail. This other reason is to illustrate why worrying about naturalness issues is useful when exploring phenomenological applications (such as quintessence cosmology). Naturalness issues are useful precisely because the cosmology of quintessence models depends in such a detailed way on the precise form of the low-energy potential. On one hand, the naturalness problem states that only very specific kinds of potentials are likely to arise as the low-energy limit of realistic microphysics. On the other hand, it has proven difficult to build a completely convincing quintessence cosmology purely by guessing different kinds of potentials. It may be true that it is only the very few potentials which can arise from real microphysics that can also provide a realistic description of cosmology as well. If so, it should be invaluable to be able to explore in detail any potential which *does* arise as the low-energy limit of real microphysics.

In the present instance the stabilization mechanism proposed suggests a qualitative new feature of the low-energy theory: it predicts that the effective low-energy couplings and masses depend logarithmically on the quintessence field, and so these all slowly run (i.e., walk) over cosmological time scales. This leads in a natural way to extended quintessence models [22], but with a specific and cosmologically-interesting type of field-dependence. Indeed the proposal of [2] was initially motivated by the earlier discovery of the attractive cosmologies which can result from quintessence models having exponential potentials with logarithmic corrections [4].

5.1. Stabilization via six-dimensional logarithms.
The models now described are based on the observation that $V(r)$ would naturally be minimized at large values for r if it were to depend logarithmically on r:

$$(5.1) \qquad V(r) = V_0 \left(\frac{\ell}{r}\right)^p \left[1 + \epsilon \log\left(\frac{r}{\ell}\right)\right] + \cdots,$$

where a and ϵ are constants and ℓ is a microscopic length scale, such as $\ell \sim 1/M_\mathrm{s}$. This ellipses indicate other terms which fall off with a higher power of ℓ/r relative to those shown. A potential of this form has a minimum at $r_0 \sim \ell \exp(1/\epsilon)$, which is exponentially larger than ℓ if ϵ is moderately smaller than one. (Values $\epsilon \sim 1/50$ are sufficient to generate the desired heirarchy in the models explored in [2].)

The key point is that a potential of precisely this form is expected if the effective six-dimensional theory at scales $E > 1/r$ contains a renormalizable (i.e., marginal) coupling, g. If so, then this coupling runs logarithmically with r and loop corrections to the radion potential involving only this coupling have the logarithmic form of (5.1), with a coefficient $\epsilon \sim g^2/(4\pi)^3$ which is naturally small. In six dimensions most interactions are not renormalizable, but such couplings can exist, such as a cubic coupling amongst six-dimensional scalar fields.

Given this generic mechanism it is clear that logarithms r are not restricted to appear in the low-energy theory only within the radion potential. Rather, they arise generically as radiative corrections to *all* couplings, and it is this logarithmic dependence which is responsible for the walking of these couplings as r evolves over cosmological time scales.

5.2. A sample quintessence cosmology. Logarithmically evolving couplings are very attractive for quintessence cosmology for two reasons. First, they introduce nonminimal couplings between the quintessence field and ordinary matter, and these turn out to provide new kinds of scaling attractor solutions. Which attractor is the endpoint for a given choice of initial conditions is determined by the parameters of the quintessence potential, and this observation is at the root of the second attractive cosmological feature of having walking couplings. A given cosmological solution can cross over from the basin of attraction of one attractor to another since the walking of the couplings moves the boundaries of these domains of attraction within field space.

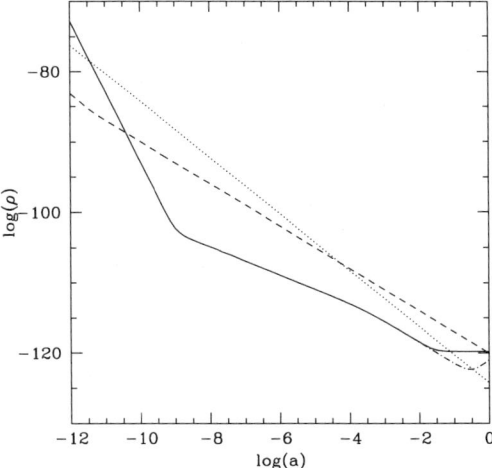

FIGURE 1. The logarithm of the energy density (in Planck units) of radiation, matter and the scalar field, as a function of the logarithm of the universal scale factor (normalized to unity at present). At the initial time the energies in decreasing order of size are: scalar, radiation and matter. The scalar kinetic and potential energies trace identical curves until the scale factor reaches 0.01, at which point the potential energy no longer changes while the kinetic energy continues to fall until beginning to rise once more.

Reference [**3**] provides a preliminary exploration of the cosmology of models built along these lines. An example of a viable cosmology which is obtained in this way is illustrated in Figure 1, which shows the cosmological evolution of the energy density in radiation, matter and the radion kinetic and potential energies. This particular model also satisfies the other two main constraints, such as the requirement that couplings to present-day matter are small enough to not introduce unacceptable new long-range forces.

A constraint which is specific to models where it is radion which is the quintessence field is the requirement that the radion not roll too far since the epoch of nucleosynthesis, since such a roll would predict an unacceptably large change in the ratio of weak and gravitational couplings, $M_w/M_p \sim 1/(M_s r)$, between then and now. This requirement is also satisfied by the model whose cosmology is illustrated in Figure 1.

Acknowledgements

I would like to thank the organizers of this workshop for providing such stimulating environs, and for their kind invitation to speak. My part of the work summarized in this article was done in collaboration with Andy Albrecht, Finn Ravndal and Constantinos Skordis.

References

1. A. Aguirre, C. P. Burgess, A. Friedland and D. Nolte, *Astrophysical constraints on modifying gravity at large distances*, Classical Quantum Gravity **18** (2001), R223–R232; hep-ph/0105083.
2. A. Albrecht, C. P. Burgess, F. Ravndal and C. Skordis, *Exponentially large extra dimensions*, Phys. Rev. D **65** (2002), 123506; hep-th/0105261.
3. ———, *Natural quintessence and large extra dimensions*, Phys. Rev. D **65** (2002), 123507; astro-ph/0107573.
4. A. Albrecht and C. Skordis, *Phenomenology of a realistic accelerating universe using only Planck-scale physics*, Phys. Rev. Lett. **84** (2000), 2076–2079; astro-ph/9908085.
5. I. Antoniadis, *A possible new dimension at a few TeV* Phys. Lett. B **246** (1990), 377–384.
6. I. Antoniadis, N. Arkani-Hamed, S. Dimopoulos and G. Dvali, *New Dimensions at a Millimeter to a Fermi and Superstrings at a TeV*, Phys. Lett. B **436** (1998), 257–263; hep-ph/9804398.
7. N. Arkani-Hamed, S. Dimopoulos and G. Dvali, *The hierarchy problem and new dimensions at a millimeter*, Phys. Lett. B **429** (1998), 263–272; hep-ph/9803315.
8. ———, *Phenomenology, astrophysics and cosmology of theories with sub-millimeter dimensions and TeV scale quantum gravity*, Phys. Rev. D **59** (1999), 086004; hep-ph/9807344.
9. N. A. Bahcall et al., *The cosmic triangle: revealing the state of the universe*, Science **284** (1999), 1481–1488.
10. K. Benakli, *Phenomenology of low quantum gravity scale models*, Phys. Rev. D **60** (1999), 104002; hep-ph/9809582.
11. C. P. Burgess, L. E. Ibañez and F. Quevedo, *Strings at the intermediate scale, or is the Fermi scale dual to the Planck scale?*, Phys. Lett. B **447** (1999), 257–265; hep-ph/9810535.
12. R. R. Caldwell and P. J. Steinhardt, *The imprint of gravitational waves in models dominated by a dynamical cosmic scalar field*, Phys. Rev. D **57** (1998), 6057–6064; astro-ph/9710062.
13. K. Coble, S. Dodelson and J. Frieman, *Dynamical lambda models of structure formation*, Phys. Rev. D **55** (1997), 1851–1859; astro-ph/9608122.
14. A. Dekel, D. Burstein and S. D. M. White, *Measuring Omega*, Critical Dialogues in Cosmology (N. Turok, ed.) (Princeton, 1996), World Scientific, Singapore, 1997, pp. 175–192; astro-ph/9611108.
15. S. Deser, R. Jackiw, and G. 't Hooft, *Three-dimensional Einstein gravity: dynamics of flat space*, Ann. Physics **152** (1984) 220–235.
16. J. Frieman, C. Hill, A. Stebbins and I. Waga, *Cosmology with ultralight pseudo Nambu-Goldstone bosons*, Phys. Rev. Lett. **75** (1995), 2077–2080; astro-ph/9505060.

17. P. Horava and E. Witten, *Eleven-dimensional supergravity on a manifold with boundary*, Nuclear Phys. B **475** (1996), 94–114; `hep-th/9603142`.
18. _____, *Heterotic and type I string dynamics from eleven dimensions*, Nuclear Phys. B **460** (1996) 506–524; `hep-th/9510209`.
19. J. D. Lykken, *Weak scale superstrings*, Phys. Rev. D **54** (1996), 3693–3697; `hep-th/9603133`.
20. P. J. E. Peebles, *Is cosmology solved? An astrophysical cosmologist's viewpoint*, Publ. Astronom. Soc. Pacific **111** (1999), 274–284.
21. S. Perlmutter et al., *Discovery of a supernova explosion at half the age of the universe and its cosmological implications*, Nature **391** (1998), 51–54; `astro-ph/9712212`.
22. F. Perrotta, C. Baccigalupi and S. Matarrese, *Extended quintessence*, Phys. Rev. D **61** (2000), 023507; `astro-ph/9906066`.
23. E. Pontón and E. Poppitz, *Casimir energy and radius stabilization in five and six dimensional orbifolds*, J. High Energy Phys. **2001**, no. 6, paper 19; `hep-ph/0105021`.
24. A. G. Riess et al., *Observational evidence from supernovae for an accelerating universe and a cosmological constant*, Astronom. J. **116** (1998) 1009–1038; `astro-ph/9805201`.
25. P. J. Steinhardt, L. M. Wang and I. Zlatev, *Cosmological tracking solutions*, Phys. Rev. D **59** (1999), 123504; `astro-ph/9812313`.
26. M. S. Turner, *Cosmology solved? Quite possibly!*, Publ. Astronom. Soc. Pacific **111** (1999), 264–273.
27. A. Vilenkin and T. Vachaspati, *Radiation of Goldstone bosons from cosmic strings*, Phys. Rev. **35** (1987), 1138–??.
28. S. Weinberg, *The cosmological constant problem*, Rev. Modern Phys. **61** (1989) 1–23.
29. C. M. Will, *The confrontation between general relativity and experiment: a 1998 update*, `gr-qc/9811036`.
30. _____, *The confrontation between general relativity and experiment*, Living Rev. Rel. **4** (2001), 4; `gr-qc/0103036`.
31. E. Witten, *Strong coupling expansion Of Calabi–Yau compactification*, Nuclear Phys. B **471** (1996) 135–158; `hep-th/9602070`.
32. I. Zlatev, L. M. Wang and P. J. Steinhardt, *Quintessence, cosmic coincidence, and the cosmological constant*, Phys. Rev. Lett. **82** (1999), 896–899; `astro-ph/9807002`.

Physics Department, McGill University, 3600 University St., Montréal, QC H3A 2T8, Canada.

E-mail address: `cliff@inca.physics.mcgill.ca`

Congruence Subgroups of PSL(2, $\mathbb{Z}$)

C. J. Cummins and S. Pauli

This paper is dedicated to the memory of Bob and Kay Sharp.

ABSTRACT. In recent years modular invariance has played an increasingly important role in many areas of mathematics and physics. In this paper we report on the computation and tabulation of congruence subgroups of PSL(2, $\mathbb{Z}$) of small genus which we have recently completed.

The group $\overline{\Gamma} = \mathrm{PSL}(2,\mathbb{Z}) = \mathrm{SL}(2,\mathbb{Z})/\{\pm 1\}$ acts on the expended upper half plane $\mathcal{H}^* = \mathcal{H} \cup \mathbb{Q} \cup \infty$ by fractional linear transformations. Although its properties have been investigated since the 19th century, recent years have seen a remarkable resurgence in the study of $\overline{\Gamma}$ and in the automorphic functions and forms of its subgroups. A particular set of subgroups contains the groups which often occur "in nature." These are the *congruence subgroups* which are defined as follows: The principal congruence subgroup of level N, $\overline{\Gamma}(N)$, is the image in PSL(2, $\mathbb{Z}$) of the group

$$\Gamma(N) = \left\{ \begin{pmatrix} a & b \\ c & d \end{pmatrix} \in \mathrm{SL}(2,\mathbb{Z}) \;\middle|\; \begin{pmatrix} a & b \\ c & d \end{pmatrix} \equiv \begin{pmatrix} 1 & 0 \\ 0 & 1 \end{pmatrix} \mod N \right\}.$$

A subgroup of finite index in $\overline{\Gamma}$ which contains some principal congruence subgroup is called a congruence group. Algorithms for determining when a given subgroup is a congruence subgroup have been described [3, 11, 12] and it appears that all subgroups which arise via RCFT are in fact congruence [3].

An important invariant attached to each subgroup G of $\overline{\Gamma}$ is its genus, which is defined to be the genus of the corresponding surface $\mathcal{H}^*/G$. The genus 0 case plays a crucial role in "Monstrous Moonshine" [1, 2, 14]—a key motivation for our work. Many years ago Rademacher conjectured that there are only finitely many genus 0 congruence subgroups. This problem was studied by Knopp and Newman [13], McQuillen [15, 16], and Dennin [8–10]. Stronger versions of the conjecture were proved by Thompson [17] and Cox and Parry [4, 5] which show that there are only finitely many congruence subgroups for any genus. Thompson's result is very

---

2000 *Mathematics Subject Classification.* 11F03, 11F22, 30F35.
Work supported by NSERC grants.
This is the final version of the paper.

©2004 American Mathematical Society

general, but does not give an explicit bound on the level or index of the subgroups. The results of Cox and Parry give the bounds:

PROPOSITION 1 (Cox and Parry). *If G is a congruence subgroup of genus g and level ℓ then*:

$$\ell \leq \begin{cases} 168, & \text{if } g = 0, \\ 12g + \frac{1}{2}(13\sqrt{48g + 121}) + 145, & \text{if } g \geq 1. \end{cases}$$

PROPOSITION 2 (Cox and Parry). *If G is a congruence subgroup of genus g and level ℓ and if p is a prime dividing ℓ then $p \leq 12g + 13$.*

Using Analytic methods derived to study the Selberg eigenvalue problem, Zograf [18] gave a bound on the index of a congruence subgroup:

PROPOSITION 3 (Zograf). *If G is a congruence subgroup of index m and genus g then*:

$$m < 128(g + 1).$$

Cox and Parry used Propositions 1 and 2 as the basis for a calculation of all the genus 0 congruence groups [4, 5]. Proposition 3 also, in principle, reduces the problem to a finite calculation—although a very large one.

We have used these three propositions to calculate all the congruence subgroups of $\mathrm{PSL}(2, \mathbb{Z})$ of genus up to genus 24. Suppressing some details, in outline the method we use is as follows: First the maximum genus, g_m, is fixed. The aim is to find all congruence subgroups with genus less than or equal to g_m. We first compute a list of "maximal" levels. A level ℓ is maximal if it satisfies the bounds of Propositions 1, 2 and 3 (with $g = g_m$) and if every multiple of ℓ fails to satisfy at least one of the bounds. For each maximal level we compute in $\mathrm{SL}(2, \mathbb{Z}/\ell\mathbb{Z})$ in a convenient permutation representation which greatly speeds the calculation. If G_1 is a subgroup of G_2, then the genus of G_1 is greater than or equal to that of G_2 and so it is sufficient to work down chains of subgroups until the genus is larger than g_m (or another constraint fails). Duplicates must, of course, be removed and the final list sorted and labelled. More details and tables for genus up to 3 are contained in [6] and complete tables and software are available online [7].

Our motivation for this tabulation were twofold. We first wanted to provide a comprehensive listing of low-genus congruence subgroups and their properties to assist both our own research and that of others. Our second motivation was the belief that generating complete tables of congruence subgroups will reveal previously unobserved patterns and so lead to new conjectures concerning their properties.

With this latter aim in mind we take a very modest step in this direction by presenting three tables. The first is a summary of the total number of congruence subgroups for each genus up to 24. The total number of subgroups is given, together with the total up to $\mathrm{PSL}(2, \mathbb{Z})$ and $\mathrm{PGL}(2, \mathbb{Z})$ conjugacy. The maximum level and index for each genus are also included. The same data is also presented for torsion-free subgroups. The final two columns list the bounds coming from Propositions 1 and 3. In Tables 2 and 3 we focus on those groups which achieve the level and index bounds given in Table 1. Thus these groups are the "extreme" cases at each genus. These groups are of interest since they give some measure of the strength of the Cox–Parry level and Zograf index bounds of Propositions 1 and 3. The inequalities are far from being saturated—at least for the range we have computed.

TABLE 1. Number of congruence subgroups of genus up to 24. The columns give the total number, the number up to $\mathrm{PSL}(2,\mathbb{Z})$ conjugacy and the number up to $\mathrm{PGL}(2,\mathbb{Z})$ conjugacy. The largest level ℓ and the largest index I at each genus is also given. The same data for torsion free subgroups is also tabulated. The columns $P1$ and $P3$ are the bounds from Propositions 1 and 3.

g	All Subgroups					Torsion-Free Subgroups					$P1$	$P3$
		PSL	PGL	ℓ	I		PSL	PGL	ℓ	I		
0	1116	132	121	48	72	254	33	33	32	60	168	128
1	2801	187	163	52	108	459	48	48	36	108	169	256
2	4107	177	145	78	108	672	49	49	64	108	192	384
3	6513	284	241	96	168	1809	108	105	72	168	214	512
4	7257	261	215	108	180	1665	87	86	81	180	235	640
5	9386	303	256	126	192	3028	133	125	75	192	256	768
6	10416	230	175	126	192	1780	55	45	121	180	275	896
7	18191	480	388	156	216	6216	213	191	128	216	295	1024
8	13726	277	212	169	220	2671	83	76	96	156	314	1152
9	21014	469	403	154	288	6711	208	203	128	288	333	1280
10	15622	304	235	168	324	4483	133	120	118	324	351	1408
11	27466	489	381	198	288	8450	195	179	147	240	370	1536
12	18095	269	198	210	330	4978	93	70	142	300	388	1664
13	33241	664	549	231	384	12447	343	303	162	384	405	1792
14	22871	268	178	252	300	4581	72	53	167	192	423	1920
15	40880	596	485	240	384	16743	289	263	179	288	441	2048
16	30809	410	294	243	364	8607	143	123	243	360	458	2176
17	54794	819	667	289	480	17453	351	317	242	480	475	2304
18	24935	273	191	264	384	4819	71	60	214	288	492	2432
19	60648	812	647	273	504	24287	411	375	256	504	509	2560
20	31137	308	203	286	408	9396	122	85	239	300	526	2688
21	66841	888	729	308	480	27542	504	450	256	480	542	2816
22	36135	365	284	361	486	11206	152	132	263	432	559	2944
23	59450	686	537	338	504	22798	312	271	274	384	576	3072
24	42289	336	212	336	546	6903	78	51	284	336	592	3200

TABLE 2. The congruence subgroups which maximize the level at each genus. The labelling of the groups is (level)(label)$^{(genus)}$—so, for example, $48A^0$ is the first group of level 48 and genus 0. The remaining columns are: I the index; Z the number of $PGL(2,\mathbb{Z})$ conjugates; L the number of $PSL(2,\mathbb{Z})$ conjugates; c_2 and c_3 the number of conjugacy classes of elements of order 2 and 3 respectively and Cusps the cusp widths in partition notation. (Note: in some cases these parameters do not distinguish the subgroups listed).

	I	Z	L	c_2	c_3	Cusps		I	Z	L	c_2	c_3	Cusps
$48A^0$	72	1	3	24	0	$24^1 48^1$	$231A^{13}$	231	2	77	27	0	231^1
							$231B^{13}$	231	2	77	27	0	231^1
$52A^1$	56	2	14	4	8	$4^1 52^1$							
							$252A^{14}$	252	2	252	30	0	252^1
$78A^2$	84	2	14	12	6	$6^1 78^1$	$240A^{15}$	240	2	40	6	12	240^1
$96A^3$	144	1	12	36	0	$48^1 96^1$	$240B^{15}$	240	2	40	6	12	240^1
$108A^4$	144	1	12	0	24	$36^1 108^1$	$243A^{16}$	324	1	108	0	0	$1^{18} 3^3 27^2 243^1$
$126A^5$	126	2	63	24	0	126^1	$289A^{17}$	306	1	306	2	0	$1^{17} 289^1$
$126B^5$	126	2	63	24	0	126^1							
							$264A^{18}$	264	2	44	18	0	264^1
$126A^6$	126	2	63	20	0	126^1	$264B^{18}$	264	2	44	18	0	264^1
							$264C^{18}$	264	2	132	18	0	264^1
$156A^7$	168	2	14	12	12	$12^1 156^1$	$264D^{18}$	264	2	132	18	0	264^1
$156B^7$	168	2	14	12	12	$12^1 156^1$							
							$273A^{19}$	294	2	98	18	3	$21^1 273^1$
$169A^8$	182	1	182	2	2	$1^{13} 169^1$	$273B^{19}$	294	2	98	18	3	$21^1 273^1$
$154A^9$	154	2	77	12	4	154^1	$286A^{20}$	308	2	154	12	8	$22^1 286^1$
$154B^9$	154	2	77	12	4	154^1							
							$308A^{21}$	308	2	308	18	2	308^1
$168A^{10}$	168	2	28	18	0	168^1	$308B^{21}$	308	2	308	18	2	308^1
$168B^{10}$	168	2	28	18	0	168^1							
$168C^{10}$	168	2	28	18	0	168^1	$361A^{22}$	380	1	380	0	2	$1^{19} 361^1$
$168D^{10}$	168	2	28	18	0	168^1							
$168E^{10}$	168	2	84	18	0	168^1	$338A^{23}$	364	1	182	0	4	$2^{13} 338^1$
$168F^{10}$	168	2	84	18	0	168^1							
$168G^{10}$	168	2	84	18	0	168^1	$336A^{24}$	336	2	56	18	0	336^1
$168H^{10}$	168	2	84	18	0	168^1	$336B^{24}$	336	2	56	18	0	336^1
							$336C^{24}$	336	2	56	18	0	336^1
$198A^{11}$	198	2	99	24	0	198^1	$336D^{24}$	336	2	56	18	0	336^1
$198B^{11}$	198	2	99	24	0	198^1	$336E^{24}$	336	2	168	18	0	336^1
							$336F^{24}$	336	2	168	18	0	336^1
$210A^{12}$	252	2	42	36	0	$42^1 210^1$	$336G^{24}$	336	2	168	18	0	336^1
							$336H^{24}$	336	2	168	18	0	336^1
							$336I^{24}$	504	2	21	72	0	$168^1 336^1$

TABLE 3. The congruence subgroups which maximize the index at each genus. The notation is as in Table 2. Note that the ordering of the groups is taken from [6, 7].

	I	Z	L	c_2	c_3	Cusps		I	Z	L	c_2	c_3	Cusps
$48A^0$	72	1	3	24	0	$24^1 48^1$	$18N^7$	216	1	6	0	0	$6^{18} 18^6$
							$18O^7$	216	1	12	0	0	$2^9 6^6 18^9$
$9H^1$	108	1	4	0	0	$3^9 9^9$	$18P^7$	216	1	24	0	9	$6^9 18^9$
$27C^1$	108	1	12	0	0	$1^9 3^6 27^3$	$36N^7$	216	1	18	0	0	$3^{12} 9^4 12^6 36^2$
							$36O^7$	216	1	36	0	0	$1^6 3^4 4^3 9^6 12^2 36^3$
$18P^2$	108	1	18	0	0	$3^6 6^6 9^2 18^2$							
$18Q^2$	108	1	36	0	0	$1^3 2^3 3^2 6^2 9^3 18^3$	$54B^7$	216	1	12	0	0	$2^{18} 6^3 54^3$
$54B^2$	108	1	36	0	0	$1^6 2^6 3^1 6^1 27^1 54^1$	$108A^7$	216	1	36	0	0	$1^{12} 3^2 4^6 12^1 27^2 108^1$
$7A^3$	168	1	1	0	0	7^{24}	$11A^8$	220	1	55	0	4	11^{20}
$49A^3$	168	1	8	0	0	$1^{21} 49^3$	$12B^9$	288	1	3	0	0	$6^{16} 12^{16}$
$10B^4$	180	1	3	0	0	$5^{12} 10^{12}$	$15C^9$	288	1	6	0	0	$3^{16} 15^{16}$
$50F^4$	180	1	18	0	0	$1^{10} 2^{10} 25^2 50^2$	$20F^9$	288	1	18	0	0	$2^8 4^8 10^8 20^8$
$8A^5$	192	1	1	0	0	8^{24}	$24AP^9$	288	1	6	0	0	$3^{16} 6^8 24^8$
$12E^5$	192	1	4	0	0	$4^{12} 12^{12}$	$30S^9$	288	1	72	0	0	$1^4 2^4 3^4 5^4 6^4 10^4 15^4 30^4$
$16M^5$	192	1	3	0	0	$4^{16} 16^8$	$36Q^9$	288	1	12	0	0	$2^{12} 4^{12} 18^4 36^4$
$16N^5$	192	1	12	0	0	$2^8 4^8 8^4 16^8$							
$16O^5$	192	1	12	0	0	$4^8 8^{12} 16^4$	$40W^9$	288	1	36	0	0	$1^8 2^4 5^8 8^4 10^4 40^4$
$21E^5$	192	1	32	0	0	$1^6 3^6 7^6 21^6$	$45G^9$	288	1	24	0	0	$1^{12} 5^{12} 9^4 45^4$
$24Z^5$	192	1	12	0	0	$2^8 6^8 8^4 24^4$	$72I^9$	288	1	24	0	0	$1^{12} 2^6 8^6 9^4 18^2 72^2$
$24AA^5$	192	1	48	0	0	$1^4 2^2 3^4 4^2 6^2 8^4 12^2 24^4$							
$24AB^5$	192	1	48	0	0	$2^4 6^6 8^2 12^6 24^2$	$9A^{10}$	324	1	1	0	0	9^{36}
$32M^5$	192	1	6	0	0	$2^{16} 8^4 32^4$	$18M^{10}$	324	1	12	0	0	$3^9 6^9 9^9 18^9$
$32N^5$	192	1	12	0	0	$2^8 4^{12} 32^4$	$27B^{10}$	324	1	4	0	0	$3^{27} 27^9$
$32O^5$	192	1	24	0	0	$1^8 2^4 4^8 8^4 32^4$							
$48I^5$	192	1	24	0	0	$1^8 3^8 4^2 12^2 16^2 48^2$	$54I^{10}$	324	1	36	0	0	$1^9 2^9 3^6 6^6 27^3 54^3$
$48J^5$	192	1	48	0	0	$1^4 2^6 3^4 6^6 16^2 48^2$							
							$81A^{10}$	324	1	12	0	0	$1^{27} 9^6 81^3$
$64C^5$	192	1	12	0	0	$1^{16} 4^4 16^2 64^2$							
$64D^5$	192	1	24	0	0	$1^8 2^{12} 16^2 64^2$	$12A^{11}$	288	1	36	8	0	12^{24}
$16C^6$	192	1	48	4	0	$8^{16} 16^4$	$20E^{11}$	288	2	36	8	0	$4^{12} 20^{12}$
$32H^6$	192	1	48	4	0	$4^{16} 32^4$	$24N^{11}$	288	1	36	8	0	$6^{16} 24^8$

Table 3 (continued)

	I	Z	L	c_2	c_3	Cusps		I	Z	L	c_2	c_3	Cusps
$40K^{11}$	288	2	36	8	0	$2^8 8^4 10^8 40^4$	$13A^{16}$	364	1	91	0	4	13^{28}
$96I^{11}$	288	1	6	48	0	$48^2 96^2$	$169A^{16}$	364	1	182	0	4	$1^{26} 169^2$
$96J^{11}$	288	1	12	48	0	$48^2 96^2$	$15B^{17}$	480	1	4	0	0	$5^{24} 15^{24}$
$96K^{11}$	288	1	36	48	0	$48^2 96^2$	$75F^{17}$	480	1	24	0	0	$1^{20} 3^{20} 25^4 75^4$
$11A^{12}$	330	1	55	6	0	11^{30}	$32A^{18}$	384	1	96	4	0	$8^{16} 16^8 32^4$
$16B^{13}$	384	1	3	0	0	$8^{32} 16^8$	$32B^{18}$	384	1	96	4	0	$8^{16} 16^8 32^4$
$16C^{13}$	384	1	6	0	0	$4^{16} 8^8 16^{16}$	$64A^{18}$	384	1	96	4	0	$4^{16} 8^8 64^4$
$24AB^{13}$	384	1	12	0	0	$4^{16} 8^4 12^{16} 24^4$	$64B^{18}$	384	1	96	4	0	$4^{16} 8^8 64^4$
$24AC^{13}$	384	1	24	0	0	$2^8 4^4 6^8 8^8 12^4 24^8$	$14A^{19}$	504	1	3	0	0	$7^{24} 14^{24}$
$32T^{13}$	384	1	3	0	0	$4^{32} 32^8$	$98A^{19}$	504	1	24	0	0	$1^{21} 2^{21} 49^3 98^3$
$32U^{13}$	384	1	12	0	0	$2^{16} 4^8 8^8 32^8$	$17A^{20}$	408	1	408	8	3	17^{24}
$48AG^{13}$	384	1	12	0	0	$2^{16} 6^{16} 16^4 48^4$	$33C^{21}$	480	1	48	0	0	$1^{10} 3^{10} 11^{10} 33^{10}$
$48AH^{13}$	384	1	48	0	0	$1^8 2^4 3^8 4^4 6^4 12^4 16^4 48^4$	$18A^{22}$	486	1	81	6	0	$9^{18} 18^{18}$
$64V^{13}$	384	1	6	0	0	$2^{32} 16^4 64^4$	$27C^{22}$	486	1	162	6	0	$9^{27} 27^9$
$64W^{13}$	384	1	24	0	0	$1^{16} 2^8 4^8 16^4 64^4$	$14A^{23}$	504	1	63	8	0	14^{36}
$96I^{13}$	384	1	24	0	0	$1^{16} 3^{16} 8^2 24^2 32^2 96^2$	$28E^{23}$	504	1	63	8	0	$7^{24} 28^{12}$
$128E^{13}$	384	1	12	0	0	$1^{32} 8^4 32^2 128^2$	$13A^{24}$	546	1	91	6	0	13^{42}
$50A^{14}$	300	1	15	20	0	$10^{10} 50^4$	$169A^{24}$	546	1	182	6	0	$1^{39} 169^3$
$50B^{14}$	300	1	30	20	0	$2^5 10^4 50^5$							
$16A^{15}$	384	1	48	8	0	$8^{16} 16^{16}$							

References

1. R. E. Borcherds, *Monstrous moonshine and monstrous Lie superalgebras* Invent. Math. **109** (1992), 405–444.
2. J. H. Conway and S. P. Norton, *Monstrous moonshine*, Bull. London Math. Soc. **11** (1979), 308–339.
3. A. Coste and T. Gannon, *Congruence subgroups and rational conformal field theory*, 1999 (preprint).
4. D. A. Cox and W. R. Parry, *Genera of congruence subgroups in Q-quaternion algebras*, J. Reine Angew. Math. **351** (1984), 66–112.
5. _____, *Genera of congruence subgroups in Q-quaternion algebras* (unabridged version).
6. C. J. Cummins and S .Pauli, *Congruence subgroups of* $PSL(2, \mathbb{Z})$ *of genus less than or equal to* 24, Experiment. Math. (to appear).
7. _____, *Full tables and source code for congruence subgroups of genus less than or equal to* 24 are available at http://www.math.tu-berlin.de/~pauli/congruence and http://www.math-stat.concordia.ca/faculty/cummins/congruence.
8. J. B. Dennin, Jr., *Fields of modular functions of genus* 0, Illinois J. Math. **15** (1971), 442–455.

9. _____, *Subfields of $K(2^n)$ of genus 0*, Illinois J. Math. **16** (1972), 502–518.
10. _____, *The genus of subfields of $K(p^n)$*, Illinois J. Math. **18** (1974), 246–264.
11. T. Hsu, *Identifying congruence subgroups of the modular group*, Proc. Amer. Math. Soc. **124** (1996), no. 5, 1351–1359.
12. _____, *Permutation techniques for coset representations of modular subgroups*, Geometric Galois Actions, 2, London Math.Soc.Lecture Note Ser., vol. 243, Cambridge Univ. Press, Cambridge, 1997, pp. 67–77.
13. M. I. Knopp and M Newman, *Congruence subgroups of positive genus in the modular group*, Illinois J Math **9** (1965), 577–583.
14. J. McKay, 1978 (unpublished letter to J. G. Thompson).
15. D. L. McQuillan, *Some results on the linear fractional group*, Illinois J. Math. **10** (1966) 24–38.
16. _____, *On the genus of fields of elliptic modular functions*, Illinois J. Math. **10** (1966), 479–487.
17. J. G. Thompson, *A finiteness theorem for subgroups of $\mathrm{PSL}(2,\mathbb{R})$ which are commensurable with $\mathrm{PSL}(2,\mathbb{Z})$*, Santa Cruz Conference on Finite Groups (Santa Cruz, CA, 1979, Proc. Sympos. Pure. Math., vol. 37, Amer. Math. Soc., Providence RI, 1980, pp. 533–555.
18. P. Zograf, *A spectral proof of Rademacher's conjecture for congruence subgroups of the modular group*, J. Reine Angew. Math. **414** (1991), 113–116.

DEPARTMENT OF MATHEMATICS AND STATISTICS, CONCORDIA UNIVERSITY, MONTRÉAL, QC H3G 1M8, CANADA
E-mail address: `cummins@cicma.concordia.ca`

INSTITUT FÜR MATHEMATIK, TECHNISCHE UNIVERSITÄT BERLIN, STRASSE DES 17. JUNI 136, 10623 BERLIN, GERMANY
E-mail address: `pauli@math.tu-berlin.de`

Asymptotic SU(2) and SU(3) Wigner Functions from the Weight Diagram

Hubert de Guise, David J. Rowe, and Barry C. Sanders

ABSTRACT. By comparing portions of the SU(2) weight diagram with physical systems, I will show how one can infer the essential features of the asymptotic form of the SU(2) Wigner functions. The same method will be used to deduce some features of asymptotic SU(3) Wigner functions.

1. Introduction

In these note, I want to expand the results of an earlier collaborative work [4] with David Rowe and Barry Sanders. My objective is to show how one can often infer, from the position of the initial and final weights $|\psi\rangle$ and $|\phi\rangle$ on the SU(2) or SU(3) weight diagram, the essential asymptotic features of the Wigner functions, describing respectively matrix elements of finite SU(2) and SU(3) transformations between $|\psi\rangle$ and $|\phi\rangle$. The basic idea is to compare portions of the weight diagram to physical systems, and use this to deduce how the Wigner functions should behave.

The examples I will discuss are restricted to cases where $|\psi\rangle$ and $|\phi\rangle$ are "close" to one another on the weight diagram (i.e., one can go from one to the other by laddering finitely many times). In these cases, one can verify, at least for SU(2), that the results agree with the well-known contraction limits. It is possible to obtain asymptotic expressions for Wigner functions in more general cases, where $|\psi\rangle$ and $|\phi\rangle$ are "far" from one another on the weight diagram; these cases do not appear related to contractions, and it has thus far not been possible to find physical systems useful for understanding the asymptotics of these cases.

A lot of work on asymptotic SU(2) Wigner function is already known from classical function theory, where limits of Jacobi polynomials were investigated back in the 19th century. Besides this intrinsic interest in asymptotics of group transformations, the major motivation for studying large quantum number limits of systems comes from our need to understand the passage from quantum to classical mechanics.

The growing interest in N-level systems, the natural algebraic formulation of transitions between these levels in terms of generators of the Lie algebra $u(N)$,

---

2000 *Mathematics Subject Classification.* 22E70, 81R05, 41A60.
This is the final form of the paper.

and the use of finite U(N) transformations to (partially) diagonalize appropriate interaction Hamiltonians justify the extension of known results of SU(2) Wigner functions.

In this contribution, some asymptotic limits of SU(3) Wigner functions will be presented. The results are limited to unitary irreducible representations (unirreps) of highest weight $(\lambda, 0)$. These representations only have trivial weight multiplicities. It would be nice to eventually obtain expressions for general representations; for instance, a density matrix formulation of a system of λ identical three-level atoms involves unirreps of the type (p, p), which arise in the decomposition of the product $(0, \lambda) \otimes (\lambda, 0)$. Results on these more complicated representations remain elusive for the moment, but the hope is that a good physical intuition about the simpler representations with no weight multiplicity will pave the way to understanding asymptotic properties of the more general cases.

The presentation of this contribution will be intuitive and "informal," emphazing the physical picture rather than more formal derivations, which can be found in [1].

2. Asymptotic SU(2) Wigner functions

The complex extension of the su(2) Lie algebra is spanned by the three operators $\{L_+, L_-, L_0\}$, with non-zero commutation relations

$$(2.1) \qquad [L_+, L_-] = 2L_0, \quad [L_0, L_\pm] = \pm L_\pm.$$

A finite SU(2) transformation T is parametrized in terms of three Euler angles α, β, γ such that

$$(2.2) \qquad T(\alpha, \beta, \gamma) = \exp(i\alpha L_0) \exp(\beta(L_+ - L_-)/2) \exp(i\gamma L_0).$$

If the Cartan subalgebra is spanned by L_0, then weight states for a finite-dimensional unitary irreducible representation are denoted by $|jm\rangle$, with

$$(2.3) \qquad L_0|jm\rangle = m|jm\rangle, \quad L_+|jj\rangle = 0,$$

where $2j \in \mathbb{Z}^+$. The last equality defines the highest weight of an irreducible SU(2) module. An SU(2) Wigner function is defined as the overlap

$$(2.4) \qquad \mathcal{D}^j_{m',m}(\alpha, \beta, \gamma) = \langle jm'|T(\alpha, \beta, \gamma)|jm\rangle$$
$$= e^{i\alpha m'} d^j_{m'm}(\beta) e^{i\gamma m},$$

with

$$(2.5) \qquad d^j_{m'm}(\beta) = \langle jm'| \exp[\beta(L_+ - L_-)/2]|jm\rangle.$$

It is the asymptotic behaviour of this last function, known as the reduced Wigner function, that is the primary interest of this section.

2.1. Case 1: input and output states near one edge of the weight diagram.
Suppose $j \to \infty$, with $|m|, |m'| \approx j$. The location of both $|jm'\rangle$ and $|jm\rangle$ is close to the end of the weight diagram. Locally, as illustrated in Figure 1, the weight diagram looks like the spectrum of a one-dimensional harmonic oscillator that has been placed horizontally rather than vertically.

Inspired by this observation, we consider the Schwinger realization of su(2):

$$(2.6) \qquad L_+ \mapsto a_1^\dagger a_2, \quad L_0 \mapsto \tfrac{1}{2}(a_1^\dagger a_1 - a_2^\dagger a_2), \quad L_- \mapsto a_1 a_2^\dagger,$$

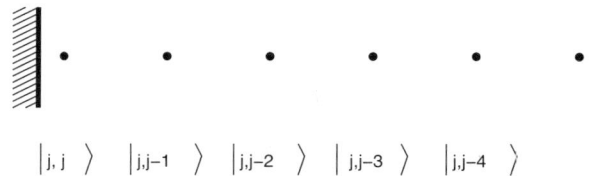

FIGURE 1. A piece of the su(2) weight diagram in the region where $|m|, |m'| \approx j$.

acting on eigenstates of a two-dimensional harmonic oscillator, such that $|jm\rangle \mapsto |n_1 n_2\rangle$, where $n_1 = j + m$, $n_2 = j - m$.

To recover the one-dimensional oscillator, we observe that, for a state $|2j-p, p\rangle$, $p \approx 0$ near the edge of the weight diagram and the matrix element

$$(2.7) \quad a_1^\dagger a_2 |2j-p, p\rangle = \sqrt{(2j-p+1)p}|2j-p+1, p-1\rangle \sim \sqrt{2jp}|2j-p+1, p-1\rangle$$

is, up to an overall constant factor $\sqrt{2j}$, effectively independent of $n_1 \approx 2j$ in $|n_1, n_2\rangle$. Physically, if $2j$ is large while p is small, a small change in the number of quanta $n_1 = 2j - p$ in the first oscillator will not significantly alter the system, whereas a small change in the number of quanta $n_2 = p$ in the second will have a comparatively much larger effect. In other words, our claim is that the properties of the reduced Wigner function, under the conditions given above, are effectively determined by the number of quanta in the second oscillator.

If this holds, then it must be possible to approximate the reduced Wigner function by

$$(2.8) \quad d^j_{mm'}(\beta) \mapsto \langle n_1 n_2| \exp[\beta(a_2 a_1^\dagger - a_2^\dagger a_1)]|n_1' n_2'\rangle$$
$$\sim \langle n_2| \exp[\chi\beta(a_2 - a_2^\dagger)/2]|n_2'\rangle,$$

where χ is a parameter which describes the "spectator" status of the first mode in the problem.

The right-hand side of (2.8) can be evaluated with the standard harmonic oscillator techniques to give (in the case where $n_2' > n_2$)

$$(2.9) \quad d^j_{mm'}(\beta) \sim \sqrt{\frac{n_2!}{n_2'!}}(-\chi\beta)^{n_2'-n_2} e^{-\chi^2\beta^2/2} L_{n_2}^{n_2'-n_2}(\chi^2\beta^2),$$

where L is an associated Laguerre polynomial. This result is known from the theory of special functions, and corresponds to the contraction $su(2) \to hw(1)$.

The parameter χ, which should depend only on quantities related to the mode 1 (the spectator mode), can be shown [1] to be $\chi = \sqrt{\frac{1}{2}(n_1 + n_1')}$; in this picture, the input and output states of the spectator modes, containing respectively n_1 and n_1' quanta with $n_1 \approx n_1'$, can be interchanged at will without changing in any essential way the physics of this approximation.

Comparisons between the exact Wigner function and its approximation by (2.9) are presented in Figures 2 and 3.

In Figure 2, both curves effectively overlap for $0 \leq \beta \leq \pi/8$, with only minor quantitative differences all the way to $\beta = 3\pi/8$.

In Figure 3, $j = 20$. Even for this low value of j, the qualitative features of the exact curve are very well reproduced by the approximation, with small numerical disagreements increasing with the angle.

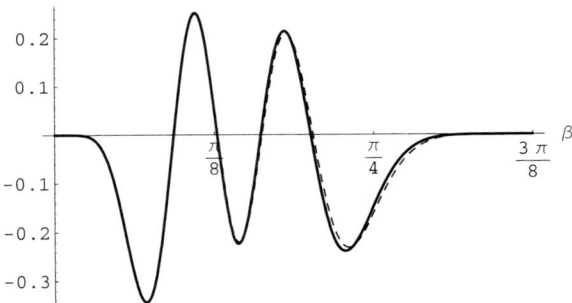

FIGURE 2. The exact Wigner function $d^{100}_{96,89}(\beta)$ (full line) and its approximation (dashed line) by (2.9).

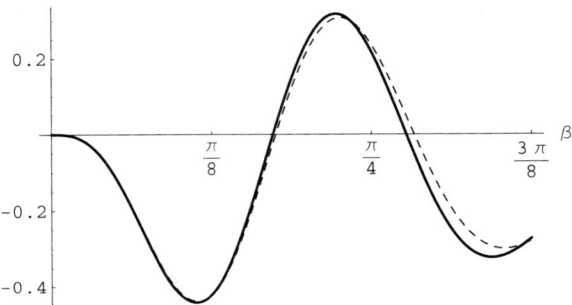

FIGURE 3. The exact Wigner function $d^{20}_{18,15}(\beta)$ (full line) and its approximation (dashed line) by (2.9).

2.2. Case 2: input and output states near the middle of the weight diagram. Suppose now $j \to \infty$, with $|m|, |m'| \ll j$ so that the location of both $|jm'\rangle$ and $|jm\rangle$ is far from either edge of the weight diagram. Locally, as illustrated in Figure 4, the weight diagram looks like a translationally invariant lattice.

Translation by one site to the right, corresponding to the action of L_+, cannot change the state by more than a phase. Thus, we expect that

$$(2.10) \qquad L_+ |jm\rangle \sim e^{i\theta} \xi |j, m+1\rangle,$$

with solution $|jm\rangle \mapsto e^{im\theta}/\sqrt{2\pi}$ in θ-space, where ξ is a real *constant* parameter that accounts for the approximately constant magnitude of the matrix element of L_+ in the neighborhood of $|jm\rangle$.

For a Hermitian realization, we have $L_- \mapsto e^{-i\theta} \xi$ so that

$$(2.11) \qquad d^j_{mm'}(\beta) \sim \frac{1}{2\pi} \int_0^{2\pi} d\theta \, e^{-im'\theta} e^{i\beta\xi \sin(\theta)} e^{im\theta}$$
$$= J_{m'-m}(\xi\beta),$$

• • • • • • • • • •

FIGURE 4. A piece of the su(2) weight diagram in the region where $|jm\rangle, |jm'\rangle$ are both far from the edges.

where J_p is a Bessel function [1,3]. Detailed calculations [1] show that the parameter ξ is given by

(2.12) $$\xi = \sqrt{j^2 - \tfrac{1}{4}(m+m')^2}.$$

ξ depends on m and m' only to second order in the expansion parameter $1/j^2$: it cannot be otherwise if the matrix element is to be approximately constant in the neighborhood of the state $|jm\rangle$ and $|jm'\rangle$. The symmetry between m and m' in the expression for (2.12) can be better understood if rewritten as $\xi = j \sin\theta_{\rm cl}$, where $\theta_{\rm cl}$ is the classical angle between two vectors $\vec{j}_1$ and $\vec{j}_2$ of equal magnitude $\sqrt{j_1(j_1+1)}$, the first having $j_{1z} = j$ and the second $j_{2z} = \tfrac{1}{2}(m+m')$. In other words, $\theta_{\rm cl}$ is the classical angle between the angular momentum state $|jj\rangle$ and the "average" angular momentum state $|j, \tfrac{1}{2}(m+m')\rangle$, although, formally, the weight $\tfrac{1}{2}(m+m')$ may not be in the irrep.

There are two interesting points to (2.11). First, this limit corresponds to the well-known contraction of ${\rm su}(2) \to {\rm e}(2)$, and the corresponding contraction of the generalized Legendre polynomials to Bessel functions. Second, the calculation shows a correction term, $\sin\theta_{\rm cl}$, that is absent from the usual expressions, and which allows us to extend the range of validity of our approximation.

Comparaisons between the exact expression and the approximate expression of (2.11) are presented in Figures 5 and 6. In Figure 5, inclusion of the $\sin\theta_{\rm cl}$ factor gives the dashed curve, which is a substantial improvement over the result from classical function theory (dotted curve) especially for large angles. This last result correctly captures the essentially oscillatory behaviour of the Wigner function, but oscillates too fast so that this estimate rapidly looses accuracy. In Figure 6, both input and output states are very close to the center of the weight diagram, so that the correction factor $\sin\theta_{\rm cl} \approx 1$: the result is that both approximations to the exact Wigner functions are practically indistinguishable.

3. Extension to SU(3) irreps of the type $(\lambda, 0)$

A realization of the complex extension of the su(3) Lie algebra is obtained by using the six number-preserving ladder operators $C_{ij} = a_i^\dagger a_j$, $i \neq j$, $i,j = 1,2,3$ together with the two traceless operators $h_1 = a_1^\dagger a_1 - a_2^\dagger a_2$ and $h_2 = a_2^\dagger a_2 - a_3^\dagger a_3$.

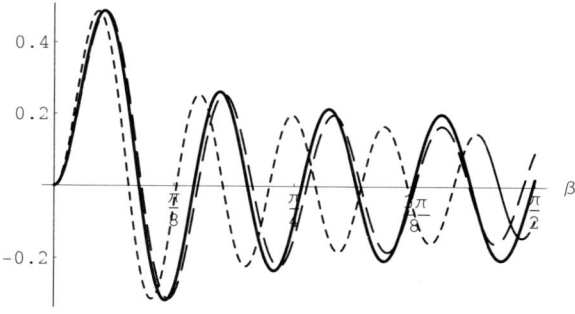

FIGURE 5. The exact Wigner function $d^{21}_{12,10}(\beta)$ (full line), its approximation (dashed line) by (2.12), and its approximation (dotted line) by a simple Bessel function with the correction factor of $\sin\theta_{\rm cl}$.

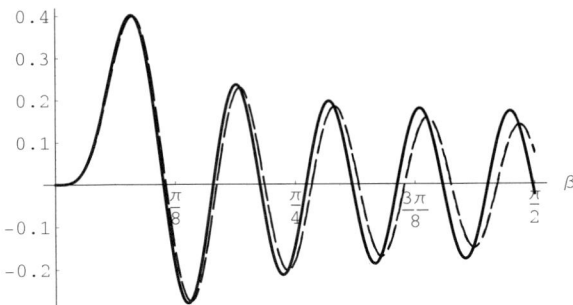

FIGURE 6. The exact Wigner function $d^{21}_{3,-1}(\beta)$ (full line) its approximation (dashed line) by (2.12) and its approximation (dotted line) by a simple Bessel function without the correction factor $\sin\theta_{\rm cl}$.

One can extract from this set three su(2) subalgebras, and it has been shown [5] that a general SU(3) transformation T can be written as a product of three SU(2) transformations:

(3.1) $\qquad T = R_{23}(\alpha_1,\beta_1,\gamma_1) \cdot R_{12}(\alpha_2,\beta_2,\alpha_2) \cdot R_{23}(\alpha_3,\beta_3,\gamma_3),$

where, for instance, $R_{23}(\alpha,\beta,\gamma)$ is a transformation of the form

$$\begin{pmatrix} 1 & 0 & 0 \\ 0 & & \\ 0 & & R(\alpha,\beta,\gamma) \end{pmatrix},$$

with $R(\alpha,\beta,\gamma)$ is a SU(2) transformation parametrized by the three Euler angles (α,β,γ).

It will be convenient throughout to label basis states $|n_1,n_2,n_3\rangle$ of an unirrep with highest weight $(\lambda,0)$ by three non-negative integers n_1, n_2, n_3, such that, for instance,

(3.2) $\qquad a_1^\dagger a_2 |n_1,n_2,n_3\rangle = \sqrt{(n_1+1)n_2}|n_1+1,n_2-1,n_3\rangle,$

with $n_1+n_2+n_3 = \lambda \in \mathbb{Z}^+$. The highest weight is $|\lambda,0,0\rangle$.

The SU(3) Wigner function is then defined as the overlap

(3.3) $\qquad \mathcal{D}^\lambda_{nn'}(\alpha_1,\beta_1,\gamma_1,\alpha_2,\beta_2,\alpha_3,\beta_3,\gamma_3) \equiv \langle n_1 n_2 n_3 | T | n'_1 n'_2 n'_3 \rangle.$

Taking advantage of the factorization scheme of (3.1) for an SU(3) element, we can rewrite this as a sum of products of SU(2) Wigner functions:

(3.4) $\quad \mathcal{D}^\lambda_{nn'}(\alpha_1,\beta_1,\gamma_1,\alpha_2,\beta_2,\alpha_3,\beta_3,\gamma_3)$
$$= \sum_{p=0}^{\lambda-n_1} D^{j'_1}_{m'm'_p}(\alpha_1,\beta_1,\gamma_1) D^{j_p}_{\mu'_p\mu_p}(\alpha_2,\beta_2,\alpha_2) D^{j_1}_{m_p m}(\alpha_3,\beta_3,\gamma_3),$$

where $D^j_{ab}(\alpha,\beta,\gamma)$ is an SU(2) Wigner function,

(3.5) $\qquad j_1 = \tfrac{1}{2}(\lambda-n_1), \quad j'_1 = \tfrac{1}{2}(\lambda-n_1), \quad j_p = \tfrac{1}{2}(\lambda-p),$

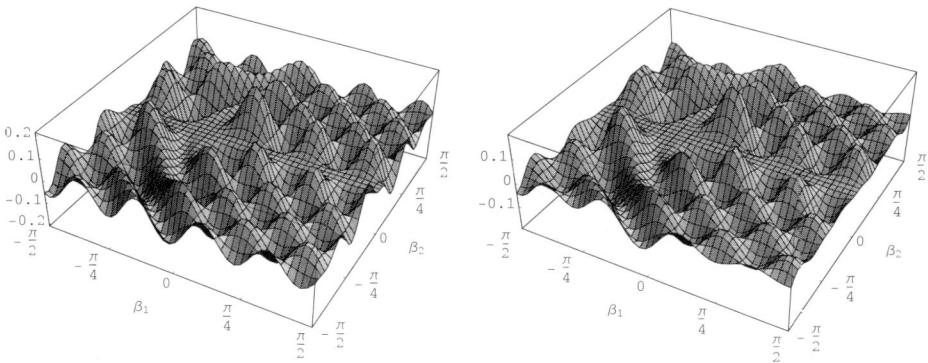

FIGURE 7. A 3-D plot of the exact Wigner function $\mathcal{D}^\lambda_{nn'}(0,\beta_1,0,0,\beta_2,0,\pi/7,0)$ with $n=(11,12,7)$, $n'=(8,12,10)$, $\lambda=30$ (left) and the same function approximated by the of Bessel functions of (3.7).

and where

$$\begin{aligned} m &= \tfrac{1}{2}(\lambda-n_1) & \mu_p &= n_1 - \tfrac{1}{2}(\lambda-p) & m_p &= \tfrac{1}{2}(\lambda-n_1)-p, \\ m' &= \tfrac{1}{2}(\lambda-n_1') & \mu_p' &= n_1' - \tfrac{1}{2}(\lambda-p) & m_p' &= \tfrac{1}{2}(\lambda-n_1')-p. \end{aligned} \tag{3.6}$$

For comparison and display purposes, we choose to eliminate the phases from the expression of $\mathcal{D}$, so we will set from now on $\alpha_1 = \gamma_1 = \alpha_2 = \alpha_3 = \gamma_3 = 0$.

Suppose the input and output states are located near the center of the weight diagram. We can approach the problem of finding the large λ limit for a Wigner function connecting these states from two perspectives.

3.1. An approximation using a sum.
First, we note that, in the sum of (3.4), the majority of the terms will involve SU(2) Wigner functions containing states near the center of the weight diagram. Thus, a first result is to simply replace every SU(2) Wigner function in (3.4) by a Bessel function and carry out the sum:

$$\mathcal{D}^\lambda_{nn'} \sim \sum_p J_{\nu_3-p}(\chi_{23}\beta_1)J_{\nu_1'-\nu_1}(\chi_{12}\beta_2)J_{p-\nu_3'}(\chi_{23}'\beta_{23}), \tag{3.7}$$

where

$$\begin{aligned} \chi_{23} &= \tfrac{1}{2}\sqrt{(2\nu_2+\nu_3-p)(\nu_3+p)}, \\ \chi_{12} &= \tfrac{1}{2}\sqrt{(\nu_1+\nu_1')(2\lambda-\nu_1-\nu_1'-2p)}, \\ \chi_{23}' &= \tfrac{1}{2}\sqrt{(2\nu_2'+\nu_3'-p)(\nu_3'+p)}. \end{aligned} \tag{3.8}$$

A three-dimensional view of the exact Wigner function $\mathcal{D}^\lambda_{nn'}(0,\beta_1,0,0,\beta_2,0,\pi/7,0)$ with $\lambda=30$, $n=(11,12,7)$, $n'=(8,12,10)$ is given on the left of Figure 7. Its approximation by a sum of Bessel function is given on the right. The qualitative agreement is obvious.

In order to make a more detailed comparison, we also present some cuts of the surfaces of Figure 7 in Figure 8. The exact result (thick lines) and approximate result of (3.7) (dashed lines) are superimposed on the same graph. There are four

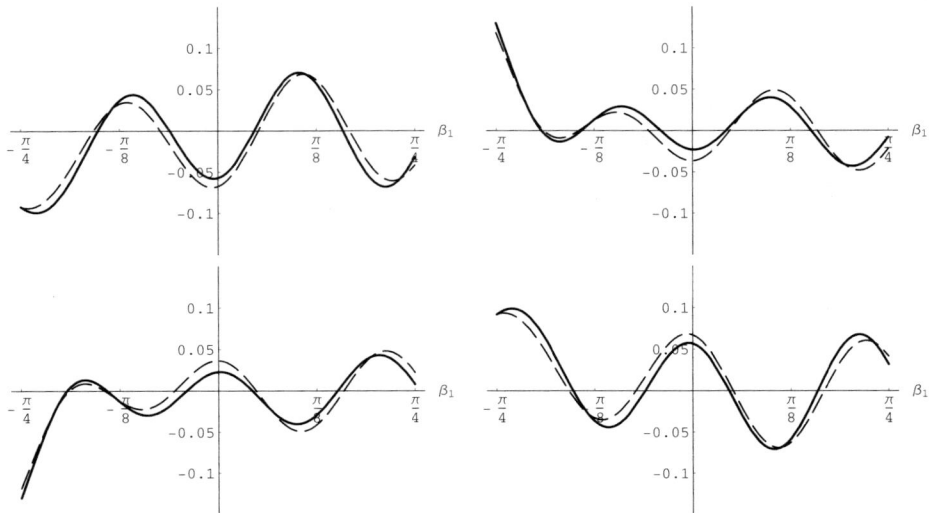

FIGURE 8. Slices in the surface plot of Figure 7. The slices for the exact (full lines) and approximate (dashed lines) expressions for $\mathcal{D}^\lambda_{nn'}(0,\beta_1,0,0,\beta_2,0,\pi/7,0)$ with $\lambda = 30$, $n = (11,12,7)$, $n' = (8,12,10)$ are presented superimposed for ease of comparaison. The top two slices are made at $\beta_2 = -\pi/3$, $-\pi/6$, and the bottom two graphs are cuts at $\beta_2 = \pi/6$ and $\pi/3$.

slices, corresponding respectively to cuts at $\beta_2 = -\pi/3$, $-\pi/6$ for the top graphs, and $\beta_2 = \pi/6$ and $\pi/3$ for the lower two graphs. In each case, $\beta_3 = \pi/7$.

3.2. An approximation using no sum. If we restrict ourselves to a neighborhood away from the edges, the weight system looks like a two-dimensional triangular lattice. We can easily generalize the argument which lead to (2.10) and (2.11). Introducing the auxiliary phases θ_1 and θ_2 in lattice space, we find

$$(3.9) \qquad |n_1 n_2 n_3\rangle \mapsto \frac{1}{2\pi} e^{i(n_1-n_2)\theta_1} e^{i(n_2-n_3)\theta_2}.$$

The approximate expressions for the su(3) generators can be inferred by comparing the way in which they act on weights with translation of these weights in the auxiliary lattice space. Hence, for instance [**2**],

$$(3.10) \qquad a_1^\dagger a_2 \mapsto \zeta_{12} e^{i(2\theta_1-\theta_2)}, \qquad a_2^\dagger a_3 \mapsto \zeta_{23} e^{i(-\theta_1+2\theta_2)}.$$

Using this and (3.9) in (3.3), we obtain the following integral as an approximation to (3.3):

$$(3.11) \quad D^\lambda_{nn'}(0,\beta_1,0,0,\beta_2,0,\beta_3,0)$$
$$= \frac{1}{4\pi^2} \int_0^{2\pi} \int_0^{2\pi} e^{-i(n_1-n_2)\theta_1} e^{-i(n_2-n_3)} e^{i(\zeta_{23}\beta_1+\zeta'_{23}\beta_3)\sin(2\theta_1-\theta_2)}$$
$$\times e^{i\zeta_{12}\beta_2 \sin(2\theta_2-\theta_1)} e^{i(n'_1-n'_2)\theta_1} e^{i(n'_2-n'_3)}$$

This is a complicated double integral that must be integrated numerically. It cannot be written in terms of known functions or factored into a product of two

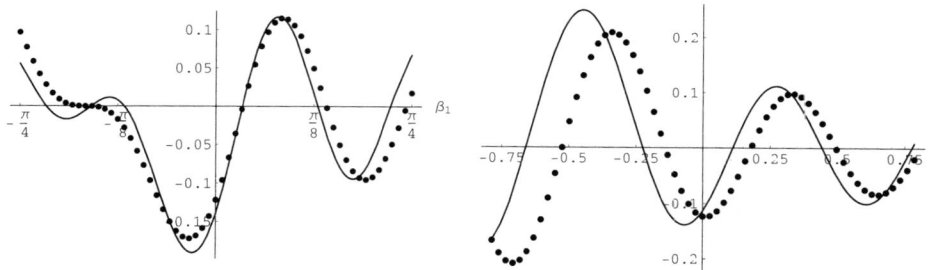

FIGURE 9. Two slices of SU(3) Wigner functions approximated by the integral of (3.11). On the left: the exact SU(3) Wigner function $D^{30}_{(8,12,10),(11,12,7)}(0,\beta_1,0,0,\pi/7,0,\pi/6,0)$ (full line); the dots are computed using the double integral of (3.11). On the right, the full line is the exact Wigner function $D^{30}_{(8,12,10),(12,9,9)}(0,\beta_1,0,0,\pi/7,0,\pi/6,0)$, and dots are calculated by (3.11).

independent integrals because the basic translation vectors are along the simple roots of su(3) and not orthogonal. Some results are shown in Figure 9. Because of the computing time involved in the numerical integration of (3.11), the results are limited to slices taken at $\beta_2 = \pi/7$ and $\beta_3 = \pi/6$, with only some points from (3.11) presented for comparaison with the exact results.

In the left part of the figure, the agreement between the approximate and exact expressions is good. On the right, however, there is too much difference to conclude that (3.11) can provide a reasonable estimate of the exact expression.

There are several interesting features of these two graphs. First, a consequence of modeling the weight space by a two-dimensional lattice is that su(3) ladder operators, which act as translations in that space, commute with one another; in other words, we are dealing with a contraction $su(3) \to \mathbb{R}^6 \oplus u(1) \oplus u(1)$. This, in turn, implies that, in this limit, the SU(2) transformations R_{12} and R_{23} of (3.1) commute with one another.

This is a drastic statement: it means that the transformation T of (3.1) can be rewritten as

$$(3.12) \qquad T \approx R_{23}(0,\beta_1+\beta_3,0)R_{12}(0,\beta_2,0).$$

Alternatively, this implies $\zeta_{23} = \zeta'_{23}$ in (3.11).

This approximation will be especially delicate in the region where $\beta_1 + \beta_3 \approx 0$. In particular, at $\beta_1 + \beta_3 = 0$, we have the surprising result

$$(3.13) \qquad T = R_{23}(0,\beta_1,0)R_{12}(0,\beta_2,0)R_{23}(0,-\beta_1,0) \approx R_{12}(0,\beta_2,0).$$

This is illustrated in a nice way on Figure 10, which present two slides taken at fixed β_1, β_3 for variable β_2.

On the left, we are in the region where $\beta_1 + \beta_3$ is well away from 0, and the agreement is reasonable for all values of the remaining angle β_2. On the right figure, however, $\beta_1 + \beta_3 \approx 0$; (3.11) diagrees with the exact result for all values of β_2.

More general results are presented in Figure 10. On the left, the number of quanta in the second oscillator is the same at input and output; (3.11) agrees reasonably well with the exact expression. On the right, the number of quanta in

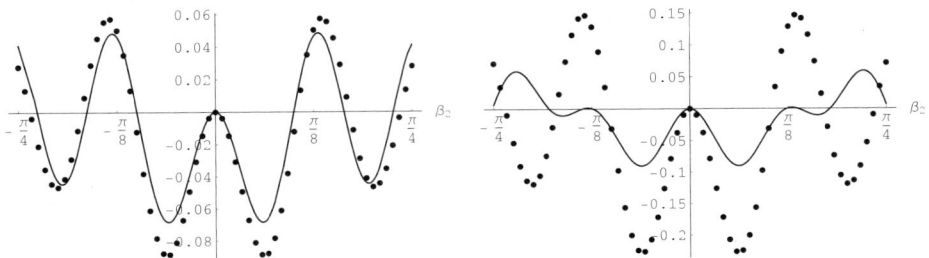

FIGURE 10. A comparaison of the exact Wigner function $\mathcal{D}^{45}_{(14,18,13),(16,15,14)}(0,\beta_1,0,0,\beta_2,0,\pi/6,0)$ with the results of (3.11). On the left, $\beta_1 = 2\pi/9$ and $\beta_1 + \beta_3 \gg 0$. On the right, $\beta_1 = -2\pi/9$ and $\beta_1 + \beta_3 < 0$.

the second oscillator is different at input and ouput. There is reasonable agreement in the region where $\beta_1 + \beta_3 \gg 0$, i.e., on the right hand side of the graph. The approximation deteriorates as we approach the point where $\beta_1 = -\beta_3$.

Exploration of numerous situation shows that the summation of Bessel functions of (3.7) is always computationally faster than the evaluation of the double integral of (3.11). Furthermore, the sum of Bessel functions never provides an approximation that is worse than the double integral. Thus, even if (3.11) is sometimes useful from a conceptual point of view, it does not represent a very practical way of approximating the SU(3) Wigner functions that we considered.

3.3. Other cases. It is also possible to investigate asymptotic SU(3) Wigner functions in cases, for instance, where both the input and output state are close to a corner of the weight diagram. In such cases, at least one of the SU(2) transformations of (3.1) cannot be approximated by an asymptotic limit: the resulting form of the Wigner function necessarily contains a sum over "finite" SU(2) Wigner functions. The same observation applies to the case where the input and output states are both close to only one edge of the weight diagram. Further details on these cases can be found in [4].

4. Conclusion

It is certainly true that some features of asymptotic SU(2) Wigner functions can be inferred only from the position of the initial and final states on the weight diagram. It is easy to construct simple models that capture these features in an insightful way when one can understand the asymptotic limit in terms of a contraction of the su(2) algebra; in this case, the SU(2) Wigner function also contracts to the Wigner function of the appropriate contracted algebra.

It would be nice if this simple method would immediately extend to SU(3). From a purely numerical point of view, it may be sufficient to use expression like (3.7) containing multiple sums to approximate asymptotic SU(3) Wigner functions. Expression like (3.7) have the advantage of containing functions that are well-known and are included in many computing packages.

On the other hand, it would be productive to extend to SU(3) the ideas developped to understand SU(2) limits. More work needs to be done in this direction. Considering that the exact SU(3) Wigner function for a general representation (λ,μ)

is complicated by the presence of weight multiplicities, it is difficult to see how the asymptotic properties of these more general functions can be understood without some guidance from physics.

References

1. M. Abramowitz and I. A Stegun (eds.), *Handbook of mathematical functions*, Dover, New York, 1966.
2. H. de Guise and M. Bertola, *Coherent state realizations of* su$(n+1)$ *on the n-torus*, J. Math. Phys. **43** (2002), no.7, 3425–3444.
3. W. Magnus, F. Oberhettinger, and R. P. Soni, *Formulas and theorems for the special functions of mathematical physics* Grundlehren Math. Wiss. vol. 52, Springer-Verlag, New York, 1966.
4. D. J. Rowe, H. de Guise, and B. C. Sanders, *Asymptotic limits of* SU(2) *and* SU(3) *Wigner functions*, J. Math. Phys. **42** (2001), no. 5, 2315–2342.
5. D. J. Rowe, B. C. Sanders and H. de Guise, *Representations of the Weyl group and Wigner functions for* SU(3), J. Math. Phys. **40** (1999), no. 7, 3604–3615.

DEPARTMENT OF PHYSICS, LAKEHEAD UNIVERSITY, THUNDER BAY, ON P7B 5E1, CANADA
E-mail address: hubert.deguise@lakeheadu.ca

DEPARTMENT OF PHYSICS, UNIVERSITY OF TORONTO, TORONTO, ON M5S 1A7, CANADA
E-mail address: rowe@physics.utoronto.ca

DEPARTMENT OF PHYSICS, MACQUARIE UNIVERSITY, SYDNEY, NSW 2109, AUSTRALIA
Current address: Department of Physics and Astronomy, University of Calgary, 2500 University Drive NW, Calgary, AB T2N 1N4, Canada
E-mail address: bsanders@phas.ucalgary.ca

Physical Applications of a Five-Dimensional Metric Formulation of Galilean Invariance

M. de Montigny, F. C. Khanna, and A. E. Santana

ABSTRACT. This presentation summarizes recent advances with a five-dimensional metric formalism of Galilean covariance. As a first example, we recover the two Galilean limits of electromagnetism investigated previously by Le Bellac and Lévy-Leblond. Then we describe the field theoretical formulation of some fluids and superfluids models: Euler equation, Takahashi Lagrangian for irrotational fluids and Thellung–Ziman Lagrangian for Helium II. Finally the non-relativistic Bhabha equations for spin 0 and 1 particles, and the Dirac equation for spin $\frac{1}{2}$ are considered.

1. Introduction

Although Galilean relativity has been superseded by Einstein's theory, there exists a wealth of low-energy systems, particularly in condensed matter physics and low-energy nuclear physics, where any new method involving Galilean invariance is likely to be useful. The Landau theory of superfluid state of ^{4}He is just an example. In fact Galilean invariance simply cannot be ignored in most many-body theories. The general program presented hereafter consists in investigating the physical applications of a metric formulation of Galilei invariance, so that one could use Galilean covariance, tensor analysis, etc., as a guiding principle to write down models in many-body problems. Hereafter we summarize the articles [1–5], where further details and references can be found. The approach follows the articles of Takahashi [17, 21, 22]. Other five-dimensional formalisms can be found in [6, 11, 12, 18, 19].

A five-dimensional *Galilei-vector* is such that a boost acts on it as

(1.1)
$$\mathbf{x}' = \mathbf{x} - \mathbf{V}t$$
$$t' = t$$
$$s' = s - \mathbf{V} \cdot \mathbf{x} + \tfrac{1}{2}\mathbf{V}^2 t$$

with **V**: relative velocity. Although it is clearly related to the phase acquired by

---

2000 *Mathematics Subject Classification.* 74A99, 78A25, 78A99.

NSERC (Canada), as well as CAPES and CNPq (Brazil) are acknowledged for financial support.

This is the final version of the paper.

the wave-function due to the non-trivial central charge of the Galilei group, an elegant interpretation of the additional coordinate s is proposed in [**11**]. Note that the units of s are L^2/T. The scalar product

$$(A \mid B) = A^\mu B_\mu \equiv \mathbf{A} \cdot \mathbf{B} - A_4 B_5 - A_5 B_4 \tag{1.2}$$

of two Galilei-vectors A and B is invariant under transformation (1.1). We define the Galilean metric by

$$g^{\mu\nu} = g_{\mu\nu} = \begin{pmatrix} 1 & 0 & 0 & 0 & 0 \\ 0 & 1 & 0 & 0 & 0 \\ 0 & 0 & 1 & 0 & 0 \\ 0 & 0 & 0 & 0 & -1 \\ 0 & 0 & 0 & -1 & 0 \end{pmatrix}. \tag{1.3}$$

This amounts to say that we consider the light-cone in 4+1 dimensions. The relation between Galilean invariance and light-cone coordinates in one more dimension has been noticed by L. Susskind in [**20**].

Equation (1.1) can be written as

$$x^{\mu'} = \Lambda^{\mu'}_\nu x^\nu \tag{1.4}$$

where $\Lambda^{\mu'}_\nu$ is the $(\mu'\nu)$-entry, or

$$\begin{pmatrix} x^{1'} \\ x^{2'} \\ x^{3'} \\ x^{4'} \\ x^{5'} \end{pmatrix} = \begin{pmatrix} 1 & 0 & 0 & -V_1 & 0 \\ 0 & 1 & 0 & -V_2 & 0 \\ 0 & 0 & 1 & -V_3 & 0 \\ 0 & 0 & 0 & 1 & 0 \\ -V_1 & -V_2 & -V_3 & \frac{1}{2}\mathbf{V}^2 & 1 \end{pmatrix} \begin{pmatrix} x^1 \\ x^2 \\ x^3 \\ x^4 \\ x^5 \end{pmatrix}. \tag{1.5}$$

For a Galilei-oneform we have:

$$x_{\mu'} = \Lambda^\nu_{\mu'} x_\nu \tag{1.6}$$

where now $\Lambda^\nu_{\mu'}$ is the $(\nu\mu')$-entry, with $\Lambda^\nu_{\mu'} x_\nu$ as in (1.5) with the change $V_j \to -V_j$.

Throughout, except in Section 2, we utilize the Galilei-vectors $(x^1, \ldots, x^5)$ with each component having units of length:

$$(x^1, \ldots, x^5) = \left(\mathbf{x}, vt, \frac{s}{v}\right) \tag{1.7}$$

where v has units of velocity. For a real field $\tilde{\phi}$, the projection is defined as

$$\tilde{\phi}(x) \equiv \phi(\mathbf{x}, t) + a_0 s \tag{1.8}$$

with a_0 a dimensionless constant ($\tilde{\phi}$, ϕ and s all have units of L^2/T). For a complex field $\tilde{\psi}$ we use the definition:

$$\tilde{\psi}(x) \equiv e^{ia_0 ms} \psi(\mathbf{x}, t) \tag{1.9}$$

with natural units, such that $\hbar = 1$. Typically we use $a_0 = \pm 1$, except in Section 3.4, where $a_0 = +1$.

If we use $(\mathbf{x}, t) \to x^\mu = (\mathbf{x}, t, s)$, then using the five-momentum $p_\mu \equiv -i\partial_\mu = (-i\mathbf{\nabla}, -i\partial_t, -i\partial_s)$ with $E = i\partial_t$ and $m = i\partial_s$, we obtain $p_\mu = (\mathbf{p}, -E, -m)$ and $p^\mu = g^{\mu\nu} p_\nu = (\mathbf{p}, m, E)$. Thereupon the mass does not enter as an external parameter, but rather as a remnant of the fifth component of the particle's momentum, starting from an apparently massless theory in five dimensions!

2. Galilean electromagnetism

Here we recover the two 'Galilean limits' of electromagnetism obtained nearly thirty years ago by Le Bellac and Lévy-Leblond [13]. For details see [2].

The Lorentz transformation of a four-vector $(u^0, \mathbf{u})$,

(2.1)
$$u^{0\prime} = \gamma\left(u^0 - \frac{1}{c}\mathbf{V}\cdot\mathbf{u}\right)$$
$$\mathbf{u}' = \mathbf{u} - \gamma\frac{\mathbf{V}}{c}u^0 + \frac{\mathbf{V}}{\mathbf{V}^2}(\gamma-1)\mathbf{V}\cdot\mathbf{u}$$

where $\gamma \equiv 1/\sqrt{1-\mathbf{V}^2/c^2}$, admits two well-defined Galilean limits [13]. One is for largely timelike vectors, with $u^{0\prime} = u^0$ and $\mathbf{u}' = \mathbf{u} - \frac{1}{c}\mathbf{V}u^0$, and corresponds to the 'electric' limit. The second limit is for largely spacelike vectors, with $u^{0\prime} = u^0 - (1/c)\mathbf{V}\cdot\mathbf{u}$ and $\mathbf{u}' = \mathbf{u}$, and is associated to the 'magnetic' limit. Throughout this section, we define the embedding of the Newtonian space-time into the de Sitter space by

(2.2)
$$(\mathbf{x}, t) \hookrightarrow x = (\mathbf{x}, t, 0)$$

so that $\partial_k = \boldsymbol{\nabla}_k$, $\partial_4 = \partial_t$ and $\partial_5 = 0$.

The electric and magnetic limits will be obtained by considering two particular embeddings of the five-potential. From (1.6):

(2.3)
$$\mathbf{A}' = \mathbf{A} + \mathbf{V}A_5$$
$$A_{4'} = A_4 + \mathbf{V}\cdot\mathbf{A} + \tfrac{1}{2}\mathbf{V}^2 A_5$$
$$A_{5'} = A_5.$$

Let us denote the components of the five dimensional electromagnetic antisymmetric tensor $F_{\mu\nu} \equiv \partial_\mu A_\nu - \partial_\nu A_\mu$ as

(2.4)
$$F_{\mu\nu} = \begin{pmatrix} 0 & b_3 & -b_2 & c_1 & d_1 \\ -b_3 & 0 & b_1 & c_2 & d_2 \\ b_2 & -b_1 & 0 & c_3 & d_3 \\ -c_1 & -c_2 & -c_3 & 0 & a \\ -d_1 & -d_2 & -d_3 & -a & 0 \end{pmatrix}.$$

They are expressed in terms of the five-potential as

(2.5)
$$\mathbf{b} = \boldsymbol{\nabla}\times\mathbf{A}$$
$$\mathbf{c} = \boldsymbol{\nabla}A_4 - \partial_4\mathbf{A}$$
$$\mathbf{d} = \boldsymbol{\nabla}A_5 - \partial_5\mathbf{A}$$
$$a = \partial_4 A_5 - \partial_5 A_4.$$

The external five-current, $j_\mu = (\mathbf{j}, j_4, j_5)$, also transforms as a five-vector:

(2.6)
$$\mathbf{j}' = \mathbf{j} + \mathbf{V}j_5$$
$$j_{4'} = j_4 + \mathbf{V}\cdot\mathbf{j} + \tfrac{1}{2}\mathbf{V}^2 j_5$$
$$j_{5'} = j_5$$

and the continuity equation takes the form

(2.7)
$$\partial^\mu j_\mu = \boldsymbol{\nabla}\cdot\mathbf{j} - \partial_4 j_5 - \partial_5 j_4 = 0.$$

In terms of the components in (2.4), the Maxwell equations,

(2.8) $$\partial_\mu F_{\alpha\beta} + \partial_\alpha F_{\beta\mu} + \partial_\beta F_{\mu\alpha} = 0$$

and

(2.9) $$\partial_\nu F^{\mu\nu} = j^\mu$$

read

(2.10) $$\begin{aligned} \boldsymbol{\nabla} \cdot \mathbf{b} &= 0 \\ \boldsymbol{\nabla} \times \mathbf{c} + \partial_4 \mathbf{b} &= \mathbf{0} \\ \boldsymbol{\nabla} \times \mathbf{d} + \partial_5 \mathbf{b} &= \mathbf{0} \\ \boldsymbol{\nabla} a - \partial_4 \mathbf{d} + \partial_5 \mathbf{c} &= \mathbf{0} \end{aligned}$$

and

(2.11) $$\begin{aligned} \boldsymbol{\nabla} \times \mathbf{b} - \partial_5 \mathbf{c} - \partial_4 \mathbf{d} &= \mathbf{j} \\ \boldsymbol{\nabla} \cdot \mathbf{c} - \partial_4 a &= -j_4 \\ \boldsymbol{\nabla} \cdot \mathbf{d} + \partial_5 a &= -j_5 \end{aligned}$$

respectively. Finally, the electromagnetic tensor transforms like $F_{\mu'\nu'} = \Lambda^\alpha_{\mu'} \Lambda^\beta_{\nu'} F_{\alpha\beta}$, so that its components transform as

(2.12) $$\begin{aligned} a' &= a + \mathbf{V} \cdot \mathbf{d} \\ \mathbf{b}' &= \mathbf{b} - \mathbf{V} \times \mathbf{d} \\ \mathbf{c}' &= \mathbf{c} + \mathbf{V} \times \mathbf{b} + \tfrac{1}{2}\mathbf{V}^2 \mathbf{d} - a\mathbf{V} - \mathbf{V}(\mathbf{V} \cdot \mathbf{d}) \\ \mathbf{d}' &= \mathbf{d}. \end{aligned}$$

2.1. Electric limit. The electric limit corresponds to the embedding

(2.13) $$(\mathbf{A}_e, \phi_e) \hookrightarrow A_e = \left(\mathbf{A}_e, 0, -\frac{1}{k_1} \phi_e \right)$$

and

(2.14) $$(\mathbf{j}_e, \rho_e) \hookrightarrow j_e = (k_2 \mathbf{j}_e, 0, -k_2 \rho_e).$$

The units of k_1 and k_2 are L^2/T^2 and ML/Q^2, respectively. (Note that k_2 has the same units as the permeability of free space, μ_0).

From (2.3) and (2.13), we find $\mathbf{A}'_e = \mathbf{A}_e - (1/k_1)\mathbf{V}\phi_e$ and $\phi'_e = \phi_e$, as in [**13**, equation (2.10)], if we choose $k_1 \equiv 1/(\mu_0 \epsilon_0)$. From (2.6) and (2.14) we obtain $\mathbf{j}'_e = \mathbf{j}_e - \mathbf{V}\rho_e$ and $\rho'_e = \rho_e$, also in agreement with [**13**], and from (2.2) and (2.7) we find

(2.15) $$\boldsymbol{\nabla} \cdot \mathbf{j} - \partial_4 j_5 - \partial_5 j_4 = \boldsymbol{\nabla} \cdot \mathbf{j}_e + \partial_t \rho_e = 0.$$

Now we define

(2.16) $$\mathbf{B}_e \equiv \mathbf{b} = \boldsymbol{\nabla} \times \mathbf{A}_e$$

and take $\mathbf{E}_e \propto \mathbf{d}$, so that

(2.17) $$\mathbf{E}_e \equiv k_1 \mathbf{d} = \frac{1}{\mu_0 \epsilon_0} \mathbf{d} = -\boldsymbol{\nabla} \phi_e$$

from (2.5). From (2.12), we find that the field components transform like

(2.18) $$\begin{aligned} \mathbf{E}'_e &= \mathbf{E}_e \\ \mathbf{B}'_e &= \mathbf{B}_e - \mu_0 \epsilon_0 \mathbf{V} \times \mathbf{E}_e \end{aligned}$$

as in [13]. From (2.10) and (2.11), with $k_2 \equiv \mu_0$, we find the wave equations

$$\boldsymbol{\nabla} \times \mathbf{E}_e = \mathbf{0}$$
$$\boldsymbol{\nabla} \cdot \mathbf{B}_e = 0$$
(2.19)
$$\boldsymbol{\nabla} \times \mathbf{B}_e - \mu_0 \epsilon_0 \partial_t \mathbf{E}_e = \mu_0 \mathbf{j}_e$$
$$\boldsymbol{\nabla} \cdot \mathbf{E}_e = \frac{1}{\epsilon_0} \rho_e$$

as in equation (2.8) of [13]!

2.2. Magnetic limit.
The magnetic limit corresponds to the embedding

(2.20) $$(\mathbf{A}_m, \phi_m) \hookrightarrow A_m = (\mathbf{A}_m, \phi_m, 0)$$

and

(2.21) $$(\mathbf{j}_m, \rho_m) \hookrightarrow j_m = (k_3 \mathbf{j}_m, -k_4 \rho_m, 0).$$

k_3 and k_4 have units ML/Q^2 and $(ML/Q^2)(L^2/T^2)$ (like μ_0 and ϵ_0^{-1}), respectively. With $k_3 \equiv \mu_0$ and $k_4 \equiv 1/\epsilon_0$ we find from (2.3) and (2.20) that now $\mathbf{A}'_m = \mathbf{A}_m$ and $\phi'_m = \phi_m - \mathbf{V} \cdot \mathbf{A}_m$. From (2.6)) and (2.21) we find $\mathbf{j}'_m = \mathbf{j}_m$ and $\rho'_m = \rho_m - \mu_0 \epsilon_0 \mathbf{V} \cdot \mathbf{j}_m$, and (2.7) gives

(2.22) $$\boldsymbol{\nabla} \cdot \mathbf{j} - \partial_4 j_5 - \partial_5 j_4 = \boldsymbol{\nabla} \cdot \mathbf{j}_m = 0$$

which is [13, equation (2.16)], and shows that the current $\mathbf{j}_m$ cannot be related to a transport of charge!

By defining

(2.23) $$\mathbf{B}_m \equiv \mathbf{b} = \boldsymbol{\nabla} \times \mathbf{A}_m$$

and taking $\mathbf{E}_m \propto \mathbf{c}$, then (2.5) shows that

(2.24) $$\mathbf{E}_m \equiv \mathbf{c} = -\boldsymbol{\nabla} \phi_m - \partial_t \mathbf{A}_m.$$

From (2.12) we get

(2.25)
$$\mathbf{E}'_m = \mathbf{E}_m + \mathbf{V} \times \mathbf{B}_m$$
$$\mathbf{B}'_m = \mathbf{B}_m.$$

Finally, (2.10) and (2.11) show that the Maxwell equations reduce to

$$\boldsymbol{\nabla} \times \mathbf{E}_m = -\partial_t \mathbf{B}_m$$
$$\boldsymbol{\nabla} \cdot \mathbf{B}_m = 0$$
(2.26)
$$\boldsymbol{\nabla} \times \mathbf{B}_m = \mu_0 \mathbf{j}_m$$
$$\boldsymbol{\nabla} \cdot \mathbf{E}_m = \frac{1}{\epsilon_0} \rho_m$$

in agreement with [13, equation (2.15)]!

3. Fluids and superfluids equations

3.1. Euler equation for fluids.
Define the functional Lagrangian as

(3.1) $$\tilde{\mathcal{L}}[\tilde{\rho}, \tilde{\phi}] = -\tfrac{1}{2} \tilde{\rho} \partial_\mu \tilde{\phi} \partial^\mu \tilde{\phi} - V(\tilde{\rho}).$$

The Euler-Lagrange equation for $\tilde{\rho}$ leads to $\tfrac{1}{2} \partial_\mu \tilde{\phi} \partial^\mu \tilde{\phi} + V'(\tilde{\rho}) = 0$. By defining the embedding as in (1.7) and (1.8), with $a_0 = -1$ and $\tilde{\rho}(x) \equiv \rho(\mathbf{x}, t)$, we find

(3.2) $$\tfrac{1}{2} \boldsymbol{\nabla} \phi \cdot \boldsymbol{\nabla} \phi + \partial_t \phi = -V'.$$

The gradient of this expression gives

(3.3) $$(\boldsymbol{\nabla}\phi \cdot \boldsymbol{\nabla})\boldsymbol{\nabla}\phi + \partial_t(\boldsymbol{\nabla}\phi) = -\boldsymbol{\nabla}(V').$$

With $\mathbf{v} = \boldsymbol{\nabla}\phi$ and $\boldsymbol{\nabla}(V') = (1/\rho)\boldsymbol{\nabla}p$ (where p is the pressure) we find the Euler equation

(3.4) $$\partial_t \mathbf{v} + (\mathbf{v} \cdot \boldsymbol{\nabla})\mathbf{v} = -\frac{1}{\rho}\boldsymbol{\nabla}p$$

which is a particular case of the Navier–Stokes equation, with viscosity and body force both equal to zero. If we perform the variation with respect to $\tilde{\phi}$, we find the continuity equation

(3.5) $$\partial_t \rho + \boldsymbol{\nabla}(\rho \boldsymbol{\nabla}\phi) = 0.$$

The Lagrangian of (3.1) can be deduced from

(3.6) $$\tilde{\mathcal{L}}[\tilde{\psi}, \tilde{\psi}^*] = k_1\big(\partial_\mu \tilde{\psi} \partial^\mu \tilde{\psi}^* - V(|\tilde{\psi}|)\big)$$

with complex field $\tilde{\psi}$, by defining the real fields $\tilde{\rho}$ and $\tilde{\phi}$ with the Madelung prescription $\tilde{\psi} \equiv \sqrt{\tilde{\rho}}e^{i\tilde{\phi}}$ so that the Lagrangian becomes

(3.7) $$\tilde{\mathcal{L}} = k_1\big(\tilde{\rho}\partial_\mu \tilde{\phi} \partial^\mu \tilde{\phi} - \overline{V}(\tilde{\rho})\big)$$

where $\overline{V} \equiv V - \frac{1}{4\tilde{\rho}}\partial_\mu \tilde{\rho} \partial^\mu \tilde{\rho}$. For further details see [**1, 3**].

3.2. Barotropic irrotational fluid. The Takahashi model for compressible irrotational barotropic fluids with pressure proportional to the square of the mass density [**17, 21, 22**] can be expressed in Galilean covariant form as

(3.8) $$\tilde{\mathcal{L}} = \frac{\rho_0}{8v_0^2}(\partial^\mu \tilde{\phi} \partial_\mu \tilde{\phi} - 2v_0^2)^2.$$

The field $\tilde{\phi}$ is related to velocity potential and v_0 has units of velocity. Variation with respect to $\tilde{\phi}$ gives the equation of motion

(3.9) $$\partial_\mu \partial^\mu \tilde{\phi} - \frac{1}{2v_0^2}(\partial_\mu \partial^\mu \tilde{\phi})(\partial_\nu \tilde{\phi} \partial^\nu \tilde{\phi}) - \frac{1}{v_0^2}(\partial^\mu \tilde{\phi})(\partial^\nu \tilde{\phi})(\partial_\mu \partial_\nu \tilde{\phi}) = 0.$$

With (1.7), (1.8) and $a_0 = -1$, (3.8) reduces to

(3.10) $$\mathcal{L} = \frac{\rho_0}{2v_0^2}(\tfrac{1}{2}\boldsymbol{\nabla}\phi \cdot \boldsymbol{\nabla}\phi + \partial_t \phi - v_0^2)^2.$$

The corresponding equation of motion is

(3.11) $$v_0^2 \nabla^2 \phi - \partial_t^2 \phi = \nabla^2 \phi(\tfrac{1}{2}\boldsymbol{\nabla}\phi \cdot \boldsymbol{\nabla}\phi + \partial_t \phi) + \tfrac{1}{2}\boldsymbol{\nabla}\phi \cdot \boldsymbol{\nabla}(\boldsymbol{\nabla}\phi \cdot \boldsymbol{\nabla}\phi) + 2\boldsymbol{\nabla}(\partial_t \phi) \cdot \boldsymbol{\nabla}\phi$$

as in [**21**, equation (5.40)]. A symmetry analysis of its one-dimensional version leads to the vector fields

(3.12)
$$\begin{aligned}
\mathbf{v}_1 &= \partial_t \\
\mathbf{v}_2 &= \partial_\phi \\
\mathbf{v}_3 &= \partial_x \\
\mathbf{v}_4 &= \left(t - \frac{\phi}{v_0^2}\right)\partial_\phi - \frac{x}{2v_0^2}\partial_x \\
\mathbf{v}_5 &= x\partial_\phi + t\partial_x \\
\mathbf{v}_6 &= x\partial_x + t\partial_t + \phi\partial_\phi.
\end{aligned}$$

The vectors $\mathbf{v}_1+c\mathbf{v}_3$ and $\mathbf{v}_3+c\mathbf{v}_2$ generate travelling-wave solutions. The subgroup generated by $\mathbf{v}_6$ lead to scale-invariant solutions $\phi(x,t) = \frac{1}{3}x^2/t + (v_0^2+k^2)t \pm i\sqrt{\frac{2}{3}}kx$. The $\mathbf{v}_5$-invariant solutions have the form $\phi(x,t) = x^2/2t - (k_1/2)\ln t + v_0^2 t - k_2/2$ and from $\mathbf{v}_4$ we find $\phi(x,t) = cx^2/(t+k) + v_0^2 t$ ($c = \frac{1}{2}$ or $\frac{1}{3}$). More details are in [1].

Now let us gauge the Lagrangian of (3.6):

$$\tilde{\mathcal{L}} \propto D_\mu \tilde{\psi} D^\mu \tilde{\psi}^* \equiv (\partial_\mu \tilde{\psi} + i\tilde{A}_\mu \tilde{\psi})(\partial^\mu \tilde{\psi}^* - i\tilde{A}^\mu \tilde{\psi}^*) \tag{3.13}$$

with $\tilde{A}_\mu = k\partial_\mu \tilde{\rho}$. Variation of this Lagrangian for $\tilde{\psi}^*$, together with (1.7) and (1.9) and $\tilde{\rho}(x) \equiv \rho(\mathbf{x},t)$ gives

$$\nabla^2 \psi - 2ia_0 \partial_t \psi + ik(\nabla \cdot \nabla \rho)\psi + 2ik\nabla\rho \cdot \nabla\psi \tag{3.14}$$
$$- 2ik(ia_0)(\partial_t \rho)\psi - k^2(\nabla\rho \cdot \nabla\rho)\psi = 0.$$

With $a_0 = -1$ and $k = m$, we find

$$i\partial_t \psi = m(\partial_t \rho)\psi - \frac{1}{2m}(\nabla + im\nabla\rho)^2 \psi \tag{3.15}$$

which is [21, equation (6.5)].

We conclude this section by expressing in covariant form yet another model suggested by Takahashi to describe the irrotational component of quantum liquids. Consider the null vector $\tilde{u}_p^\mu \equiv \left(\mathbf{u}_p, v, \frac{1}{2v}\mathbf{u}_p^2\right)$. Then the scalar product

$$\tilde{\Phi}_p = \tilde{u}_p^\mu \partial_\mu \tilde{\phi} = \mathbf{u}_p \cdot \nabla\phi + \frac{1}{2}a_0 \mathbf{u}_p^2 + \partial_t \phi \tag{3.16}$$

with $a_0 = -1$ leads to

$$\Phi_p = \partial_t \phi + \mathbf{u}_p \cdot \nabla\phi - \tfrac{1}{2}\mathbf{u}_p^2. \tag{3.17}$$

Similarly, one can construct $\Phi_m = \partial_t \phi + \mathbf{u}_m \cdot \nabla\phi - \tfrac{1}{2}\mathbf{u}_m^2$. Next define the expressions $\chi_m = \frac{i}{2}\tilde{u}_m^\mu \left[\tilde{\psi}^* \partial_\mu \tilde{\psi} - (\partial_\mu \tilde{\psi}^*)\tilde{\psi}\right]$, $\eta_m(x) \equiv \alpha \psi^*(x) + \alpha^* \psi(x)$, and $\rho_m(x) \equiv m\psi^*(x)\psi(x)$. Then the Lagrangian (3.12) of [21] takes the form

$$\mathcal{L} = \frac{\rho_0}{2v_0^2}(\Phi_p(x) - v_0^2)^2 + \eta_m(x)\Phi_p(x) + \rho_m(x)(\Phi_p(x) - \Phi_m(x)) + \chi_m(x). \tag{3.18}$$

3.3. Generalized models for non-barotropic fluids. Here we generalize (3.8) by relaxing $p \propto \rho^2$ (p: pressure, ρ: its density) to $p \propto \rho^\gamma$ ($\gamma \geq 1$). For $\gamma \neq 1$ we consider

$$\tilde{\mathcal{L}} = k_\gamma (\partial \tilde{\phi} \partial \tilde{\phi} - v_0^2)^\gamma \tag{3.19}$$

so that variation of the field $\tilde{\phi}$ gives

$$(\tfrac{1}{2}\partial_\mu \tilde{\phi} \partial^\mu \tilde{\phi} - v_0^2)\partial_\nu \partial^\nu \tilde{\phi} + (\gamma - 1)\partial_{\mu\nu}\tilde{\phi}\partial^\mu \tilde{\phi} \partial^\nu \tilde{\phi} = 0. \tag{3.20}$$

Using (1.7) with $a_0 = -1$, it becomes

$$v_0^2 \nabla^2 \phi - (\gamma - 1)\partial_t^2 \phi = \nabla^2 \phi (\tfrac{1}{2}\nabla\phi \cdot \nabla\phi + \partial_t \phi) \tag{3.21}$$
$$+ (\gamma - 1)\nabla\phi \cdot \nabla(\tfrac{1}{2}\nabla\phi \cdot \nabla\phi + 2\partial_t \phi).$$

If $\gamma = 1$, it reduces to

$$v_0^2 \nabla^2 \phi = \nabla^2 \phi (\tfrac{1}{2}\nabla\phi \cdot \nabla\phi + \partial_t \phi). \tag{3.22}$$

When $\gamma \neq 1$, we recover the Takahashi model [17, 21, 22].

Another possibility, which corresponds to $p \propto \rho$, is

$$\tilde{\mathcal{L}}[\tilde{\phi}] = k \exp(\partial \tilde{\phi} \partial \tilde{\phi} - v_0^2). \tag{3.23}$$

The associated equation of motion is

$$v_0^2 \partial_\mu \partial^\mu \tilde{\phi} + \partial_{\mu\nu} \tilde{\phi} \partial^\mu \tilde{\phi} \partial^\nu \tilde{\phi} = 0 \tag{3.24}$$

and it reduces to

$$v_0^2 \boldsymbol{\nabla}^2 \phi + \partial_{tt} \phi + \boldsymbol{\nabla}\phi \cdot \boldsymbol{\nabla}(\tfrac{1}{2}\boldsymbol{\nabla}\phi \cdot \boldsymbol{\nabla}\phi + 2\partial_t \phi) = 0. \tag{3.25}$$

Other equations relevant in condensed matter physics are obtained by generalizing (3.6). For instance, consider

$$\tilde{\mathcal{L}}[\tilde{\psi}, \tilde{\psi}^*] \propto (\partial \tilde{\psi} \partial \tilde{\psi}^* - V(|\tilde{\psi}|))^p \tag{3.26}$$

with a complex field $\tilde{\psi}$. The choices $p = 1$ and $V = \lambda |\tilde{\psi}|^4$, with the embedding in (1.7) and (1.9), give us

$$\mathcal{L} = k_1 (\boldsymbol{\nabla}\psi \cdot \boldsymbol{\nabla}\psi^* - im(\psi^* \partial_t \psi - \psi \partial_t \psi^*) - \lambda |\psi|^4). \tag{3.27}$$

The Euler–Lagrange equation, with $a_0 = -1$, leads to the non-linear Schrödinger equation

$$i\partial_t \psi = -\frac{1}{2m} \boldsymbol{\nabla}^2 \psi + \frac{\lambda}{m} |\psi|^2 \psi. \tag{3.28}$$

Similar equations are used in effective theories of superconductivity and Bose–Einstein condensation (see references in [**3**]).

3.4. Model of non-viscous fluids and liquid helium.
As a last example, let us consider (3.1) with a five-dimensional Clebsch transformation $\partial \tilde{\phi} \to \partial \tilde{\phi} + \tilde{\alpha} \partial \tilde{\beta}$:

$$\tilde{\mathcal{L}} = -\frac{\tilde{\rho}}{2v_0^2}(\partial_\mu \tilde{\phi} + \tilde{\alpha} \partial_\mu \tilde{\beta})(\partial^\mu \tilde{\phi} + \tilde{\alpha} \partial^\mu \tilde{\beta}) - V(\tilde{\rho}). \tag{3.29}$$

Next we define $\tilde{\alpha}(x) = \alpha(\mathbf{x}, t)$, $\tilde{\beta}(x) = \beta(\mathbf{x}, t)$, and $\tilde{\rho}(x) = \rho(\mathbf{x}, t)$, with (1.8) for $\tilde{\phi}(x)$ and Eq. (1.7) for the coordinates. Here we take $a_0 = +1$. Then the Lagrangian in (3.29) becomes

$$\mathcal{L} = \frac{\rho}{v_0^2}(\partial_t \phi - \tfrac{1}{2}\boldsymbol{\nabla}\phi \cdot \boldsymbol{\nabla}\phi + \alpha(\partial_t \beta - \tfrac{1}{2}\alpha \boldsymbol{\nabla}\beta \cdot \boldsymbol{\nabla}\beta) - \alpha \boldsymbol{\nabla}\phi \cdot \boldsymbol{\nabla}\beta) - V(\rho). \tag{3.30}$$

This may be expressed as

$$\mathcal{L} = \frac{\rho}{v_0^2}(\partial_t \phi + \alpha \partial_t \beta - \tfrac{1}{2}\mathbf{v}^2) - V(\rho) \tag{3.31}$$

where $\mathbf{v} = -\boldsymbol{\nabla}\phi - \alpha \boldsymbol{\nabla}\beta$. This Lagrangian was employed by Thellung and Ziman to describe the rotational component of liquid Helium (see [**3**, Section 4.3]).

4. Bhabha and Duffin–Kemmer–Petiau equations: spin zero and spin one

The Duffin–Kemmer–Petiau (DKP) equation is

$$(\beta^\mu \partial_\mu + k)\Psi = 0 \tag{4.1}$$

with matrices β satisfying the DKP algebra:

$$\beta^\mu \beta^\lambda \beta^\nu + \beta^\nu \beta^\lambda \beta^\mu = g^{\mu\lambda}\beta^\nu + g^{\nu\lambda}\beta^\mu \tag{4.2}$$

$g^{\mu\nu}$ is the Galilean metric. The adjoint of Ψ is defined as $\overline{\Psi} \equiv \Psi^\dagger \eta$, where

(4.3) $$\eta = (\beta^4 + \beta^5)^2 + \mathbf{1}.$$

In the following we use the momentum version of (4.1):

(4.4) $$(\beta^\mu p_\mu - ik)\Psi = 0.$$

For details see [4, 5].

4.1. DKP equation for spin zero. For spinless particles, the β's can be taken as

(4.5) $$\beta^1 = \begin{pmatrix} 0 & 0 & 0 & 0 & 0 & 1 \\ 0 & 0 & 0 & 0 & 0 & 0 \\ 0 & 0 & 0 & 0 & 0 & 0 \\ 0 & 0 & 0 & 0 & 0 & 0 \\ 0 & 0 & 0 & 0 & 0 & 0 \\ 1 & 0 & 0 & 0 & 0 & 0 \end{pmatrix}, \quad \beta^2 = \begin{pmatrix} 0 & 0 & 0 & 0 & 0 & 0 \\ 0 & 0 & 0 & 0 & 0 & 1 \\ 0 & 0 & 0 & 0 & 0 & 0 \\ 0 & 0 & 0 & 0 & 0 & 0 \\ 0 & 0 & 0 & 0 & 0 & 0 \\ 0 & 1 & 0 & 0 & 0 & 0 \end{pmatrix},$$

$$\beta^3 = \begin{pmatrix} 0 & 0 & 0 & 0 & 0 & 0 \\ 0 & 0 & 0 & 0 & 0 & 0 \\ 0 & 0 & 0 & 0 & 0 & 1 \\ 0 & 0 & 0 & 0 & 0 & 0 \\ 0 & 0 & 0 & 0 & 0 & 0 \\ 0 & 0 & 1 & 0 & 0 & 0 \end{pmatrix}, \quad \beta^4 = \begin{pmatrix} 0 & 0 & 0 & 0 & 0 & 0 \\ 0 & 0 & 0 & 0 & 0 & 0 \\ 0 & 0 & 0 & 0 & 0 & 0 \\ 0 & 0 & 0 & 0 & 0 & 1 \\ 0 & 0 & 0 & 0 & 0 & 0 \\ 0 & 0 & 0 & 0 & -1 & 0 \end{pmatrix},$$

$$\beta^5 = \begin{pmatrix} 0 & 0 & 0 & 0 & 0 & 0 \\ 0 & 0 & 0 & 0 & 0 & 0 \\ 0 & 0 & 0 & 0 & 0 & 0 \\ 0 & 0 & 0 & 0 & 0 & 0 \\ 0 & 0 & 0 & 0 & 0 & 1 \\ 0 & 0 & 0 & -1 & 0 & 0 \end{pmatrix}.$$

They are the generators of the Lie algebra $so(5,1)$ such that $\beta^\mu = J^{\mu 6}$, with $\mu = 1, \ldots, 5$. The matrices in (4.5) satisfy the DKP algebra, (4.2).

If we introduce a DKP spinor

(4.6) $$\Psi \equiv \begin{pmatrix} \mathbf{A} \\ \theta \\ \varphi \\ \phi \end{pmatrix}$$

with $\mathbf{A} = (A_x, A_y, A_z)$, then (4.4) gives

(4.7) $$\begin{aligned} -ik\mathbf{A} + \mathbf{p}\phi &= \mathbf{0} \\ -ik\theta + p_4\phi &= 0 \\ -ik\varphi + p_5\phi &= 0 \\ \mathbf{p} \cdot \mathbf{A} - p_5\theta - p_4\varphi - ik\phi &= 0. \end{aligned}$$

These equations can be expressed in terms of ϕ as $\mathbf{p}^2\phi - 2p_4 p_5 \phi + k^2\phi = 0$, which reduces to the Schrödinger equation

(4.8) $$E\phi = \frac{\mathbf{p}^2}{2m}\phi$$

if we impose the condition $p_4 p_5 = mE$, and absorb the constant k into the energy as $E \to E - k^2/(2m)$.

The DKP oscillator is described by performing the non-minimal substitution

(4.9) $$\mathbf{p} \to \mathbf{p} + i\omega\eta\mathbf{r}$$

with η given by (4.3). Thus we find

(4.10) $$E\phi = \left(\frac{\mathbf{p}^2}{2m} + \frac{1}{2}m\omega^2\mathbf{r}^2 - \frac{3}{2}\hbar\omega\right)\phi$$

after performing the change $\omega \to m\omega$. This equation has been obtained in [16] for the expression of the non-relativistic energy of the DKP oscillator.

4.2. DKP equation for spin one. We use the shorthand notation e_{ij} to represent a 15-by-15 matrix whose only non-zero entry is ij, defined to be 1, that is, $(e_{ij})_{mn} \equiv \delta_{im}\delta_{jn}$. Then the DKP generators are

(4.11)
$$\beta^1 = e_{13,1} + e_{14,4} + e_{12,8} - e_{11,9} - e_{9,11} + e_{8,12} + e_{1,13} + e_{4,14}$$
$$\beta^2 = e_{13,2} + e_{14,5} - e_{12,7} + e_{10,9} + e_{9,10} - e_{7,12} + e_{2,13} + e_{5,14}$$
$$\beta^3 = e_{13,3} + e_{14,6} + e_{11,7} - e_{10,8} - e_{8,10} + e_{7,11} + e_{3,13} + e_{6,14}$$
$$\beta^4 = -e_{10,4} - e_{11,5} - e_{12,6} + e_{1,10} + e_{2,11} + e_{3,12} + e_{15,14} + e_{13,15}$$
$$\beta^5 = -e_{10,1} - e_{11,2} - e_{12,3} + e_{4,10} + e_{5,11} + e_{6,12} - e_{15,13} - e_{14,15}.$$

They also can be seen as the generators $J^{\mu 6}$ of $so(5,1)$.

Here we only consider the DKP simple harmonic oscillator by first performing the non-minimal substitution, (4.9), where η is given by (4.3) and (4.11), and substitute (4.11) into (4.4), with the DKP spinor given by

(4.12) $$\Psi = \begin{pmatrix} v_1 \\ \vdots \\ v_{15} \end{pmatrix}.$$

Then (4.4) can be cast into the form

(4.13) $$E\mathbf{A} = \left[\frac{\mathbf{p}^2}{2m} + \frac{1}{2}m\omega^2\mathbf{r}^2 - \frac{3}{2}\hbar\omega - \frac{\omega}{\hbar}\mathbf{L}\cdot\mathbf{S}\right]\mathbf{A}.$$

This is the non-relativistic energy obtained in [16].

5. Dirac equation: spin $\frac{1}{2}$

The details for this section are in [4]. Some recent developments, including the interaction with an external gauge field, are described in [2]. Here we examine the non-relativistic Dirac equation, also considered by Omote et al. in [17]. Our purpose is to see how the covariant formalism allows us to recover the non-relativistic limit of the Dirac oscillator [15].

The non-relativistic Dirac equation is

(5.1) $$(\gamma^\mu\partial_\mu + k)\Psi = 0, \quad \mu = 1,\ldots,5$$

written in momentum space as (4.4) with the β's replaced by γ's. The gamma matrices satisfy

(5.2) $$\{\gamma^\mu, \gamma^\nu\} = \gamma^\mu\gamma^\nu + \gamma^\nu\gamma^\mu = 2g^{\mu\nu}$$

and can be chosen as

$$\gamma^n = \begin{pmatrix} \sigma_n & 0 \\ 0 & -\sigma_n \end{pmatrix}, \quad \gamma^4 = \begin{pmatrix} 0 & 0 \\ -\sqrt{2} & 0 \end{pmatrix}, \quad \gamma^5 = \begin{pmatrix} 0 & \sqrt{2} \\ 0 & 0 \end{pmatrix}, \tag{5.3}$$

where each entry is a two-by-two matrix and the σ_n are the spin Pauli matrices. As for the integer spins, they generate the Lie algebra so(5,1) by taking $\gamma^\mu = J^{\mu 6}$, with $\mu = 1, \ldots, 5$. The adjoint spinor is defined as $\overline{\Psi} = \Psi^\dagger \zeta$, where

$$\zeta = \frac{-i}{\sqrt{2}}(\gamma^4 + \gamma^5) = \begin{pmatrix} 0 & -i \\ i & 0 \end{pmatrix}. \tag{5.4}$$

The matrix ζ plays a role similar to η, (4.3), in DKP equations.

Now let us consider directly the harmonic oscillator. If we write the Dirac equation in momentum space, using the representation, (5.3), and then perform the non-minimal substitution analogous to 4.9, with η now replaced by ζ, for a spinor

$$\Psi = \begin{pmatrix} \varphi \\ \chi \end{pmatrix} \tag{5.5}$$

one finds Lévy-Leblond equation [**7, 14**]:

$$(\boldsymbol{\sigma} \cdot \mathbf{p} - ik)\varphi + (\omega\boldsymbol{\sigma} \cdot \mathbf{r} + \sqrt{2}p_5)\chi = 0 \tag{5.6}$$
$$(\boldsymbol{\sigma} \cdot \mathbf{p} + ik)\chi + (\sqrt{2}p_4 - \omega\boldsymbol{\sigma} \cdot \mathbf{r})\varphi = 0.$$

Defining $p_4 = p_5$ and $\chi = -i\varphi$ we find (see [**4**, Appendix B])

$$E\varphi = \left(\frac{\mathbf{p}^2}{2m} + \frac{1}{2}m\omega^2\mathbf{r}^2 - \frac{3}{2}\hbar\omega - \frac{2}{\hbar}\omega\mathbf{L}\cdot\mathbf{S}\right)\varphi \tag{5.7}$$

where $\mathbf{S} \equiv \frac{1}{2}\hbar\boldsymbol{\sigma}$. Obviously, a similar equation can be obtained in terms of the field χ. This result, (5.7), is in agreement with the non-relativistic energy of the Dirac oscillator investigated in [**15**]. The non-relativistic Dirac equation with an external gauge field is studied in [**2**].

Concluding remarks

Our purpose has been to illustrate a formalism which allows one to write down non-relativistic equations using covariant expressions similar to the well known relativistic Lorentz covariance. We are currently investigating the quantization using this scheme. Further physical applications are related to superfluidity and Bose-Einstein condensates.

This contribution is dedicated to the memory of Professor Robert T. Sharp, human computer, baseball fan and an incredibly nice person! We are grateful to the organizers, in particular Professor P. Winternitz, for such a wonderful meeting.

References

1. M. de Montigny, F. C. Khanna, and A. E. Santana, *On Galilei-covariant Lagrangian models of fluids*, J. Phys. A **34** (2001), 10921–10937.
2. _____, *Non-relativistic wave equation with external gauge field*, Internat. J. Theoret. Phys. (to appear).
3. _____, *Lorentz-like covariant equations of non-relativistic fluids*, J. Phys. A 36 (2003), no. 8, 2009–2029.
4. M. de Montigny, F. C. Khanna, A. E. Santana, and E. S. Santos, *Galilean covariance and non-relativistic Bhabha equations*, J. Phys. A **34** (2001), 8901–8917.

5. M. de Montigny, F. C. Khanna, A. E. Santana, E. S. Santos, and J. D. M. Vianna, *Galilean covariance and the Duffin–Kemmer–Petiau equation*, J. Phys. A **33** (2000), L273–L278.
6. C. Duval, G. Burdet, H. P. Künzle, and M. Perrin, *Bargmann structures and Newton–Cartan theory*, Phys. Rev. D **31** (1985), 1841–1853.
7. W. I. Fuschich and A. G. Nikitin, *Symmetries of equations in quantum mechanics*, Allerton Press, New York, 1994.
8. M. Hassaïne and P. A. Horváthy, *Field-dependent symmetries of a non-relativistic fluid model*, Ann. Phys. **282** (2000), 218–246.
9. _____, *Relativistic Chaplygin gas with field-dependent Poincaré symmetry*, Lett. Math. Phys. **57** (2001), 33–40.
10. R. Jackiw, *Lectures on fluid dynamics*, CRM Ser. Math. Phys., Springer, New York, 2002.
11. E. Kapuścik, *On the physical meaning of the Galilean space-time coordinates*, Acta Phys. Pol. B **17** (1986), 569–575.
12. H. P. Künzle and C. Duval, *Relativistic and nonrelativistic physical theories on five-dimensional space-time*, Semantical Aspects of Spacetime Theories (U. Majer and H. J. Schmidt, eds.), BI-Wissenschaftsverlag, Mannheim, 1994, pp. 113-129.
13. M. Le Bellac and J. M. Lévy-Leblond, *Galilean electromagnetism*, Nuovo Cimento B **14** (1973), 217–233.
14. J. M. Lévy-Leblond, *Nonrelativistic particles and wave equations*, Comm. Math. Phys. **6** (1967), 286–311.
15. M. Moshinsky and A. Szczepaniak, *The Dirac oscillator*, J. Phys. A **22** (1989), L817–L819.
16. Y. Nedjadi and R. C. Barrett, *The Duffin–Kemmer–Petiau oscillator*, J. Phys. A **27** (1994), 4301–4315.
17. M. Omote, S. Kamefuchi, Y. Takahashi, and Y. Ohnuki, *Galilean covariance and the Schrödinger equation*, Fortschr. Phys. **37** (1989), 933–950.
18. G. Pinski, *Galilean tensor calculus*, J. Math. Phys. **9** (1968), 1927–1930.
19. D. E. Soper *Classical field theory*, Section 7.3, Wiley and Sons, New York, 1976.
20. L. Susskind, *Model of self-dual strong interactions*, Phys. Rev. **165** (1968), 1535–1546.
21. Y. Takahashi, *Towards the many-body theory with the Galilei invariance as a guide. I*, Fortschr. Phys. **36** (1988), 63–81; II, 83–96.
22. _____, *An invitation to a Galilei invariant world*, Wandering in the Fields. (K. Kwarabayashi and A. Ukawa, eds.), World Scientific, Singapore, 1987, pp. 117–127.

FACULTÉ SAINT-JEAN, UNIVERSITY OF ALBERTA, 8406, RUE MARIE-ANNE GABOURY, EDMONTON, AB T6C 4G9, CANADA
E-mail address: montigny@phys.ualberta.ca

DEPARTMENT OF PHYSICS, UNIVERSITY OF ALBERTA, 412 AVADH BHATIA PHYSICS LABORATORY, EDMONTON, AB T6G 2J1, CANADA
E-mail address: khanna@phys.ualberta.ca

INSTITUTO DE FÍSICA, UNIVERSIDADE FEDERAL DA BAHIA, CAMPUS DE ONDINA, SALVADOR, BAHIA 40210-340, BRAZIL
E-mail address: santana@ufba.br

Variations on Dedekind's Eta

Terry Gannon

This paper is dedicated to Professor R. T. Sharp.

1. Introduction

Bob Sharp sat on my Ph.D. examining committee, and asked more and harder questions than anyone else. Perhaps it was because he really wasn't impressed—I'll never know. But I'll take the more optimistic view here that it was because he was interested in it, at least a little. In any case, in my talk and in this paper, I'll discuss a couple of developments concerning the material in my thesis.

It may seem that this paper has little to do with the other contributions to this volume. Actually, many of these papers fit into a common larger picture: on the one hand we have algebraic structures and on the other we have modular forms and functions (I'll define these shortly). These two seemingly unrelated mathematical areas are connected in the middle by *conformal field theory* (see, e.g., the book [7] and references therein), or what is the same thing, *perturbative string theory* (see, e.g., the book [13]). For example, on one side you have lattices Λ, and on the other their theta functions $\Theta_\Lambda(\tau)$ (see (2.2) below) and the closely related Dedekind eta η, and they are connected by the bosonic string compactified on the lattice. On one side you have affine Kac–Moody algebras (loop algebras), on the other their characters χ_λ, and they are connected by the WZW models. On the one side you have the Monster finite simple group, and on the other we have the j-function and other Hauptmoduls, and in the middle you have a remarkable $c = 24$ conformal field theory. This paper pertains mostly to the lattice example.

The Dedekind eta is the function

$$(1.1) \qquad \eta(\tau) = q^{1/24} \prod_{n=1}^{\infty}(1-q^n) = q^{1/24}\sum_{m\in\mathbb{Z}}(-1)^m q^{(3m^2+m)/2},$$

where τ lives in the upper half-plane $\mathbb{H} := \{\tau \in \mathbb{C} \mid \operatorname{Im}\tau > 0\}$ and $q := e^{2\pi i\tau}$ lives in the unit disc. It arises in many contexts and is probably familiar to any reader. It appears throughout string theory—for instance the 1-loop partition function of d free bosons is proportional to η^{-d}. Also, [4] uses identities it satisfies, in order to

---

2000 *Mathematics Subject Classification.* Primary 11F03; Secondary 81T30, 11E45.
This research is supported in part by NSERC.
This is the final form of the paper.

come up with very fast recursive algorithms for computing digits of π. The deep relation of $\log \eta$ to geometry is discussed in [**1**].

Most important is its role as a *modular form* for the group $\mathrm{SL}_2(\mathbb{Z})$ of 2×2 integer matrices with determinant 1. This means that it is holomorphic in $\mathbb{H}$ (also at the cusps $\mathbb{Q} \cup \{\infty\}$, but we can ignore that), and most important, it obeys the symmetry

(1.2a) $$\eta\left(\frac{a\tau+b}{c\tau+d}\right) = \xi \sqrt{c\tau+d}\, \eta(\tau), \quad \forall \begin{pmatrix} a & b \\ c & d \end{pmatrix} \in \mathrm{SL}_2(\mathbb{Z}),$$

where ξ is some 24th root of unity. For example,

(1.2b) $$\eta(\tau+1) = e^{\pi i/12}\, \eta(\tau),$$

(1.2c) $$\eta(-1/\tau) = \sqrt{\frac{\tau}{i}}\, \eta(\tau),$$

and these generate all other transformations. Other important examples of modular forms are

(1.3a) $$\theta_3(\tau) = \sum_{n \in \mathbb{Z}} q^{n^2/2} = \prod_{m=1}^{\infty}(1-q^m)(1+q^{(2m-1)/2})^2,$$

(1.3b) $$\theta_2(\tau) = \sum_{n \in \mathbb{Z}} q^{(n+1/2)^2/2} = 2q^{1/8} \prod_{m=1}^{\infty}(1-q^m)(1+q^m)^2 = \theta_3(\tau/4) - \theta_3(\tau),$$

(1.3c) $$\theta_4(\tau) = \sum_{n \in \mathbb{Z}} (-1)^n q^{n^2/2} = \prod_{m=1}^{\infty}(1-q^m)(1-q^{(2m-1)/2})^2 = 2\theta_3(4\tau) - \theta_3(\tau).$$

$\mathrm{SL}_2(\mathbb{Z})$ transforms them into each other; the most important relations are

(1.3d) $$\theta_3(-1/\tau) = \sqrt{\frac{\tau}{i}}\, \theta_3(\tau),$$

(1.3e) $$\theta_2(-1/\tau) = \sqrt{\frac{\tau}{i}}\, \theta_4(\tau).$$

If you've never seriously encountered modular forms before, they must look a little arbitrary. One way to see their importance is to consider *Riemann surfaces*, i.e., a (real) surface with a complex structure on it. As the reader surely knows, any smooth surface S, with or without punctures, is diffeomorphic (using C^∞-maps) to a sphere (genus $g = 0$), a torus ($g = 1$), a double-torus ($g = 2$), ..., with a number n of punctures. That is, a smooth surface is completely characterised (up to C^∞ equivalence) by the genus g and number n of punctures. But if we consider this 2-dimensional *real surface* to be a one-dimensional *complex curve*—i.e., to be a Riemann surface—then there is a continuum of different curves (up to complex analytic equivalence) corresponding to the same real surface, i.e., the same pair (g,n). More precisely, this space of different surfaces is called the *moduli space* $\mathcal{M}_{g,n}$, and for all but the smallest pairs (g,n) this space has real dimension $6g-6+2n$. But apart from a few small (g,n) exceptions, all Riemann surfaces are (complex analytic) equivalent to one of the form $\mathbb{H}/G$, where G is a discrete subgroup of the Lie group $\mathrm{SL}_2(\mathbb{R})$ of symmetries of the hyperbolic plane = upper half-plane $\mathbb{H}$, which act on $\mathbb{H}$ by fractional linear transformations $\begin{pmatrix} a & b \\ c & d \end{pmatrix}.\tau = (a\tau+b)/(c\tau+d)$.

For any such surface $S = \mathbb{H}/G$, a meromorphic function $f \colon S \to \mathbb{C}$ living on S will correspond to a meromorphic function $f \colon \mathbb{H} \to \mathbb{C}$ with symmetry G:

$f((a\tau+b)/(c\tau+d)) = f(\tau)$. These f are called *modular functions* for the group G. Differential k-forms $f\,\mathrm{d}\tau^k$ on S would correspond to functions obeying

$$f\left(\frac{a\tau+b}{c\tau+d}\right)\left(\frac{\mathrm{d}}{\mathrm{d}\tau}\frac{a\tau+b}{c\tau+d}\right)^k \mathrm{d}\tau^k = f(\tau)\,\mathrm{d}\tau^k,$$

i.e., $f((a\tau+b)/(c\tau+d)) = (c\tau+d)^{2k} f(\tau)$. As the reader knows, differential k-forms (for $k = 1, 2$) are the integrands of contour and area integrals, respectively, on S. Any such f, for any k, is called a *modular form* of level $2k$ for G.

The simplest and most important possibility for a group G is $\mathrm{SL}_2(\mathbb{Z})$; its orbit space $\mathbb{H}/\mathrm{SL}_2(\mathbb{Z})$ (which can be naturally identified with the moduli spaces $\mathcal{M}_{1,0}$ and $\mathcal{M}_{1,1}$ of tori) corresponds topologically to the plane $\mathbb{R}^2 \cong \mathbb{C}$. Its modular functions are generated by (i.e., are rational functions of) the so-called j-function, which is the partition function of the $c = 24$ Moonshine CFT. Both $\eta(\tau)^{24}$ and $\theta_2(\tau)^8 + \theta_3(\tau)^8 + \theta_4(\tau)^8$ are modular forms for $\mathrm{SL}_2(\mathbb{Z})$, of weights 12 and 4 respectively; the j-function equals

$$j(\tau) = \frac{(\theta_2(\tau)^8 + \theta_3(\tau)^8 + \theta_4(\tau)^8)^3}{8\eta(\tau)^{24}},$$

and so has weight 0, as it must. Another important choice for G is the subgroup of $\mathrm{SL}_2(\mathbb{Z})$ generated by $\tau \mapsto \tau+2$ and $\tau \mapsto -1/\tau$: θ_3 is a modular form (of weight $k = \frac{1}{2}$, with a "multiplier") for this G; the corresponding surface $\mathbb{H}/G$ is a cylinder.

We will discuss two topics. The main one concerns a 1-parameter deformation of the η-function which appeared recently in string theory. We will address this in Section 3. The other, which we give next section, connects directly with my thesis, and with Bob Sharp.

2. Eta and theta identities

Modular form and function identities have a wide range of uses. For example, the identity

(2.1) $$\theta_3(\tau)^4 = \theta_2(\tau)^4 + \theta_4(\tau)^4$$

can be used to prove that every integer $n > 0$ is a sum of 4 squares $n = a^2+b^2+c^2+d^2$, as well as analytically embed tori explicitly into the complex projective 3-space $\mathbb{P}^3(\mathbb{C})$. $\eta(\tau)$ identities have been used to accurately compute π [4], while identities ("modular equations") involving the j-function and other Hauptmoduls plays a central role [6] in the proof of the Monstrous Moonshine conjectures. The $\sum = \prod$ identity (1.3a) is related to "bosonisation" in string theory (see [13, p. 155]). A polynomial identity relating $\theta_3(\tau)/\eta(\tau)$ to $\theta_3(5\tau)/\eta(5\tau)$ can be used to solve any quintic polynomial (as is well known, radicals alone cannot solve general polynomials of degree ≥ 5). The great Indian mathematician Ramanujan came up with several identities involving theta and eta functions (see, e.g., [3]), as does string theory to this day.

Part of my thesis, as well as the papers [9, 10] coauthored with my thesis supervisor C.S. Lam, developed the following systematic way to generate identities. Start with any n-dimensional lattice Λ. This discrete periodic set can be thought of as the span over the integers of a basis $\{\mathbf{v}_1, \ldots, \mathbf{v}_n\}$ of $\mathbb{R}^n$. The standard reference for lattice lore is [5]. We are interested in *rational* lattices, i.e. we require that all dot-products $\mathbf{u} \cdot \mathbf{v}$ in Λ be rational numbers. To any such lattice define its *theta function* Θ_Λ by

$$\Theta_\Lambda(\tau) := \sum_{\mathbf{v} \in \Lambda} q^{\mathbf{v}^2/2}. \tag{2.2}$$

This theta function will be a modular form for some subgroup of $\mathrm{SL}_2(\mathbb{Z})$, as will the theta function of any translate $\Lambda + \mathbf{t}$, for $\mathbf{t} \in \mathbb{Q}\Lambda$. For example, the only rational 1-dimensional lattices are $\sqrt{r}\mathbb{Z}$; their theta functions are $\theta_3(r\tau)$, while that of the translate $\mathbb{Z} + r$ equals $\psi_{r^{-1}}(\tau)$, where

$$\psi_r(\tau) := \sum_{n \in \mathbb{Z}} q^{(n+1/r)^2/2}. \tag{2.3a}$$

Note the obvious equalities

$$\psi_r(\tau) = \psi_{r/(rn\pm 1)}(\tau), \qquad \forall n \in \mathbb{Z}, \tag{2.3b}$$

$$\theta_3(n\tau) = \sum_{k=1}^{n} \psi_{n/k}(\tau), \qquad \forall n = 1, 2, 3, \ldots. \tag{2.3c}$$

Together, (2.3b) and (2.3c) exhaust all linear identities obeyed by the ψ's [9].

The Gram–Schmidt orthogonalisation process implies that any n-dimensional rational lattice Λ has several n-dimensional orthogonal sublattices Λ_0. An orthogonal lattice is a lattice with an orthogonal basis—up to rotation it looks like the orthogonal direct sum $\sqrt{r_1}\mathbb{Z} \oplus \cdots \oplus \sqrt{r_n}\mathbb{Z}$, and has theta function $\theta_3(r_1\tau) \cdots \theta_3(r_n\tau)$. The quotient Λ/Λ_0 will be a finite abelian group of size $\sqrt{|\Lambda_0|/|\Lambda|}$, where $|\Lambda|$ denotes the determinant of Λ (see [5]; geometrically, $|\Lambda|$ is the volume-squared of a fundamental region of Λ). Hence we can write $\Lambda/\Lambda_0 \cong \mathbb{Z}_{k_1} \times \mathbb{Z}_{k_2} \times \cdots \times \mathbb{Z}_{k_n}$, where $\mathbb{Z}_k$ is the cyclic group $\mathbb{Z}/k\mathbb{Z}$, and where each k_i divides k_{i+1}. The largest of these, k_n, is called the *exponent* of the quotient Λ/Λ_0; the order of any class $[\mathbf{v}] \in \Lambda/\Lambda_0$ will divide it.

Each choice of n-dimensional orthogonal sublattice $\Lambda_0 \subset \Lambda$ enables us to write Θ_Λ as a sum and product of the 1-dimensional theta functions ψ_r in (2.3a). In fact, Θ_Λ will be a homogeneous polynomial of degree n in these ψ's. This is because we get the decomposition $\Lambda = \bigcup_{[\mathbf{v}] \in \Lambda/\Lambda_0} [\mathbf{v}]$ of Λ into translates of Λ_0. The numerators of the r's involved in this expression will divide the exponent k_n of Λ/Λ_0. Comparing these expressions for Θ_Λ, for different choices of Λ_0, we obtain polynomial identities for the ψ's.

Now, certain of the ψ's can be expressed in terms of θ_3: the complete list is

$$\psi_1(\tau) = \theta_3(\tau),$$
$$\psi_2(\tau) = \theta_2(\tau) = \theta_3(\tau/4) - \theta_3(\tau),$$
$$\psi_3(\tau) = \tfrac{1}{2}\{\theta_3(\tau/9) - \theta_3(\tau)\},$$
$$\psi_4(\tau) = \tfrac{1}{2}\theta_2(\tau/4),$$
$$\psi_6(\tau) = \tfrac{1}{2}\{\theta_2(\tau/9) - \theta_2(\tau)\},$$

together with the corresponding expressions coming from (2.3b) ($\psi_{3/2} = \psi_3$ etc). These expressions are immediate consequences of (2.3b) and (2.3c). If the exponent k_n of the quotient group Λ/Λ_0 equals 1, 2, 3, 4, or 6, then the function Θ_Λ can be expressed solely in terms of θ_3; two such Λ_0 will yield a degree-n polynomial identity for θ_3. This is the key observation of [9].

For the simplest example, take the square lattice $\Lambda = \mathbb{Z} \oplus \mathbb{Z}$. The obvious choice for orthogonal sublattice Λ_0 is Λ itself, and we obtain $\Theta_\Lambda(\tau) = \theta_3(\tau)^2$. A second choice though is $\Lambda_0 = \text{span}\{(1,1),(1,-1)\} \cong \sqrt{2}\mathbb{Z} \oplus \sqrt{2}\mathbb{Z}$, and for it we obtain $\Theta_\Lambda(\tau) = \psi_1(2\tau)^2 + \psi_2(2\tau)^2$. Thus we get the identity

$$\theta_3(\tau)^2 = \theta_3(2\tau)^2 + \theta_2(2\tau)^2.$$

In [9] about 30 different quadratic identities for θ_3 were found, all but 3 of which were new. All known quadratic θ_3 identities are in the list of [9] (or trivially derived from identities there, by applying modular transformations). Conjecturally, all polynomial identities for θ_3 can be derived in this way from lattices. Strong additional support for this conjecture was provided in [10], where it was proved that if the identity has full variable dependence (so a polynomial identity of degree n would have n complex variables z_i in addition to τ) then that identity comes from a lattice in the above sense. Conversely, any ψ identity produced by this lattice method can be "lifted" into an identity with full variable dependence.

Reference [11] pushes this a little further. It begins with the observation that ψ_{12} and $\psi_{12/5}$ can be expressed in terms of θ_3 and η: in particular

$$\theta_3(\tau/144) - \theta_3(\tau/9) - \theta_2(\tau/9) - \theta_2(\tau/4) \pm 2\eta(\tau/12)$$

equals $4\psi_{12}(\tau)$ and $4\psi_{12/5}(\tau)$, respectively. Thus if we simply widen the possibilities for the orthogonal sublattice Λ_0 by allowing the quotients Λ/Λ_0 to have exponent 12, then the above method will produce identities involving θ_3 and η. Again, conjecturally this method produces all polynomial identities involving θ_3 and η.

Of course, many such identities are already known—e.g.

$$\eta(2\tau)\theta_4(2\tau) = \eta(\tau)^2,$$
$$\theta_2(\tau)\theta_3(\tau)\theta_4(\tau) = 2\eta(\tau)^3.$$

In [11] we find over 100 different new quadratic identities in η and θ_3 (hence θ_2 and θ_4). Conjecturally our list is complete.

It can be asked: what meaning can be given to the ψ identities when the exponents of the quotients Λ/Λ_0 are other than 1, 2, 3, 4, 6, 12? [11] also provides the answer there: these identities can be re-expressed as identities between various Dirichlet twists of θ_3. The Dedekind eta is an example of one of these twists.

3. Deformations of the Dedekind eta

For any real number $m \geq 0$, define as in [2] the function

(3.1a) $$\eta^{(m)}(\tau) = q^{-\Delta_m}\sqrt{1-q^m}\prod_{n=1}^{\infty}(1-q^{\sqrt{m^2+n^2}}),$$

where

$$\Delta_m = -\frac{1}{(2\pi)^2}\sum_{n=1}^{\infty}\int_0^{\infty} e^{-n^2 s - \pi^2 m^2/s}\,ds = -\frac{1}{8\pi}\int_0^{\infty} e^{-\pi m^2/s}(\theta_3(is)-1)\,ds.$$

Note that

(3.1b) $$\eta^{(m)}(\tau) \to \sqrt{\frac{\tau}{i}}\sqrt{2\pi m}\,\eta(\tau) \quad \text{as } m \to 0,$$

so we should regard $\eta^{(m)}$ as a deformation of η and m as the deformation parameter. Then

(3.1c) $$\eta^{(mt)}(\mathrm{i}/t) = \eta^{(m)}(\mathrm{i}t), \quad \forall t > 0.$$

In the $m \to 0$ limit this recovers equation (1.2c).

Note that for each fixed complex m with $\mathrm{Re}(m) > 0$, $\eta^{(m)}(\tau)$ exists and is holomorphic throughout

$$\mathbb{H}^{(m)} := \{\tau \in \mathbb{H} \mid \mathrm{Im}(m\tau) > 0\}.$$

$\mathbb{H}^{(m)}$ will be an open wedge containing the positive imaginary axis. Similarly, for fixed $\tau \in \mathbb{H}$, $\eta^{(m)}(\tau)$ is a holomorphic function of m in the domain

$$\mathbb{H}_{(\tau)} := \{m \in \mathbb{C} \mid \mathrm{Re}(m) > 0 \text{ and } \mathrm{Im}(m\tau) > 0\}.$$

Thus (3.1c) generalises to

(3.1d) $$\eta^{(m\tau/\mathrm{i})}(-1/\tau) = \eta^{(m)}(\tau), \quad \forall \tau \in \mathbb{H}^{(m)}, \ \mathrm{Re}(m) > 0.$$

To see this, first establish (3.1c) for any $\mathrm{Re}(m) > 0$ by considering the function $\eta^{(mt)}(\mathrm{i}/t) - \eta^{(m)}(\mathrm{i}t)$ for fixed t: it is holomorphic throughout $\mathrm{Re}(m) > 0$, and vanishes along the positive real axis, so it vanishes everywhere. Then (3.1d) is established by considering the function $\eta^{(m\tau/\mathrm{i})}(-1/\tau) - \eta^{(m)}(\tau)$, for fixed m with $\Re(m) > 0$.

Incidentally, there is an alternative for extending the domain of $\eta^{(m)}(\mathrm{i}t)$ and similar functions, and is given in [16], generalising a result of [15]. This extension will not be meromorphic (unlike $\eta^{(m)}(\tau)$), but it will also be covariant with respect to translations $\tau + n$ (also unlike $\eta^{(m)}(\tau)$). So something is lost and something gained, and it is quite possible that modularity is more desirable than holomorphicity here (as we'll discuss later). In any case it would be interesting to interpret our results along the lines of [16]—the functions $Z_{a,b}^{(m)}$ of [16] correspond to the square of what we will call $F_{\psi_{1/b}(\tau),\theta_3(\tau+a)}^{(m)}$ below.

The function $\eta^{(m)}(\mathrm{i}t)$ arises physically, in a perturbative type IIB string theory, as follows. The masslike parameter m is a measure of how curved is the 10-dimensional background space-time. $\eta^{(m)}$ arises in the one-loop vacuum-to-vacuum amplitude of an open string with endpoints on a pair of D-branes (i.e., a cylindrical worldsheet). "t" parametrises the different cylinders. The modular transformation $t \mapsto 1/t$ interchanges time and space on the worldsheet, and takes the open string to a closed one exchanged between the two D-branes. Δ_m is the Casimir energy of a single boson of mass m on the cylinder. See [2] for details; related functions can also be found in [8, 15, 16]. (It is interesting that the relation of string theory to modular forms is only apparent perturbatively.)

Of course it is easy to construct functions invariant (or covariant) under any finite group. Here we can take *any* function $f: \mathbb{H} \to \mathbb{C}$, then of course the averaged function $\tau \mapsto f(\tau) + f(-1/\tau)$ will be invariant under $\tau \mapsto -1/\tau$. This isn't the point however. In a similar way, it is easy to find, e.g., modular forms of $\mathrm{SL}_2(\mathbb{Z})$ (in fact the so-called Eisenstein functions are trivially modular and generate them all), but that doesn't trivialise the theory of modular forms—quite the contrary. The point is that there are functions arising naturally in interesting contexts, which possess an unexpected symmetry.

The transformation properties of η and the theta functions with respect to translations $\tau \mapsto \tau + n$ are obvious and elementary. The behaviour with respect to $\tau \mapsto -1/\tau$ remains mysterious. For instance this (and its straightforward generalisation to the so-called Siegel modular forms) was the theme of the major paper [17]. Weil's opening paragraph is pure moonshine:

> By force of habit, the fact that theta series define modular forms has nearly ceased to amaze us. But the appearance of the symplectic group [i.e., $\mathrm{Sp}_{2g}(\mathbb{Z})$, e.g., $\mathrm{Sp}_2(\mathbb{Z}) \cong \mathrm{SL}_2(\mathbb{Z})$] as a *deus ex machina* in the famous work of Siegel on quadratic forms has still lost none of its mysterious character.

Weil "sheds a little light" on this mystery by relating the modularity to the representation theory of the Heisenberg group. On the other hand, the deformations $\eta^{(m)}$ probe this mystery in a different way by focusing on the most important modular transformation: $\tau \mapsto -1/\tau$.

The standard way to prove, e.g., (1.3d) is to use Poisson summation, but this doesn't work here (at least not directly). Instead, we will express $\eta^{(m)}$ as an integral involving two θ_3's, and then apply (3d). The Taylor expansion of '$\log(1-x)$' gives

$$\log \eta^{(m)}(\mathrm{i}t) = 2\pi t \Delta_m - \sum_{\ell=1}^{\infty} \frac{1}{2\ell} q^{\ell m} - \sum_{n=1}^{\infty} \sum_{\ell=1}^{\infty} \frac{1}{\ell} q^{\ell \sqrt{m^2+n^2}}.$$

Rewrite $q^{\ell\sqrt{m^2+n^2}}$ using the integral identity ([12, number 3.325])

$$\exp(-2\sqrt{ab}) = 2\sqrt{\frac{a}{\pi}} \int_0^\infty \exp(-ax^2 - bx^{-2})\, \mathrm{d}x$$

with $a = \ell^2 \pi t, b = (m^2+n^2)\pi t$, and making the change-of-variables $x^2 = s$, we obtain the desired expression:

(3.1e) $$\log \eta^{(m)}(\mathrm{i}t) = 2\pi t \Delta_m + \frac{2\pi}{t}\Delta_{mt}$$
$$- \frac{\sqrt{t}}{4} \int_0^\infty s^{-1/2} e^{-\pi t m^2/s} (\theta_3(\mathrm{i}st) - 1)\left(\theta_3(\mathrm{i}t/s) - \sqrt{\frac{s}{t}}\right) \mathrm{d}s.$$

Now apply (1.3d) to the second θ_3; we get

$$\log \eta^{(m)}(\mathrm{i}t) = 2\pi t \Delta_m + \frac{2\pi}{t}\Delta_{mt} - \frac{1}{4}\int_0^\infty e^{-\pi t m^2/s}(\theta_3(\mathrm{i}st)-1)(\theta_3(\mathrm{i}s/t)-1)\,\mathrm{d}s,$$

which is manifestly invariant under the simultaneous transformations $t \mapsto 1/t$, $m \mapsto mt$. A more explicit derivation of (3.1c) is given in [2], but perhaps it leaves somewhat obscure the transformation law (3.1c).

Similar deformations exist for the theta functions $\theta_2, \theta_3, \theta_4$:

(3.2a) $$\theta_2^{(m)}(\tau) = q^{-3\Delta_m}\sqrt{1-q^m}(1+q^m)\prod_{n=1}^{\infty}(1-q^{\sqrt{m^2+n^2}})(1+q^{\sqrt{m^2+n^2}})^2,$$

(3.2b) $$\theta_3^{(m)}(\tau) = q^{-\overline{\Delta}_m}\sqrt{1-q^m}\prod_{n=1}^{\infty}(1-q^{\sqrt{m^2+n^2}})(1+q^{\sqrt{m^2+(n+1/2)^2}})^2,$$

(3.2c) $$\theta_4^{(m)}(\tau) = q^{-\overline{\Delta}_m}\sqrt{1-q^m}\prod_{n=1}^{\infty}(1-q^{\sqrt{m^2+n^2}})(1-q^{\sqrt{m^2+(n+1/2)^2}})^2,$$

where

$$\overline{\Delta}_m = -\frac{1}{8\pi}\int_0^\infty e^{-\pi m^2/s}(\theta_3(\mathrm{i}s) + 2\theta_4(\mathrm{i}s) - 3)\,\mathrm{d}s.$$

For $m \to 0$, $\theta_i(\tau) \to \sqrt{2\pi m\tau/\mathrm{i}}\theta_i(\tau)$ as before, for $i = 2, 3, 4$, so again they are deformations. As before, they are holomorphic throughout $\tau \in \mathbb{H}^{(m)}$, for any fixed complex m with $\mathrm{Re}(m) > 0$. Using the previous arguments, it can be verified that

$$\theta_2^{(mt)}(\mathrm{i}/t) = \theta_4^{(m)}(\mathrm{i}t),$$
$$\theta_3^{(mt)}(\mathrm{i}/t) = \theta_3^{(m)}(\mathrm{i}t),$$
$$\theta_4^{(mt)}(\mathrm{i}/t) = \theta_2^{(m)}(\mathrm{i}t),$$

and as before the analogues of (3.1d) hold.

We see that the key to (3.1c) is the appearance in (3.1e) of the two theta functions. An obvious generalisation of this deformation then is to replace θ_3 with any modular form. In particular, let f, g, f', g' be functions obeying

$$f(\mathrm{i}/t) = t^k f'(\mathrm{i}t), \qquad g(\mathrm{i}/t) = t^k g'(\mathrm{i}t), \quad \forall t > 0,$$

for some constant k. For example, f, g could be modular forms with multipliers, for some group G containing $\begin{pmatrix} 0 & 1 \\ -1 & 0 \end{pmatrix}$. Suppose the limit $\lim_{t\to\infty}$ of f, g, f', g' is a_0, b_0, a'_0, b'_0, respectively, and write $f(\mathrm{i}t) = \sum_{\ell \geq 0} a_\ell q^\ell$ and $g(\mathrm{i}t) = \sum_{n \geq 0} b_n q^n$ (ℓ, n need not be integers, but this should be a discrete sum and in the most interesting cases n, ℓ will be rational). Define a function $F_{f,g}^{(m)}$ by

(3.3a)

$$\log F_{f,g}^{(m)}(\mathrm{i}t) = 2\pi t^{k+1/2}\Delta_m(f,g) + 2\pi t^{-k-1/2}\Delta_{mt}(g',f')$$
$$- \frac{t^k}{4}\int_0^\infty s^{-1/2}e^{-\pi tm^2/s}(g(\mathrm{i}st) - b_0)\left(f(\mathrm{i}t/s) - a'_0\left(\frac{s}{t}\right)^k\right)\mathrm{d}s$$

(3.3b)
$$= -2\pi t^{k+1/2}\Delta_{f,g}^{(m)} - \frac{t^{k-1/2}}{4}\sum_{\ell>0}\sum_{n\geq 0}\frac{a_n b_\ell}{\sqrt{2\ell}}q^{\sqrt{2\ell(2n+m^2)}},$$

where

$$\Delta_m(f,g) = \frac{-b_0}{8\pi}\int_0^\infty s^{k-1/2}e^{-\pi m^2/s}(f'(\mathrm{i}s) - a'_0)\,\mathrm{d}s$$

(we assume these sums and integrals all exist). Then from (3.3a)

(3.3c) $$F_{f,g}^{(m)}(\mathrm{i}t) = F_{g',f'}^{(mt)}(\mathrm{i}/t).$$

The choice of $f = g = \theta_3$ recovers $\eta^{(m)}$, so $F_{f,g}^{(m)}$ can be regarded as a fairly large generalisation of $\eta^{(m)}$. Similarly, the functions $F_{\theta_i,\theta_j}^{(m)}$ for $(i,j) = (3,4), (2,4), (2,3)$, respectively, equal the functions $f_2^{(m)}, f_3^{(m)}, f_4^{(m)}$ appearing in [**2**].

Note that $\log F_{f,g}^{(m)}$ is bilinear in f, g: for any constants A, B, C, D,

(3.4a) $F_{Af_1+Bf_2, Cg_1+Dg_2}^{(m)}(\mathrm{i}t)$
$$= (F_{f_1,g_1}^{(m)}(\mathrm{i}t))^{AC}(F_{f_1,g_2}^{(m)}(\mathrm{i}t))^{AD}(F_{f_2,g_1}^{(m)}(\mathrm{i}t))^{BC}(F_{f_2,g_2}^{(m)}(\mathrm{i}t))^{BD}.$$

For example, $F^{(m)}_{\theta_3,\theta_3+2\theta_4} = \theta_2^{(m)}$ and $F^{(m)}_{\theta_3+2\theta_2,\theta_3} = \theta_4^{(m)}$. Rescaling the arguments of f,g also doesn't give anything new: for any constants $\alpha,\beta > 0$, let $\tilde{f}(it) = f(i\alpha t)$ and $\tilde{g}(it) = g(i\beta t)$; then

(3.4b) $$F^{(m)}_{\tilde{f},\tilde{g}}(\mathrm{i}t) = (F^{(m\sqrt{\alpha})}_{f,g}(\mathrm{i}\sqrt{\alpha\beta}t))^a,$$

where $a = \alpha^{-k/2+1/4}\beta^{-k/2-1/4}$.

It would be nice to identify what (if anything) $F^{(m)}_{f,g}$ is the deformation of, or more precisely, what the limit $m \to 0$ of $F^{(m)}_{f,g}$ equals. First, note that $\Delta_0(f,g)$ equals $-b_0/(8\pi)$ times the Mellin transform of $f' - a'_0$, evaluated at $k+1/2$. We are most interested in the case where $\Delta_0(f,g)$ is rational (as it is for $\eta^{(m)}$). Equation (3.3b) gives us the q-expansion. The sum there over ℓ when $n = 0$ will typically diverge to $-\infty$ (when $a_0 \neq 0$), which tells us that (like $\eta^{(m)}$) $F^{(m)}_{f,g}$ will vanish as $m \to 0$. Raising $F^{(m)}_{f,g}$ to the power $t^{1/2-k}$ (i.e., dividing this from its log), and removing that vanishing factor, we see that the resulting function will have a chance to be invariant under some translation $\mathrm{i}t \mapsto \mathrm{i}t + N$ provided f and g have the property that a_ℓ, b_n vanish unless $\sqrt{\ell}, \sqrt{n} \in \mathbb{Q}$. Of course this is true for $\eta^{(m)}$, but it also is true for many other choices of f,g: e.g., $\theta_4, \theta_2, \psi_r$.

Thus we can expect that the most interesting of these "deformations" $F^{(m)}_{f,g}$ will be for choices such as $f,\, g \in \{\theta_3, \psi_r\}$, for $r \in \mathbb{Q}$, and also for any linear combination thereof (e.g., Dirichlet twists and rational translates such as η and $\theta_3(\tau+r)$). For all these choices, $k = \frac{1}{2}$. In these cases we will also have holomorphicity etc., as for $\eta^{(m)}$. The value of the integral Δ_0 for $f = \psi_r$ is the dilogarithm Li_2 evaluated at a root of unity:

$$\Delta_0(g,\psi_r) = \frac{b_0}{4\pi^2}\sum_{k=1}^{\infty}\frac{\cos(2k\pi/r)}{k^2} = b_0\left(\frac{1}{24} - r^{-1} + r^{-2}\right),$$

which will be rational (provided b_0 is), where we used [**12**, equation 1.443]:

$$\sum_{k=1}^{\infty}\frac{\cos(kx)}{k^2} = \frac{\pi^2}{6} - \pi\frac{x}{2} + \frac{x^2}{4}.$$

Suppose $f(\tau) = \sum_i f_i \psi_{r_i}(m_i)$, $g(\tau) = \sum_j g_j \psi_{s_j}(n_j\tau)$ (both finite sums), where each $r_i, s_j, \sqrt{m_i}, \sqrt{n_j} \in \mathbb{Q}$, and $f_i, g_j \in \mathbb{Z}$. Write $k = a_0 \sum_j g_j/2$. Then $F_0(\mathrm{i}t) := \lim_{m \to 0}(mt)^{-k}F^{(m)}_{f,g}(\mathrm{i}t)$ exists, has meromorphic extension throughout $\mathbb{H}$, obeys $F_0(-1/\tau) = (\tau/i)^k F_0(\tau)$, and is invariant under some translation $\tau \mapsto \tau+N$. This follows from (3.3b) by evaluating the $m \to 0$ asymptotics of the $n=0$ term of (3.3b) for $g = \psi_r$, which is dominated by $-\frac{1}{2}\log(1-q^m)$ for any r. Thus these $F^{(m)}_{f,g}$'s are indeed quite interesting deformations of certain (meromorphic) modular forms of arbitrary level.

Of course we can't expect that these $F^{(m)}_{f,g}$ will always have a nice product form, even though $\eta^{(m)}$ does, since most modular forms don't have such a form. However, any $F_{f,\theta_3(\tau+r)}$ will have a product form. Moreover, we'll see next how the $\eta^{(m)}$ product generalises to the situation where $\mathbb{Z}$ there can be replaced by any n-dimensional lattice.

Choose any pair Λ, Λ' of n-dimensional (positive-definite) rational lattices, as well as any positive real number m. Let $\mathcal{P}$ be the set of all 'primitive vectors' in Λ', i.e., all vectors $\mathbf{v}$ in Λ' with the property that the only scalar multiples $x\mathbf{v}$ which are in Λ', are integer multiples. E.g., if $\Lambda' = \mathbb{Z}$, then $\mathcal{P}$ consists of the 2 "vectors" ± 1. If however the dimension of Λ' is greater than 1, then $\mathcal{P}$ will contain infinitely many vectors. Note that if $\mathbf{v}$ is in $\mathcal{P}$, so will be $-\mathbf{v}$; by '$\mathcal{P}/\pm$' below we mean to take one representative from each pair $\pm\mathbf{v} \in \mathcal{P}$—it doesn't matter which ones are chosen. For later convenience, write $Q(\mathbf{v})$ for the quantity $\mathbf{v}^2 + m^2$, and let $|\mathbf{v}|$ denote the length $\sqrt{\mathbf{v}^2}$ of the vector $\mathbf{v}$.

Define the function $E_{\Lambda,\Lambda'}^{(m)}(\mathrm{i}t)$, for $q = e^{-\pi t}$, by

$$(3.5\mathrm{a}) \qquad E_{\Lambda,\Lambda'}^{(m)}(\mathrm{i}t) := q^{-t^{(n-1)/2}\Delta_m(\Lambda,\Lambda')} \prod_{\mathbf{w}\in\mathcal{P}/\pm} \prod_{\mathbf{v}\in\Lambda} (1 - q^{|\mathbf{w}|Q(\mathbf{v})})^{t^{(n-1)/2}/(2|\mathbf{w}|)},$$

where

$$\Delta_m(\Lambda,\Lambda') = \frac{-1}{8\pi\sqrt{|\Lambda|}} \int_0^\infty s^{(n-1)/2} e^{-\pi m^2/s} (\Theta_{\Lambda^*}(\mathrm{i}s) - 1)\,\mathrm{d}s.$$

This equals $F_{f,g}^{(m)}$ for the choice $f = \Theta_\Lambda$, $g = \Theta_{\Lambda'}$, where we are using this generalisation of (1.3d):

$$\Theta_\Lambda(-1/\tau) = \frac{1}{\sqrt{|\Lambda|}} \left(\frac{\tau}{\mathrm{i}}\right)^{n/2} \Theta_{\Lambda^*}(\tau),$$

where Λ^* denotes the dual lattice of Λ [5]. Note that $E_{\Lambda,\Lambda'}^{(m)}$ reduces to $\eta^{(m)}$ when $\Lambda = \Lambda' = \mathbb{Z}$, and that

$$(3.5\mathrm{b}) \qquad E_{\Lambda,\Lambda'}^{(m)}(\mathrm{i}t) = E_{\Lambda'^*,\Lambda^*}^{(mt)}(\mathrm{i}/t).$$

These functions $E_{\Lambda,\Lambda'}^{(m)}(\mathrm{i}t)$, and more generally most of the $F_{f,g}^{(m)}(\mathrm{i}t)$, won't extend to holomorphic functions of τ. That we lose holomorphicity is not significant—automorphic forms (a natural generalisation of modular forms) aren't holomorphic in general either. In this context though, it would be interesting to see if these $F_{f,g}^{(m)}$ satisfy some reasonably simple (partial) differential equation—typically this is what replaces holomorphicity in the more general setting. Nevertheless we should perhaps take this as further evidence that $F_{f,g}^{(m)}$ is really meant for the $k = \frac{1}{2}$ functions f, g such as θ_i and ψ_r. For other k, f, g, we may need to replace the factor $e^{-\pi tm^2/s}$.

The strange product over $\mathcal{P}/\pm$ in (3.5a) suggests a generalisation of logarithm to any lattice:

$$\mathrm{Log}_\Lambda(x) := \sum_{\substack{\mathbf{v}\in\Lambda \\ \mathbf{v}\neq 0}} \frac{1}{|\mathbf{v}|} x^{|\mathbf{v}|} = \sum_{\mathbf{v}\in\mathcal{P}/\pm} \log(1 - x^{|\mathbf{v}|}).$$

Of course, $\mathrm{Log}_\mathbb{Z}(x)$ equals the usual $\log(1-x)$. We need this generalisation of logarithm here in order to compensate for the higher-dimensional lattice occurring in Q. It would be interesting to see what properties if any this logarithm possesses.

Certainly other related generalisations exist (for easy examples, try replacing $e^{-\pi tm^2/s}$ in (3.1e) with other functions of m), and the big question is which ones to focus on. Presumably the deformations $\eta^{(m)}$ are of interest, if only through their connection with string theory. String theory can help by giving us more examples of similar deformations (see [8]), and—better yet—finding special properties their deformations will have (e.g., a differential equation they will satisfy, which will help explain the presence of $e^{-\pi t m^2/s}$). Incidentally, the factor $s^{-1/2}$ in the integrand of (3.3a) is needed to yield the q-expansion (3.3b).

Our deformations (3.3a) are simpler for $k = \frac{1}{2}$, and simplest of all for choices

(3.6) $\quad f(\tau), g(\tau) \in \{\theta_2(\tau+r), \theta_3(\tau+r), \theta_4(\tau+r), \eta(\tau+r), \psi_s(\tau+r)\}, \quad \forall r, s \in \mathbb{Q}.$

Because of the elementary identity (2.3b), we can restrict to $s > 2$ in (3.6). Equations (3.4) tell us there will be several identities among these—e.g., $\theta_2^{(m)} = F_{\theta_3, \theta_3+2\theta_4}^{(m)}$ can be expressed in terms of $\eta^{(m)}$ alone, and each $r \neq 0$ can be reduced to several $r = 0$'s—but infinitely many of our functions will be independent. In this authors opinion these are well worth studying in more detail. After all, 4 of these $F_{\theta_i, \theta_j}^{(m)}$ have already appeared in string theory [2]. These generalisations $F_{f,g}^{(m)}$ suggest that there is indeed a mathematical theory behind the deformation (3.1a), as we have constructed infinite families of related functions.

What is the significance of the factor $e^{-\pi t m^2/s}$ in (3.1e)? This is how the deformation parameter m enters into $\eta^{(m)}$, but perhaps other choices will enable us to find nice deformations for $k \neq \frac{1}{2}$. Of course, simply by redefining the parameter m we could replace $e^{-\pi t m^2/s}$ in (3.1e) with $e^{-\pi t m/s}$ or even $m^{t/s}$, so we have in mind here nontrivial changes of $e^{-\pi t m^2/s}$. Can we deform any modular form? So far we can deform any modular form constructed out of η and $\theta_2, \theta_3, \theta_4$, but how about an arbitrary lattice theta function? For this reason it would be desirable to explicitly compute the $m \to 0$ limit of, e.g., $F_{\psi_r, \psi_s}^{(m)}$. Another direction of study would be to see what the deformation (3.2b), say, means at the level of the Riemann zeta function $\zeta(s)$. After all, θ_3 and ζ are related by the Mellin transform, and the familiar transformation (1.3d) corresponds to the fundamental functional relation

$$\zeta(s) = 2(2\pi)^{s-1} \sin(s\pi/2) \Gamma(1-s) \zeta(1-s)$$

connecting $\zeta(1-s)$ with $\zeta(s)$. Along these lines, can we deform a Dirichlet series with Euler product, and have it mean something at the level of its Mellin transform (which will be a modular form)?

Another potential application starts from the thought that quasicrystals are a sort of deformed lattice. Is it possible that our deformations, or ones like them, are directly related to quasicrystals? This is made more plausible by the fundamental role in quasicrystal theory of a sort of Poisson summation (see [14] for details).

Acknowledgements

The author wishes to thank Matthias Gaberdiel and Mark Walton for helpful comments, questions and corrections which helped shape this little paper. Also, he thanks the organisers of the Sharp conference, for organising such a stimulating weekend.

References

1. M. Atiyah, *The logarithm of the Dedekind η-function*, Math. Ann. **278** (1987), 335–380.
2. O. Bergman, M. R. Gaberdiel, and M. B. Green, *D-brane interactions in type IIB plane-wave background*, hep-th/0205183.
3. B. C. Berndt and L.-C. Zhang, *Ramanujan's identities for eta-functions*, Math. Ann. **292** (1992), 561–573.
4. J. M. Borwein and F. G. Garvan, *Approximations to π via Dedekind eta function*, Organic Mathematics, (J. Borwein et al., eds.), CMS Conf. Proc., vol.20, Amer. Math. Soc., Providence, RI, 1997, pp. 89–115.

5. J. H. Conway and N. J. A. Sloane, *Sphere packings, lattices and groups*, 3rd ed., Grundlehren Math. Wiss., vol. 290, Springer-Verlag, New York, 1999.
6. C. Cummins and T. Gannon, *Modular equations and the genus zero property of moonshine functions*, Invent. Math. **129** (1997), 413–443.
7. Ph. Di Francesco, P. Mathieu, and D. Sénéchal, *Conformal field theory*, Grad. Texts Contemp. Phys., Springer-Verlag, New York, 1997.
8. M. R. Gaberdiel and M. B. Green, *The D-instanton and other supersymmetric D-branes in IIB planewave string theory*, `hep-th/0211122`.
9. T. Gannon and C. S. Lam, *Lattices and Θ-function identities. I. Theta constants*, J. Math. Phys. **33** (1992), 854–870.
10. ———, *Lattices and Θ-function identities. II. Theta series*, J. Math. Phys. **33** (1992), 871–887.
11. T. Gannon and R. Kooistra, *Dedekind eta and Jacobi theta function identities* (in preparation).
12. I. S. Gradshteyn and I. M. Ryzhik, *Table of integrals, series, and products*, 5th ed., Academic Press, Boston, 1994.
13. M. B. Green, J. H. Schwarz, and E. Witten, *Superstring theory*, Vol. 1, Cambridge Monogr. Math. Phys., Cambridge Univ. Press, Cambridge, 1988.
14. J. C. Lagarias, *Mathematical quasicrystals and the problem of diffraction*, Directions in Mathematical Quasicrystals, CRM Monograph Ser., vol. 13, Amer. Math. Soc., Providence, RI, 2000, pp. 61–93.
15. H. Saleur and C. Itzykson, *Two-dimensional field theories close to criticality*, J. Stat. Phys. **48** (1987), 449–475.
16. T. Takayanagi, *Modular invariance of strings on pp-waves with RR-flux*, `hep-th/0206010`.
17. A. Weil, *Sur certains groupes d'opérateurs unitaires*, Acta Math. **111** (1964), 143–211.

MATHEMATICS DEPARTMENT, UNIVERSITY OF ALBERTA, EDMONTON, AB T6G 2G1, CANADA
E-mail address: `tgannon@math.ualberta.ca`

Examples of Berezin–Toeplitz Quantization: Finite Sets and Unit Interval

J.-P. Gazeau, T. Garidi, E. Huguet, M. Lachièze-Rey, and J. Renaud

In memory of Bob Sharp

ABSTRACT. We present a quantization scheme of an arbitrary measure space based on overcomplete families of states and generalizing the Klauder and the Berezin–Toeplitz approaches. This scheme could reveal itself as an efficient tool for quantizing physical systems for which more traditional methods like geometric quantization are uneasy to implement. The procedure is illustrated by (mostly two-dimensional) elementary examples in which the measure space is a N-element set and the unit interval. Spaces of states for the N-element set and the unit interval are the 2-dimensional Euclidean $\mathbb{R}^2$ and Hermitian $\mathbb{C}^2$ planes.

1. Quantum processing of a measure space

Quantum Physics and Signal Analysis have many aspects in common. As a departure point of their respective formalism, one finds a *raw* set $X = \{x\}$ of basic parameters or data. This set may be a classical phase space in the former case whereas it might be a temporal line or a time-frequency half-plane in the latter one. In reality it can be any set of data accessible to observation. The minimal significant structure one requires of it is the existence of a measure $\mu(dx)$, together with a σ-algebra of measurable subsets. As a measure space, X will be given the name of an *observation* set in the present context, and the existence of a measure provides us with a statistical reading of the set of measurable real or complex valued functions $f(x)$ on X: computing for instance average values on subsets with bounded measure. Actually, both approaches deal with quadratic mean values and correlation/convolution involving signal pairs, and the natural frameworks of studies are the real (Euclidean) or complex (Hilbert) spaces, $L^2(X,\mu) \equiv L^2_\mathbb{R}(X,\mu)$ or $L^2_\mathbb{C}(X,\mu)$ of square integrable functions $f(x)$ on the observation set X: $\int_X |f(x)|^2\, \mu(dx) < \infty$. One will speak of *finite-energy* signal in Signal Analysis and of quantum state in Quantum Mechanics. However, it is precisely at this stage that "quantum processing" of X differs from signal processing on at least three points:

2000 *Mathematics Subject Classification.* 81R30, 81R60, 81S30, 81S10.
This is the final form of the paper.

(1) not all square integrable functions are eligible as quantum states,
(2) a quantum state is defined up to a nonzero factor,
(3) those ones among functions $f(x)$ that are eligible as quantum states with unit norm, $\int_X |f(x)|^2 \mu(dx) = 1$, give rise to a probability interpretation: $X \supset \Delta \to \int_\Delta |f(x)|^2 \mu(dx)$ is a probability measure interpretable in terms of localisation in the measurable Δ. This is inherent to the computing of mean values of quantum observables, (essentially) self-adjoint operators with domain included in the set of quantum states.

The first point lies at the heart of the *quantization* problem: what is the more or less canonical procedure allowing to select quantum states among simple signals? In other words, how to select the right (projective) Hilbert space $\mathcal{H}$, a closed subspace of $L^2(X, \mu)$, or equivalently the corresponding orthogonal projector $\mathbb{I}_\mathcal{H}$?

In various cicumstances, this question may be answered through the selection, among elements of $L^2(X, \mu)$, of an orthonormal set $\mathcal{S}_N = \{\phi_n(x)\}_{n=1}^N$, N being finite or infinite, which spans, by definition, the separable Hilbert subspace $\mathcal{H} \equiv \mathcal{H}_N$. Furthermore, and this is a crucial assumption [1, 2, 8], we require that

$$(1.1) \qquad \mathcal{N}(x) \equiv \sum_n |\phi_n(x)|^2 < \infty \quad \text{almost everywhere.}$$

Of course, if N is finite the above condition is trivially checked.

We then consider the family of states $\{|x\rangle\}_{x \in X}$ through the following linear superpositions:

$$(1.2) \qquad |x\rangle \equiv \frac{1}{\sqrt{\mathcal{N}(x)}} \sum_n \phi_n(x) |n\rangle,$$

in which the ket $|n\rangle$ designates the element $\phi_n(x)$ in a "Fock" notation. This defines an injective map

$$(1.3) \qquad X \ni x \to |x\rangle \in \mathcal{H}_N \in \mathcal{H}_N$$

(in Dirac notations), and it is not difficult to check that states (1.2) are *coherent* in the sense that they obey the following two conditions:

- **Normalisation**

$$(1.4) \qquad \langle x \mid x \rangle = 1,$$

- **Resolution of the unity in $\mathcal{H}_N$**

$$(1.5) \qquad \int_X |x\rangle\langle x| \, \nu(dx) = \mathbb{I}_{\mathcal{H}_N},$$

where $\nu(dx) = \mathcal{N}(x) \mu(dx)$ is another measure on X, absolutely continuous with respect to $\mu(dx)$. The coherent states (1.2) form in general an overcomplete (continuous) basis of $\mathcal{H}$.

The resolution of the unity in $\mathcal{H}_N$ can alternatively been understood in terms of the scalar product $\langle x \mid x' \rangle$ of two states of the family. Indeed, (1.5) implies that, to any vector $|\phi\rangle$ in $\mathcal{H}_N$ one can (anti-)isometrically associate the function

$$(1.6) \qquad \phi^*(x) \equiv \sqrt{\mathcal{N}(x)} \langle x \mid \phi \rangle$$

in $L^2(X, \mu)$, and this function obeys

$$(1.7) \qquad \phi^*(x) = \int_X \sqrt{\mathcal{N}(x)\mathcal{N}(x')} \langle x \mid x' \rangle \phi^*(x') \, \mu(dx').$$

Hence, $\mathcal{H}_N$ is (anti-)isometric to a reproducing Hilbert space with kernel

(1.8) $$\mathcal{K}(x,x') = \sqrt{\mathcal{N}(\S)\mathcal{N}(\S')}\langle x \mid x'\rangle,$$

and the latter assumes finite diagonal values (a.e.), $\mathcal{K}(x,x) = \mathcal{N}(x)$, by construction.

A *classical* observable is a function $f(x)$ on X having specific properties in relationship with some supplementary structure allocated to X, topology, geometry or something else. Its quantization [3, 7] simply consists in associating to $f(x)$ the operator

(1.9) $$A_f := \int_X f(x)|x\rangle\langle x|\,\nu(dx).$$

In this context, $f(x)$ is said upper (or contravariant) symbol of the operator A_f and denoted by $f = \hat{A}_f$, whereas the mean value $\langle x|f(x)|x\rangle$ is said lower (or covariant) symbol of A_f [3, 5] and denoted by $\check{A}_f$. Through this approach, one can say that a quantization of the observation set is in one-to-one correspondence with the choice of a frame in the sense of (1.4) and (1.5). To a certain extent, a quantization scheme consists in adopting a certain point of view in dealing with X. This frame can be discrete, continuous, depending on the topology furthermore allocated to the set X, and it can be overcomplete, of course. The validity of a precise frame choice is asserted by comparing spectral characteristics of quantum observables A_f with data issued from a predefined experimental protocole. Of course, operators acting in $\mathcal{H}_N$ are not all of them of the "diagonal" type A_f, and many different classical $f(x)$'s can give rise to the *same* operator A_f. The frame should be complete or rich enough in order to meet all experimental possibilities determined by the protocole.

Let us illustrate the above construction with the well-known Klauder–Glauber–Sudarshan coherent states [5] and the subsequent so-called canonical quantization. The observation set X is the classical phase space $\mathbb{R}^2 \simeq \mathbb{C} = \{x \equiv z = (1/\sqrt{2})\times(q+ip)\}$ (in complex notations) of a particle with one degree of freedom. The measure on X is Gaussian, $\mu(dx) = (1/\pi)e^{-|z|^2}d^2z$ where d^2z is the Lebesgue measure of the plane. The functions $\phi_n(x)$ are the normalised powers of the complex variable z, $\phi_n(x) \equiv z^n/\sqrt{n!}$, so that the Hilbert subspace $\mathcal{H}$ is the so-called Fock–Bargmann space of all entire functions that are square integrable with respect to the Gaussian measure. Since $\sum_n |z|^2/n! = e^{|z|^2}$, the coherent states read

(1.10) $$|z\rangle = e^{-|z|^2/2}\sum_n \frac{z^n}{\sqrt{n!}}|n\rangle,$$

and one easily checks the normalisation and unity resolution:

(1.11) $$\langle z \mid z\rangle = 1, \quad \frac{1}{\pi}\int_\mathbb{C}|z\rangle\langle z|\,d^2z = \mathbb{I}_\mathcal{H},$$

Note that the reproducing kernel is simply given by $e^{\bar{z}z'}$. The quantization of the observation set is hence achieved by selecting in the original Hilbert space $L^2(\mathbb{C},(1/\pi)e^{-|z|^2}d^2z)$ all holomorphic entire functions, which geometric quantization specialists would call a choice of polarization. Quantum operators acting on $\mathcal{H}$ are yielded by using (1.9). We thus have for the most basic one,

(1.12) $$\frac{1}{\pi}\int_\mathbb{C} z|z\rangle\langle z|\,d^2z = \sum_n \sqrt{n+1}|n\rangle\langle n+1| \equiv a,$$

which is the lowering operator, $a|n\rangle = \sqrt{n}|n-1\rangle$. Its adjoint $a^\dagger$ is obtained by replacing z by $\bar{z}$ in (1.12), and we get the factorisation $N = a^\dagger a$ for the number operator, together with the commutation rule $[a, a^\dagger] = \mathbb{I}_\mathcal{H}$. Also note that $a^\dagger$ and a realize on $\mathcal{H}$ as multiplication operator and derivation operator respectively, $a^\dagger f(z) = zf(z)$, $af(z) = df(z)/dz$. From $q = \frac{1}{2}(z+\bar{z})$ et $p = \frac{1}{2i}(z-\bar{z})$, one easily infers by linearity that q and p are upper symbols for $\frac{1}{2}(a+a^\dagger) \equiv Q$ and $(1/2i)(a-a^\dagger) \equiv P$ respectively. In consequence, the self-adjoint operators Q and P obey the canonical commutation rule $[Q, P] = i\mathbb{I}_\mathcal{H}$, and for this reason fully deserve the name of position and momentum operators of the usual (Galilean) quantum mechanics, together with all localisation properties specific to the latter.

The next examples which are presented in this paper are, although elementary, rather unusual. In particular, we start with observation sets which are not necessarily phase spaces, and such sets are far from having any physical meaning in the common sense. We first consider a two-dimensional quantization of a N-element set which leads, for $N \geq 4$, to a Pauli algebra of observables. We then study two-dimensional (and higher-dimensional) quantizations of the unit segment. In the conclusion, we shall mention some questions of physical interest which are currently under investigation.

2. Quantum processing of an N-element set

An elementary (but not trivial!) exercise for illustrating the quantization scheme introduced in the previous section involves an arbitrary N-element set $X = \{x_i\}$ as observation set. An arbitrary non-degenerate measure on it is given by a sum of Dirac measures:

$$(2.1) \qquad \mu(dx) = \sum_{i=1}^N a_i \delta_{\{x_i\}}, \quad a_i > 0.$$

The Hilbert space $L^2(X, \mu)$ is simply isomorphic to $\mathbb{C}^N$. An obvious orthonormal basis is given by $\{1/\sqrt{a_i}\chi_{\{x_i\}}(x), i = 1, \ldots, N\}$, where $\chi_{\{a\}}$ is the characteristic function of the singleton $\{a\}$. We now consider the two-element orthonormal set $\{\phi_1 \equiv \phi_\alpha \equiv |\boldsymbol{\alpha}\rangle, \phi_2 \equiv \phi_\beta \equiv |\boldsymbol{\beta}\rangle\}$ defined in the most generic way by:

$$(2.2) \qquad \phi_\alpha(x) = \sum_{i=1}^N \alpha_i \frac{1}{\sqrt{a_i}} \chi_{\{x_i\}}(x), \quad \phi_\beta(x) = \sum_{i=1}^N \beta_i \frac{1}{\sqrt{a_i}} \chi_{\{x_i\}}(x),$$

where complex coefficients α_i and β_i obey

$$(2.3) \qquad \sum_{i=1}^N |\alpha_i|^2 = 1 = \sum_{i=1}^N |\beta_i|^2, \quad \sum_{i=1}^N \alpha_i \overline{\beta_i} = 0.$$

In a Hermitian geometry language, our choice of $\{\phi_\alpha, \phi_\beta\}$ amounts to selecting in $\mathbb{C}^N$ the two orthonormal vectors $\boldsymbol{\alpha} = \{\alpha_i\}$, $\boldsymbol{\beta} = \{\beta_i\}$, and this justifies our notations for indices.

It follows the expression for the coherent states:

$$(2.4) \qquad |x\rangle = \frac{1}{\sqrt{\mathcal{N}(x)}}[\phi_\alpha(x)|\boldsymbol{\alpha}\rangle + \phi_\beta(x)|\boldsymbol{\beta}\rangle],$$

in which $\mathcal{N}(x)$ is given by

$$\mathcal{N}(x) = \sum_{i=1}^{N} \frac{|\alpha_i|^2 + |\beta_i|^2}{a_i} \chi_{\{x_i\}}(x). \tag{2.5}$$

The resolution of unity (1.5) here reads as:

$$\mathbb{I} = \sum_{i=1}^{N} \left(|\alpha_j|^2 + |\beta_j|^2\right) |x_i\rangle\langle x_i| \tag{2.6}$$

The overlap between two coherent states is given by the following kernel:

$$\langle x_i \mid x_j \rangle = \frac{\overline{\alpha_i}\alpha_j + \overline{\beta_i}\beta_j}{\sqrt{|\alpha_i|^2 + |\beta_i|^2}\sqrt{|\alpha_j|^2 + |\beta_j|^2}}. \tag{2.7}$$

To any real-valued function $f(x)$ on X, i.e., to any vector $\boldsymbol{f} \equiv (f(x_i))$ in $\mathbb{R}^N$, there corresponds the following Hermitian operator A_f in $\mathbb{C}^2$, expressed in matrix form with respect to the orthonormal basis (2.2):

$$A_f = \int_X \mu(dx)\,\mathcal{N}(x)\,f(x)|x\rangle\langle x| = \begin{pmatrix} \sum_{i=1}^{N}|\alpha_i|^2 f(x_i) & \sum_{i=1}^{N} \alpha_i\overline{\beta_i}f(x_i) \\ \sum_{i=1}^{N} \overline{\alpha_i}\beta_i f(x_i) & \sum_{i=1}^{N}|\beta_i|^2 f(x_i) \end{pmatrix} \tag{2.8}$$

$$\equiv \begin{pmatrix} \langle \boldsymbol{F} \rangle_\alpha & \langle \boldsymbol{\beta}|\boldsymbol{F}|\boldsymbol{\alpha}\rangle \\ \langle \boldsymbol{\alpha}|\boldsymbol{F}|\boldsymbol{\beta}\rangle & \langle \boldsymbol{F} \rangle_\beta \end{pmatrix},$$

where $\boldsymbol{F}$ holds for the diagonal matrix $\{(f(x_i))\}$. It is clear that, for a generic choice of the complex α_i's and β_i's, all possible Hermitian 2×2-matrices can be obtained in this way if $N \geq 4$. By *generic* we mean that the following $4 \times N$-real matrix

$$\mathcal{C} = \begin{pmatrix} |\alpha_1|^2 & |\alpha_2|^2 & \cdots & |\alpha_N|^2 \\ |\beta_1|^2 & |\beta_2|^2 & \cdots & |\beta_N|^2 \\ \mathrm{Re}(\alpha_1\overline{\beta_1}) & \mathrm{Re}(\alpha_2\overline{\beta_2}) & \cdots & \mathrm{Re}(\alpha_N\overline{\beta_N}) \\ \mathrm{Im}(\alpha_1\overline{\beta_1}) & \mathrm{Im}(\alpha_2\overline{\beta_2}) & \cdots & \mathrm{Im}(\alpha_N\overline{\beta_N}) \end{pmatrix} \tag{2.9}$$

has rank equal to 4. The case $N = 4$ with $\det \mathcal{C} \neq 0$ is particularly interesting since then one has uniqueness of upper symbols of Pauli matrices $\sigma_1 = \begin{pmatrix} 0 & 1 \\ 1 & 0 \end{pmatrix}$, $\sigma_2 = \begin{pmatrix} 0 & -i \\ i & 0 \end{pmatrix}$, $\sigma_3 = \begin{pmatrix} 1 & 0 \\ 0 & -1 \end{pmatrix}$, $\sigma_0 = \mathbb{I}$, which form a basis of the four-dimensional Lie algebra of complex Hermitian 2×2-matrices. As a matter of fact, the operator (2.8) decomposes with respect to this basis as:

$$A_f = \langle f \rangle_+ \sigma_0 + \langle f \rangle_- \sigma_3 + \mathrm{Re}(\langle \boldsymbol{\beta}|\boldsymbol{F}|\boldsymbol{\alpha}\rangle)\sigma_1 - \mathrm{Im}(\langle \boldsymbol{\beta}|\boldsymbol{F}|\boldsymbol{\alpha}\rangle)\sigma_2, \tag{2.10}$$

where the symbols $\langle f \rangle_\pm$ stand for the following averagings:

$$\langle f \rangle_\pm = \frac{1}{2}\sum_{i=1}^{N}(|\alpha_i|^2 \pm |\beta_i|^2)f(x_i) = \frac{1}{2}(\langle \boldsymbol{F} \rangle_\alpha \pm \langle \boldsymbol{F} \rangle_\beta). \tag{2.11}$$

Note that $\langle f \rangle_+$ alone has a meanvalue status, precisely with respect to the probability distribution

$$p_i = \tfrac{1}{2}(|\alpha_i|^2 + |\beta_i|^2). \tag{2.12}$$

Also note the appearance of these averagings in the spectral values of the quantum observable A_f:

$$\mathrm{Sp}(f) = \{\langle f \rangle_+ \pm \sqrt{(\langle f \rangle_-)^2 + |\langle \boldsymbol{\beta}|\boldsymbol{F}|\boldsymbol{\alpha}\rangle|^2}\}. \tag{2.13}$$

Just remark that if vector $\boldsymbol{\alpha} = (1, 0, \ldots, 0)$ is part of the canonical basis and $\boldsymbol{\beta} = (0, \beta_2, \ldots, \beta_n)$ is unit vector orthogonal to $\boldsymbol{\alpha}$, then A_f is diagonal and $\mathrm{Sp}(f)$ is trivially reduced to $(f(x_1), \langle \boldsymbol{F} \rangle_{\boldsymbol{\beta}})$. The upper symbols for Pauli matrices read in vector form as

$$(2.14) \quad \hat{\boldsymbol{\sigma}}_0 = \begin{pmatrix} 1 \\ 1 \\ 1 \\ 1 \end{pmatrix}, \quad \hat{\boldsymbol{\sigma}}_1 = \mathcal{C}^{-1} \begin{pmatrix} 0 \\ 0 \\ 1 \\ 0 \end{pmatrix}, \quad \hat{\boldsymbol{\sigma}}_2 = \mathcal{C}^{-1} \begin{pmatrix} 0 \\ 0 \\ 0 \\ -1 \end{pmatrix}, \quad \hat{\boldsymbol{\sigma}}_3 = \mathcal{C}^{-1} \begin{pmatrix} 1 \\ -1 \\ 0 \\ 0 \end{pmatrix}.$$

On the other hand, and for any N, components of the lower symbol of A_f are given in terms of another probability distribution in which the importance of each one is precisely doubled relatively to its counterpart in (2.12):

$$(2.15) \quad \langle x_l | A_f | x_l \rangle = \check{A}_f(x_l) = \sum_{i=1}^{N} \varpi_{li} f(x_i),$$

with

$$(2.16) \quad \varpi_{ll} = |\alpha_l|^2 + |\beta_l|^2, \quad \varpi_{li} = \frac{|\overline{\alpha_l}\alpha_i + \overline{\beta_l}\beta_i|^2}{|\alpha_l|^2 + |\beta_l|^2}, \quad i \neq l.$$

Note that the matrix (ϖ_{li}) is stochastic. As a matter of fact, components of lower symbols of Pauli matrices are given by:

$$(2.17) \quad \check{\sigma}_0(x_l) = 1, \qquad \check{\sigma}_1(x_l) = \frac{2\,\mathrm{Re}(\overline{\alpha_l}\beta_l)}{|\alpha_l|^2 + |\beta_l|^2},$$

$$(2.18) \quad \check{\sigma}_2(x_l) = \frac{2\,\mathrm{Im}(\overline{\alpha_l}\beta_l)}{|\alpha_l|^2 + |\beta_l|^2}, \qquad \check{\sigma}_3(x_l) = \frac{|\alpha_l|^2 - |\beta_l|^2}{|\alpha_l|^2 + |\beta_l|^2}.$$

Hidden behind this formal game lies an interpretation resorting to Hermitian geometry probability. For instance, consider $X = \{x_i\}$ as a set of N real numbers. One then can view the real-valued function f defined by $f(x_i) = x_i$ as the *position* observable, the measurement of which on the quantum level determined by the choice of $\boldsymbol{\alpha} = \{\alpha_i\}$, $\boldsymbol{\beta} = \{\beta_i\}$ has the two possible outcomes given by (2.13). Moreover, the *position* x_l is privileged to a certain (quantitative) extent in the expression of the average value of the *position* operator when computed in state $|x_l\rangle$.

Before ending this section, let us examine the lower-dimensional cases $N = 2$ and $N = 3$. When $N = 2$ the basis change (2.2) reduces to a U(2) transformation with SU(2) parameters $\alpha = \alpha_1$, $\beta = -\overline{\beta}_1$, $|\alpha|^2 - |\beta|^2 = 1$, and some global phase factor. The operator (2.8) simplifies as

$$(2.19) \quad A_f = f_+ \mathbb{I} + f_- \begin{pmatrix} |\alpha|^2 - |\beta|^2 & -2\alpha\beta \\ -2\overline{\alpha}\overline{\beta} & |\beta|^2 - |\alpha|^2 \end{pmatrix},$$

with $f_\pm := \bigl(f(x_1) \pm f(x_2)\bigr)/2$. We now have a two-dimensional commutative algebra of "observables" A_f, generated by the identity matrix $\mathbb{I} = \sigma_0$ and the SU(2) transform of σ_3: $\sigma_3 \to g\sigma_3 g^\dagger$ with $g = \begin{pmatrix} \alpha & \beta \\ -\overline{\beta} & \overline{\alpha} \end{pmatrix} \in \mathrm{SU}(2)$. As is easily expected in this case, lower symbols reduce to components:

$$(2.20) \quad \langle x_l | A_f | x_l \rangle = \check{A}_f(x_l) = f(x_l), \quad l = 1, 2.$$

Finally, it is interesting to consider the $N = 3$ case when all considered vector spaces are real. The basis change (2.2) involves four real independent parameters, say α_1, α_2, β_1, and β_2, all with modulus < 1. The counterpart of (2.9) reads here as

$$(2.21) \qquad \mathcal{C}_3 = \begin{pmatrix} (\alpha_1)^2 & (\alpha_2)^2 & 1-(\alpha_1)^2-(\alpha_2)^2 \\ (\beta_1)^2 & (\beta_2)^2 & 1-(\beta_1)^2-(\beta_2)^2 \\ \alpha_1\beta_1 & \alpha_2\beta_2 & -\alpha_1\beta_1-\alpha_2\beta_2 \end{pmatrix}$$

If $\det \mathcal{C}_3 = (\alpha_1\beta_2 - \alpha_2\beta_1)(\beta_1\beta_2 - \alpha_1\alpha_2) \neq 0$, then one has uniqueness of upper symbols of Pauli matrices σ_1, σ_3, and $\sigma_0 = \mathbb{I}$ which form a basis of the three-dimensional Jordan algebra of real symmetric 2×2-matrices. These upper symbols read in vector form as

$$(2.22) \qquad \hat{\sigma}_0 = \begin{pmatrix} 1 \\ 1 \\ 1 \end{pmatrix}, \quad \hat{\sigma}_1 = \mathcal{C}_3^{-1} \begin{pmatrix} 0 \\ 0 \\ 1 \end{pmatrix}, \quad \hat{\sigma}_3 = \mathcal{C}_3^{-1} \begin{pmatrix} 1 \\ -1 \\ 0 \end{pmatrix}.$$

Finally, the extension of this quantization formalism to N'-dimensional subspaces of the original $L^2(X,\mu) \simeq \mathbb{C}^N$ appears as being straighforward on a technical if not interpretational level.

3. Quantum processing of the unit interval

3.1. Quantization with finite subfamilies of Haar wavelets. Further simple examples of quantization are provided when we deal with the unit interval $X = [0,1]$ of the real line and its associated Hilbert space $L^2[0,1]$.

Let us start out by simply selecting the two first elements of the orthonormal Haar basis [4], namely the characteristic function $\mathbf{1}(x)$ of the unit interval and the Haar wavelet:

$$(3.1) \qquad \phi_1(x) = \mathbf{1}(x), \quad \phi_2(x) = \mathbf{1}(2x) - \mathbf{1}(2x-1).$$

Then we have,

$$(3.2) \qquad \mathcal{N}(x) = \sum_{n=1}^{2} |\phi_n(x)|^2 = 2 \quad \text{a.e..}$$

The corresponding coherent states read as

$$(3.3) \qquad |x\rangle = \frac{1}{\sqrt{2}}[\phi_1(x)|1\rangle + \phi_2(x)|2\rangle].$$

To any integrable function $f(x)$ on the interval there corresponds the linear operator A_f on $\mathbb{R}^2$ or $\mathbb{C}^2$:

$$(3.4) \qquad A_f = 2\int_0^1 dx\, f(x)|x\rangle\langle x|$$
$$= \left[\int_0^1 dx\, f(x)\right][|1\rangle\langle 1| + |2\rangle\langle 2|] + \left[\int_0^1 dx\, f(x)\phi_2(x)\right][|1\rangle\langle 2| + |2\rangle\langle 1|],$$

or, in matrix form with respect to the orthonormal basis (3.1),

$$(3.5) \qquad A_f = \begin{pmatrix} \int_0^1 dx\, f(x) & \int_0^1 dx\, f(x)\phi_2(x) \\ \int_0^1 dx\, f(x)\phi_2(x) & \int_0^1 dx\, f(x) \end{pmatrix}.$$

In particular, with the choice $f = \phi_1$ we recover the identity whereas for $f = \phi_2$, $A_{\phi_2} = \begin{pmatrix} 0 & 1 \\ 1 & 0 \end{pmatrix} = \sigma_1$, the first Pauli matrix. With the choice $f(x) = x^p$, $\text{Re}\, p > -1$,

$$(3.6) \qquad A_{x^p} = \frac{1}{p+1} \begin{pmatrix} 1 & 2^{-p} - 1 \\ 2^{-p} - 1 & 1 \end{pmatrix}.$$

For an arbitrary coherent state $|x_0\rangle$, $x_0 \in [0,1]$, it is interesting to evaluate the average values (lower symbols) of A_{x^p}. This gives

$$(3.7) \qquad \langle x_0 | A_{x^p} | x_0 \rangle = \begin{cases} 2^{-p}/(p+1) & 0 \leq x_0 \leq \frac{1}{2}, \\ (2 - 2^{-p})/(p+1) & \frac{1}{2} \leq x_0 \leq 1, \end{cases}$$

the two possible values being precisely the eigenvalues of the above matrix. Note the average values of the "position" operator: $\langle x_0 | A_x | x_0 \rangle = \frac{1}{4}$ if $0 \leq x_0 \leq \frac{1}{2}$ and $\frac{3}{4}$ if $\frac{1}{2} \leq x_0 \leq 1$.

Clearly, like in the $N = 2$ case of the previous section, all operators A_f commute, since they are linear combinations of the identity matrix and the Pauli matrix σ_1. The procedure is easily generalized to higher dimensions. Let us add to the previous set $\{\phi_1, \phi_2\}$ other elements of the Haar basis, say up to "scale" J:

$$(3.8) \quad \{\phi_1(x), \phi_2(x), \phi_3(x) = \sqrt{2}\phi_2(2x), \phi_4(x) = \sqrt{2}\phi_2(2x - 1),$$
$$\ldots, \phi_s(x) = 2^{j/2}\phi_2(2x - k), \phi_N(x) = 2^{J/2}\phi_2(2x - 2^J + 1)\},$$

where, at given $j = 1, 2, \ldots, J$, the integer k assumes its values in the range $0 \leq k \leq 2^j - 1$. The total number of elements of this orthonormal system is $N = 2^{J+1}$. The expression of (1.1) is also given by $\mathcal{N}(x) = 2^{J+1}$, and this clearly diverges at the limit $J \to \infty$. Then, it is remarkable if not expected that spectral values as well as average values of the "position" operator are given by $\langle x_0 | A_{x^p} | x_0 \rangle = (2k+1)/2^{J+1}$ for $k/2^J \leq x_0 \leq (k+1)/2^J$ where $0 \leq k \leq 2^J - 1$. Our quantization scheme in the present case achieves a dyadic discretization of the localization in the unit interval.

3.2. A two-dimensional non-commutative quantization of the unit interval.
Now we choose another orthonormal system, in the form of the two first elements of the trigonometric Fourier basis,

$$(3.9) \qquad \phi_1(x) = \mathbf{1}(x), \quad \phi_2(x) = \sqrt{2} \sin 2\pi x.$$

Then we have,

$$(3.10) \qquad \mathcal{N}(x) = \sum_{n=1}^{2} |\phi_n(x)|^2 = 1 + 2\sin^2 2\pi x,$$

and corresponding coherent states read as

$$(3.11) \qquad |x\rangle = \frac{1}{\sqrt{1 + 2\sin^2 2\pi x}}[|1\rangle + \sqrt{2} \sin 2\pi x |2\rangle].$$

To any integrable function $f(x)$ on the interval, corresponds the linear operator A_f on $\mathbb{R}^2$ or $\mathbb{C}^2$ (in its matrix form),

$$(3.12) \qquad A_f = \begin{pmatrix} \int_0^1 dx\, f(x) & \sqrt{2} \int_0^1 dx\, f(x) \sin 2\pi x \\ \sqrt{2} \int_0^1 dx\, f(x) \sin 2\pi x & 2\int_0^1 dx\, f(x) \sin^2 2\pi x \end{pmatrix}.$$

Like in the previous case, with the choice $f = \phi_1$ we recover the identity whereas for $f = \phi_2$, $A_{\phi_2} = \sigma_1$, the first Pauli matrix.

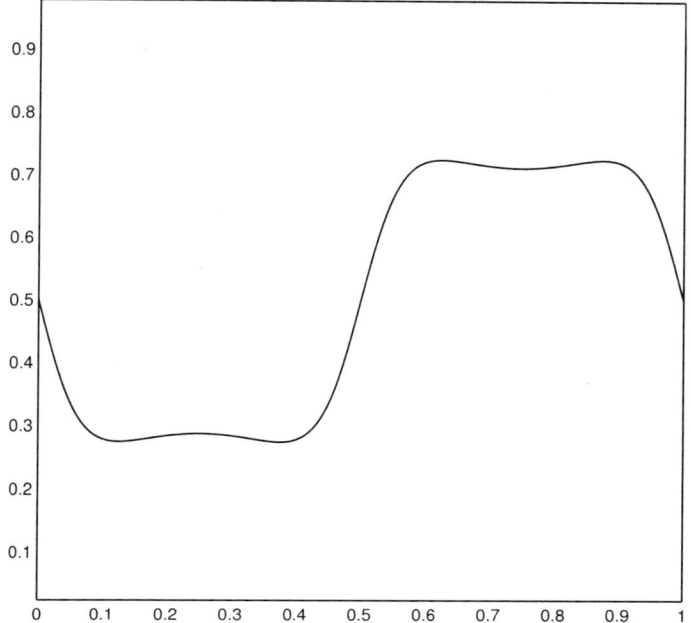

FIGURE 1. Average value $\langle x_0|A_x|x_0\rangle$ of *position* operator A_x versus x_0, (compare with eigenvalues of A_x).

We now have to deal with a non-commutative Jordan algebra of operators A_f, like in the $N = 3$ real case of the previous section. It is generated by the identity matrix and the two real Pauli matrices σ_1 and σ_3.

In this context, the *position* operator is given by:

$$A_x = \begin{pmatrix} \frac{1}{2} & -1/(\sqrt{2}\pi) \\ -1/(\sqrt{2}\pi) & \frac{1}{2} \end{pmatrix},$$

with eigenvalues $\frac{1}{2} \pm 1/(\sqrt{2}\pi)$ Note its average values in function of the coherent state parameter $x_0 \in [0,1]$:

$$\langle x_0|A_x|x_0\rangle = \frac{1}{2} - \frac{2}{\pi}\frac{\sin 2\pi x_0}{1 + 2\sin^2 2\pi x_0}$$

In Figure 1 we give the curve of $\langle x_0|A_x|x_0\rangle$ in function of x_0. It is interesting to compare with the two-dimensional Haar quantization presented in the previous section.

4. Conclusion

The examples we have given in this contribution are mainly of pedagogical nature. Other examples, specially devoted to Euclidean and pseudo-Euclidean spheres will be presented elsewhere, having in view possible connections with objects of noncommutative geometry (like fuzzy spheres, see for instance [**6, 9**]). They show the extreme freedom we have in analyzing a set X of *data* or *possibilities* just equipped with a measure by following a quantumlike procedure. The crucial step lies in the choice of a countable orthonormal subset in $L^2(X,\mu)$ obeying (1.1). A $\mathbb{C}^N$

(or l^2 if $N = \infty$) unitary transform of this original subset would actually lead to the same specific quantization, and the latter could as well be obtained by using unitarily equivalent *continuous* orthonormal distributions defined within the framework of some Gel'fand triplet. Of course, further structure like symplectic manifold combined with spectral constraints imposed to some specific observables will considerably restrict that freedom and will lead hopefully to a unique solution, like Weyl quantization, deformation quantization, or geometric quantization are able to achieve in specific situations. Nevertheless, we believe that the generalization of Berezin quantization which has been described here, and which goes far beyond the context of Classical and Quantum Mechanics, not only will shed light on the specific nature of the latter, but also will help to solve in a simpler way some quantization problems.

References

1. S. T. Ali, J.-P. Antoine, and J.-P. Gazeau, *Continuous frames in Hilbert spaces*, Ann. Physics **222** (1993), 1–37.
2. _____, *Coherent states, wavelets and their generalizations*, Grad. Texts Contemp. Phys., Springer-Verlag, New York, 2000.
3. F. A. Berezin, *General concept of quantization*, Comm. Math. Phys. **40** (1975), 153–174.
4. I. Daubechies, *Ten lectures on wavelets*, CBMS-NSF Regional Conf. Ser. in Appl. Math., vol. 61, SIAM, Philadelphia, PA, 1992.
5. D. H. Feng, J. R. Klauder, and M. R Strayer (eds.), *Coherent states—past, present and future* (Oak Ridge, 1993), World Scientific, Singapore, 1994.
6. L. Friedel and K. Krasnov, *The fuzzy sphere ∗-product and spin networks* J. Math. Phys. **43** (2002), 1737–1754.
7. J. R. Klauder, *Continuous-representation theory. I. Postulates of continuous-representation theory*, J. Math. Phys. **4** (1963), 1055–1058; II. *Generalized relation between quantum and classical dynamics*, 1058–1073.
8. _____, *Quantization without Quantization* Ann. Physics **237** (1995), 147–160.
9. M. Lachièze-Rey, J.-P. Gazeau, E. Huguet, J. Renaud, and T. Gardi, *Quantization of the sphere with coherent states*, Internat. J. Theoret. Phys. (to appear).

Laboratoire de Physique Théorique de la Matière Condensée and Fédération de Recherches Astroparticules et Cosmologie, Boîte 7020, Université Paris 7 Denis-Diderot, 75251 Paris Cedex 05, France
  E-mail address, J.-P. Gazeau: `gazeau@ccr.jussieu.fr`
  E-mail address, T. Garidi: `garidi@ccr.jussieu.fr`
  E-mail address, J. Renaud: `renaud@ccr.jussieu.fr`

GEPI, Observatoire de Paris, 5, place J.-Janssen, 92195 Meudon Cedex, France and Fédération de Recherches Astroparticules et Cosmologie, Boîte 7020, Université Paris 7 Denis-Diderot, 75251 Paris Cedex 05, France
  E-mail address: `eric.huguet@obspm.fr`

Service d'Astrophysique du CEA et Fédération de Recherches Astroparticules et Cosmologie, CEA/Saclay, 91191 Gif-sur-Yvette Cedex, France
  E-mail address: `marclr@cea.fr`

A Modified Weierstrass Representation for CMC-Surfaces in Multi-Dimensional Euclidean Spaces

A. M. Grundland and W. J. Zakrzewski

ABSTRACT. An extension of the generalized Weierstrass representation for conformally parameterized surfaces in multi-dimensional Euclidean spaces is presented. This representation is associated with the CP^{N-1} sigma model which permits the study of constant mean curvature surfaces immersed in $(N^2 - 1)$-dimensional Euclidean space. It is demonstrated that there exists a one-to-one correspondence between the modified version of the Weierstrass system and the equations of the CP^{N-1} sigma model. This proof is based on the linear spectral problem of the CP^{N-1} model which gives a set of conserved quantities. A possible geometric interpretation of the CP^{N-1} harmonic maps is discussed.

1. Introduction

Over the last century, the problem of surfaces and their deformations under various types of dynamics has become an increasingly important field of research. In particular, surfaces with constant mean curvature (CMC) have been shown to play an essential role in geometry and to have numerous applications in several areas in physics (e.g., two-dimensional gravity [6], quantum field theory [1], statistical physics [14], fluid dynamics [17], the Canham–Helfrich membrane model in biophysics [16], etc). A formula for inducing minimal surfaces immersed in 3-dimensional Euclidean space has been introduced by Enneper [5] and Weierstrass [18] one and a half century ago for the purpose of solving Gauss–Weingarten and Gauss–Codazzi equations. Since then a number of attempts to generalize their approach and its various applications have been made (see, e.g., [11, 15]).

Most recently, the study of the generalized Weierstrass representations for surfaces immersed into multi-dimensional spaces was initiated and extensively developed by B. Konopelchenko et al. (see, e.g., [12] and references therein). Our analysis

---

2000 *Mathematics Subject Classification.* 53A05.

The first author would like to thank the University of Durham and the Grey College for the Alan Richards fellowship that allowed him to spend two terms in Durham during the academic years 2001 and 2002.

The first author's research has been supported also by research grants from NSERC of Canada and FCAR du Québec.

This is the final form of the paper.

©2004 American Mathematical Society

is based on an alternative methodological approach, namely on a completely integrable Euclidean CP^{N-1} sigma model in two dimensions and provides an extension of their results.

The dynamics of CP^{N-1} sigma models involving maps from $\mathbb{R}^2$ or S^2 is given by the Lagrangian density (see, e.g.,[19])

$$L = \tfrac{1}{4}(D_\mu z)^\dagger (D_\mu z), \tag{1.1}$$

together with the constraint

$$z^\dagger z = 1,$$

where $\Omega \subset \mathbb{C} \ni \zeta = x + iy \to z = (z^1, z^2, \ldots, z^N) \in \mathbb{C}^N$ are the homogeneous coordinates, $\dagger$ denotes a Hermitian conjugation, D_μ are the covariant derivatives which act on $z \colon S^2 \to CP^{N-1}$ according to the formula

$$D_\mu z = \partial_\mu - (z^\dagger \partial_\mu z) z. \tag{1.2}$$

Here the index $\mu = 1, 2$ denotes x, y and ∂_μ denotes partial differentiation with respect to x or y. Note that CP^{N-1} model possesses an Abelian $U(1)$ symmetry and its composite gauge field

$$A_\mu = z^\dagger \partial_\mu z, \quad A_\mu^\dagger = -A_\mu \tag{1.3}$$

is a pure imaginary function. The action of the CP^{N-1} model is defined as

$$\mathcal{L} = \int L \, d\zeta \, d\bar\zeta \tag{1.4}$$

and if the model is defined over S^2 then we require that $\mathcal{L}$ is finite.

From a computational point of view it is convenient to define a new dependent variable

$$z = \frac{f}{|f|}, \quad |f| = (f^\dagger \cdot f)^{1/2}. \tag{1.5}$$

The Lagrangian density (1.1) in terms of f becomes

$$L = \frac{1}{4} \partial_\mu \left(\frac{f^\dagger}{|f|}\right) \cdot \Pi \cdot \partial_\mu \left(\frac{f}{|f|}\right), \tag{1.6}$$

where

$$\Pi = \mathbf{1} - \frac{f \otimes f^\dagger}{|f|^2} \tag{1.7}$$

is a projector. The corresponding Euler-Lagrange equations take the form

$$\Pi \cdot \left[\partial\bar\partial f - \frac{(f^\dagger \bar\partial f)}{|f|^2} \partial f - \frac{(f^\dagger \partial f)}{|f|^2} \bar\partial f\right] = 0, \tag{1.8}$$

where the derivatives are abbreviated $\partial = \partial/\partial\zeta$ and $\bar\partial/\partial\bar\zeta$. Note that operator Π projects onto a space perpendicular to f and (when applied as a right hand side operator) to $f^\dagger$. So for any $\alpha \in \mathbb{C}$, equation (1.8) can be written in an equivalent form

$$\partial\bar\partial f - \frac{(f^\dagger \bar\partial f)}{|f|^2} \partial f - \frac{(f^\dagger \partial f)}{|f|^2} \bar\partial f = \alpha f.$$

The simplest CP^{N-1} model corresponds to $N = 2$. It is well known [19] that this model CP^1 is equivalent to O(3) sigma model. The Euler–Lagrange equations derived from (1.8) take the form

$$\partial\bar\partial w - \frac{2\bar w}{1+|w|^2}\partial w \bar\partial w = 0 \tag{1.9}$$

and its respective complex conjugate equations. Here $w = f_1/f_2$ where f_1 and f_2 are two components of the vector f. In [3] it was shown that there exists a one-to-one correspondence between CP^1 maps w and (φ,ψ) of the generalized Weierstrass problem given by

$$\partial\psi = (|\psi|^2+|\varphi|^2)\varphi, \quad \bar\partial\varphi = -(|\psi|^2+|\varphi|^2)\psi. \tag{1.10}$$

This system describes the constant mean curvature surfaces in the conformal parameterization ζ and $\bar\zeta$. The connection between equations (1.9) and (1.10) is given [3] by a change of dependent variables

$$\psi = \epsilon w\frac{(\bar\partial\bar w)^{1/2}}{1+|w|^2}, \quad \varphi = \epsilon\frac{(\partial w)^{1/2}}{1+|w|^2}, \quad \epsilon = \pm 1, \tag{1.11}$$

where

$$w = \frac{\psi}{\varphi}. \tag{1.12}$$

Different analytic descriptions of CMC-surfaces immersed in $\mathbb{R}^3$ had been formulated and subsequently discussed in [2, 12, 15] (see also references therein).

The objective of this paper is to establish a generalization of the Weierstrass system equivalent to the CP^{N-1} sigma model ($N \geq 2$). The paper is organized as follows. In Section 2, we derive a modified version of the Weierstrass system and demonstrate that there is a connection between this system and the equations of the completely integrable Euclidean CP^{N-1} sigma model. In Section 3, exploiting this link, we describe in detail the Weierstrass representation for CMC-surfaces in (N^2-1)-dimensional Euclidean space. Next, in the context of the CP^{N-1} sigma model we discuss some geometric aspects of CP^{N-1} harmonic maps. Section 4 contains final remarks.

2. The modified Weierstrass system

As shown in [13], the linear spectral problem for CP^{N-1} equations (1.8) is of the form

$$\partial\psi = \frac{2}{1+\lambda}[\partial P, P]\psi, \quad \bar\partial\psi = \frac{2}{1-\lambda}[\bar\partial P, P]\psi, \tag{2.1}$$

where λ is a spectral parameter, ψ is an N-component auxiliary vector and the $N\times N$ matrix P is a projector given in terms of the vector f by

$$P = \frac{1}{A}f\otimes f^\dagger, \quad A = f^\dagger\cdot f. \tag{2.2}$$

The compatibility conditions for the linear spectral system (2.1) coincide with the CP^{N-1} equations (1.8) and can be written in the form of a conservation law (CL)

$$\partial K - \bar\partial K^\dagger = 0, \tag{2.3}$$

where the traceless matrix K, expressed in terms of components of the vector f, has the form

(2.4) $$K_{ij} = \bar{f}_j \bar{\phi}_i^2 - f_i \overline{\varphi}_j^2, \quad i,j = 1, \ldots, N.$$

For computational purposes it is useful to introduce the following functions

(2.5) $$\begin{aligned} \varphi_i^2 &= A^{-2} \bar{f}_k F_{ki}, \quad F_{ij} = f_i \partial f_j - f_j \partial f_i \\ \phi_i^2 &= A^{-2} f_k \bar{G}_{ki}, \quad G_{ij} = f_i \bar\partial f_j - f_j \bar\partial f_i, \quad i,j = 1, \ldots, N \end{aligned}$$

From now on we adopt the summation convention over repeated indices unless otherwise stated. From the equations (2.5) we are able to deduce that functions φ_i^2 and ϕ_i^2, $i = 1, \ldots, N$, are linearly dependent, in each set respectively. Indeed, it is due to the fact that these functions satisfy two algebraic constraints

(2.6) $$\bar{f}_k \varphi_k^2 = 0, \quad f_k \phi_k^2 = 0.$$

As a consequence of (2.6) we can choose $2(N-1)$ functions $\varphi_2^2, \ldots, \varphi_N^2$ and $\phi_2^2, \ldots, \phi_N^2$ as linearly independent functions and express φ_1^2 and ϕ_1^2 in terms of these remaining functions, respectively. Using the scaling symmetry of CP^{N-1} equations (1.8), $f \to f' = \alpha f$, $\alpha \in \mathbb{C}$, we can, without loss of generality set the first component of f to one, $f_1 = 1$. Moreover, the functions φ_i^2 and ϕ_i^2 are expressible in terms of f and its first derivatives as follows

(2.7) $$\begin{aligned} \varphi_i^2 &= A^{-1} \partial f_i - A^{-2}(\bar{f}_k \partial f_k) f_i, \quad i = 2, \ldots, N, \\ \phi_i^2 &= A^{-1} \bar\partial f_i - A^{-2}(\bar{f}_k \bar\partial f_k) f_i, \quad A = 1 + |f_2|^2 + \cdots + |f_N|^2. \end{aligned}$$

Note that in expression (2.7) the indices i and k run from 2 up to N, not from 1, which reflects the fact that only φ_i^2 and ϕ_i^2 ($i = 2, \ldots, N$) are linearly independent. Note also that when the terms with the indices i and k coincide ($k = i$) in expressions (2.7) we obtain a cancellation of some terms which allows us to invert expression (2.7) and find explicitly ∂f_i in terms of f_i and φ_i^2

(2.8) $$\partial f_i = A[\varphi_i^2 + (\bar{f}_k \cdot \varphi_k^2) f_i], \quad i = 2, \ldots, N$$

Of course the similar fact holds for the function ϕ_i^2 and $\bar\partial f_i$.

We now proceed to construct a modified version of the Weierstrass system, in the M-dimensional Euclidean space, which is an extension of the system (1.10) associated with CP^1 equations (1.9). By analogy with the CP^1 case (1.12) it is convenient to define a set of $2(N-1)$ new dependent complex-valued variables ψ_i and φ_i such that their ratios (no summation)

(2.9) $$f_i = \frac{\psi_i}{\overline{\varphi}_i}, \quad i = 2, \ldots, N$$

satisfy the CP^{N-1} equations (1.8). Now, we can express first derivatives $\bar\partial \varphi_i$ and $\partial \psi_i$ in terms of functions f_i and their first derivatives. Carrying out the differentiation (2.9) with respect to ∂ gives (no summation)

(2.10) $$\partial \psi_i = \partial f_i \overline{\varphi}_i + f_i \partial \overline{\varphi}_i, \quad i = 2, \ldots, N$$

Making use of the first derivatives of φ_i^2 in expression (2.7) we get

(2.11) $$\begin{aligned} \bar\partial \varphi_i^2 = &-2A^{-3}(f^\dagger \bar\partial f + \bar\partial f^\dagger f) \varphi_i^2 \\ &+ A^{-2}[(1+|f|^2) \partial\bar\partial f_i + (f^\dagger \bar\partial f) \partial f_i - (f^\dagger \partial f) \bar\partial f_i - (f^\dagger \partial \bar\partial f) f_i]. \end{aligned}$$

We can eliminate the second derivatives $\partial\bar{\partial} f_i$ in expression (2.11) by using the $\mathbb{CP}^{N-1}$ equations (1.8)

(2.12) $\quad \partial\bar{\partial} f_i = A^{-1}[(f^\dagger \bar{\partial} f)\partial f_i + (f^\dagger \partial f)\bar{\partial} f_i + (f^\dagger \partial\bar{\partial} f) f_i] - 2A^{-2}(f^\dagger \partial f)(f^\dagger \bar{\partial} f) f_i$

From equations (2.10), (2.11) and (2.12) we can express the derivatives $\bar{\partial}\varphi_i$ and $\partial\psi_i$ in terms of functions φ_i and ψ_i only

(2.13)
$$\bar{\partial}\varphi_i = -\frac{1}{2}\left[A(\varphi^\dagger \psi)\varphi_i + \frac{|\varphi^2|^2}{\overline{\varphi}_i} f_{(i)} + (\varphi^\dagger \psi)(\psi^\dagger \varphi)\frac{\psi_i}{\overline{\varphi}_{(i)}\varphi_i}\right], \quad i = 2,\ldots,N$$
$$\partial\psi_i = A[\varphi_{(i)}^2 \overline{\varphi}_i + \tfrac{1}{2}(\psi^\dagger \varphi)\psi_i] - \frac{1}{2}\left[|\varphi|^2 \frac{\psi_i}{\overline{\varphi}_i^2} + (\varphi^\dagger \psi)(\psi^\dagger \varphi)\frac{\overline{\psi}_{(i)}\psi_i}{\overline{\varphi}_{(i)}^2 \varphi_i}\right]$$

where $|\varphi^2|^2 \equiv \overline{\varphi}_2^2 \varphi_2^2 + \cdots + \overline{\varphi}_N^2 \varphi_N^2$. The first order system so obtained we call the modified version of Weierstrass (MW) system. Thus we have

PROPOSITION 1. *There exists a one-to-one correspondence between MW system (2.13) and the equations of the $\mathbb{CP}^{N-1}$ sigma model (1.8).*

PROOF. Suppose that a set of functions ψ_i and φ_i is a solution of the first order system (2.13). Then the functions f_i defined by (2.9) are a solution of the $\mathbb{CP}^{N-1}$ sigma model equations (1.8).

Conversely, if a set of functions f_i, $i = 2,\ldots,N$, is a solution to the $\mathbb{CP}^{N-1}$ equations (1.8), then the functions ψ_i and φ_i defined by (2.7) and (2.9) satisfy the MW system (2.13). □

Note that writing the MW system (2.13) in terms of functions f_i and φ_i^2 through (2.7) and (2.8) leads to its simplified form

(2.14)
$$\bar{\partial}\varphi^2 = -A(\varphi^2 \otimes \varphi^{2\dagger}) f - [(\varphi^2 \otimes \varphi^{2\dagger}):(\mathbf{1} + f \otimes f^\dagger)] f,$$
$$\partial f = A(\mathbf{1} + f \otimes f^\dagger)\varphi^2,$$
$$A = 1 + \bar{f} f,$$

where all the vectors have components $k = 2,\ldots,N$.

We finish this section with some basic facts about the MW system (2.13) which possesses several conserved quantities. One of them is the conservation of the current which can be written in terms of f as follows

(2.15) $\quad\quad\quad\quad\quad\quad\quad J = A^{-1} \partial f^\dagger \cdot \Pi \cdot \partial f.$

Differentiation of J gives

(2.16) $\quad\quad\quad\quad\quad\quad\quad \bar{\partial} J = 0,$

whenever $\mathbb{CP}^{N-1}$ equations (1.8) are satisfied. This means that J is any holomorphic function. Equivalently, we can write representation of J in terms of functions ψ_i and φ_i

(2.17) $\quad\quad\quad\quad\quad\quad\quad J = \varphi_i \partial \bar{\psi}_i - \bar{\psi}_i \partial \varphi_i,$

and (2.16) holds, whenever the MW system (2.13) is satisfied.

Note that under the requirement of finiteness of action (1.4) all solutions of the $\mathbb{CP}^{N-1}$ equations (1.8), which describe fields defined on the sphere S^N, are well known [4]. They split into three separate classes. The first two are described by analytic fields $f_i = f_i(\zeta)$ and antianalytic ones $f_i = f_i(\bar{\zeta})$ and for the analytic case J vanishes. The third class is called the mixed one, and it can be determined from

N arbitrary holomorphic or antiholomorphic functions by the following procedure. We introduce a new set of N arbitrary holomorphic functions $h_i = h_i(\zeta)$ and define new complex valued functions (now all indices take values $1, \ldots, N$)

$$g_i = \bar{f}_k F_{ki}, \tag{2.18}$$

where

$$F_{ij} = h_i \partial h_j - h_j \partial h_i, \quad \bar{\partial} h_i = 0, \quad i, j = 1, \ldots, N. \tag{2.19}$$

As it was shown in [4], a class of mixed solutions of the CP^{N-1} equations (1.8) can be expressed by taking the ratios of the components g_i

$$f_l = \frac{g_l}{g_1}, \quad l = 2, \ldots, N. \tag{2.20}$$

Other classes can be obtained by the repetition of this procedure; with the successive replacement of h_i by g_is. Alternatively, we can start with N antiholomorphic functions $h_i = h_i(\bar{z})$ and construct g_i in the same way as above but using $\bar{\partial}$ instead of ∂ in equations (2.19). These procedures lead to large classes of solutions of the MW system (2.13).

Note that the system of equations (2.13) for $N = 2$ reduces to the Weierstrass system associated with the CP^1 model. If we introduce the following notation $\varphi_2 = \varphi$, $\psi_2 = \psi$ then system (2.13) takes the form

$$\partial \psi = (|\varphi|^2 + |\psi|^2)\varphi, \quad \bar{\partial}\varphi = -(|\varphi|^2 + |\psi|^2)\psi, \tag{2.21}$$

while for $N = 3$ equations (2.13) reduce to the Weierstrass system associated with the CP^2 model (no summation)

$$\begin{aligned}
\partial \psi_i &= A[\bar{\varphi}_i \varphi_i^2 + \tfrac{1}{2}(\bar{\psi}_1 \varphi_1 + \bar{\psi}_2 \varphi_2)\psi_i] \\
&\quad - \tfrac{1}{2}[(\bar{\psi}_1 \varphi_1 + \bar{\psi}_2 \varphi_2)(\psi_1 \bar{\varphi}_1 + \psi_2 \bar{\varphi}_2) - (\bar{\varphi}_1^2 \varphi_1^2 + \bar{\varphi}_2^2 \varphi_2^2)]\frac{\bar{\psi}_i \psi_i}{\bar{\varphi}_i^2 \varphi_i}, \\
\bar{\partial}\varphi_i &= -\frac{1}{2}\Big\{A(\bar{\varphi}_1 \psi_1 + \bar{\varphi}_2 \psi_2)\varphi_i \\
&\quad + [(\bar{\psi}_1 \varphi_1 + \bar{\psi}_2 \varphi_2)(\psi_1 \bar{\varphi}_1 + \psi_2 \bar{\varphi}_2) + (\bar{\varphi}_1^2 \varphi_1^2 + \bar{\varphi}_2^2 \varphi_2^2)]\frac{\psi_i}{\bar{\varphi}_i \varphi_i}\Big\}, \\
&\quad i = 1, 2,
\end{aligned} \tag{2.22}$$

where the following notation has been used $\varphi_1 = \varphi_2$, $\varphi_2 = \varphi_3$, $\psi_1 = \psi_2$, $\psi_2 = \psi_3$. According to (2.15), the current J associated with MW system (2.13) for $N = 3$ has the form

$$J = \varphi_1 \partial \bar{\psi}_1 - \bar{\psi}_1 \partial \varphi_1 + \varphi_2 \partial \bar{\psi}_2 - \bar{\psi}_2 \partial \varphi_2, \quad \partial \bar{J} = 0 \tag{2.23}$$

whenever (2.13) holds.

3. The surfaces in $\mathbb{R}^M$

We now define a set of M real valued functions $Z_l(z, \bar{z})$ constructed out of the Weierstrass data ψ_i's and φ_i's or equivalently, f_i's and φ_i^2's which are determined by the MW system (2.13) or (2.14), respectively. Let Σ be a piece of a smooth orientable surface embedded in $\mathbb{R}^M$, which is diffeomorphic to the unit disk $D^2 = \{\zeta = x + iy \in \mathbb{C} : |\zeta| < 1\}$. Let $Z: D^2 \to \mathbb{R}^M$ be a parameterization of Σ. We will show that if f_i's and φ_i^2's obey the MW system (2.14) then for $M = N^2 - 1$ we can find the geometric characteristics of surfaces having a simple and compact form. To construct an appropriate coordinate system of a surface it is convenient to exploit the CL (2.3). To simplify our analysis we will first consider the case of conformally parameterized surfaces in $\mathbb{R}^M$. It can be shown that in this case the terms ϕ_j^2 in expressions (2.3) and (2.4) for matrices K and $K^\dagger$ can be dropped and still the CL associated with $\mathbb{C}P^{N-1}$ equations is valid. In [8], a similar fact has been discussed in detail for the isometric surfaces in $\mathbb{R}^3$. In the multi-dimensional case we can define new traceless matrices

$$(3.1) \qquad K'_{ij} = f_i \overline{\varphi}_j^2,$$

and

$$(3.2) \qquad (K'_{ij})^\dagger = \varphi_i^2 \bar{f}_j,$$

whenever the MW system (2.14) is satisfied. As a consequence of such CL (3.2) we can define N^2 real valued quantities $Z_l(z, \bar{z})$ in terms of the Weierstrass variables f_i and φ_i^2. However, $N^2 - 1$ of these quantities Z_l are linearly independent, since matrices K and $K^\dagger$ are traceless. So, we can write the vector Z in the form (no summation over repeated indices)

$$(3.3) \qquad Z = (Z_{kk} - \delta_{kN} \sum_{a=1}^{N-1} z_{aa}, X_{kl}, Y_{kl}) \in \mathbb{R}^{N^2-1}, \quad k,l = 1, \ldots, N$$

The components Z_{ij} are constructed by taking diagonal and off-diagonal entries of the matrices K and $K^\dagger$ and have [9] the form (no summation)

$$(3.4) \qquad Z_{ll} = \int_\gamma (\bar{f}_l \varphi_l^2 \, d\zeta + f_l \overline{\varphi}_l^2 \, d\bar{\zeta}), \quad k = l$$

and

$$(3.5) \quad Z_{kl} = X_{kl} + iY_{kl} = \int_\gamma (\alpha \bar{f}_l \varphi_k^2 + \bar{\alpha} \bar{f}_k \varphi_l^2) \, d\zeta + \int_\gamma (\bar{\alpha} f_l \overline{\varphi}_k^2 + \alpha f_k \overline{\varphi}_l^2) \, d\bar{\zeta}, \quad k \neq l$$

where $\alpha = (1+i)/2$ and γ is any curve from a fixed point to ζ in $\mathbb{C}$. Using the tracelessness of matrices K and $K^\dagger$ we find that the diagonal terms in (3.4) satisfy

$$(3.6) \qquad \sum_{j=1}^N Z_{jj} = 0.$$

In our expressions we take all $k, l = 1, \ldots N$ and for index $k = 1$ or $l = 1$ we use our algebraic constraints (2.6) to write all our expressions in terms of linearly independent functions φ_i^2 $i = 2, \ldots, N$. For our real variables Z_i we take the components of Z_{aa}, X_{kl} and Y_{kl}. Hence, we can treat these components as the coordinates of the position vector Z of a surface Σ in Euclidean space $\mathbb{R}^{N^2-1}$. Note that the conservation laws (2.3) guarantee that Z_{ll} and Z_{kl} do not depend on the choice of the contour γ but only on its endpoints in the complex plane $\mathbb{C}$. This is due to the fact that all components of Z can be written in the form

$$(3.7) \qquad Z = \int_\gamma F(z, \bar{z}) \, d\zeta + \bar{F}(z, \bar{z}) \, d\bar{\zeta},$$

where F is a conserved quantity which satisfies the condition

$$(3.8) \qquad \partial \overline{F} - \bar{\partial} F = 0.$$

It shows that the integrands of (3.7) are total derivatives of a real valued function F. Note also that all our expressions for Z_{ll} and Z_{kl} can be written in terms of the Weierstrass variables φ_i and ψ_i and their complex conjugates. They formally include φ_1 and ψ_1 with $\psi_1 = f_1\overline{\varphi}_1$. However, both these quantities ψ_1 and φ_1 should be eliminated by using the algebraic constraint (2.6) with $f_1 = 1$. In particular, for $N = 2$, it is easy to check that in the CP^1 case this process of elimination leads to the integrands which are quadratic in terms of φ_2 and ψ_2 and their conjugates

$$Z_{11} = -2\int_\gamma \bar{\psi}_2\varphi_2\, d\zeta' + \psi_2\overline{\varphi}_2\, d\overline{\zeta'}, \quad X_{12} = \int_\gamma (\bar{\psi}_2^2 - \varphi_2^2)\, d\zeta' + (\psi_2^2 - \overline{\varphi}_2^2)\, d\overline{\zeta'},$$

(3.9)

$$Y_{12} = \int_\gamma (\psi_2^2 + \varphi_2^2)\, d\zeta' - (\psi_2^2 + \varphi_2^2)\, d\overline{\zeta'},$$

Thus we recover the result obtained in [**12**].

Once the Weierstrass representations (2.13) are adopted then we can calculate some geometric characteristics of a surface in our multi-dimensional space. For instance, the induced metric on Σ is given by the formula

(3.10) $$g_{\alpha\beta} = \sum_{i,j=1}^N \frac{\partial Z_{ij}}{\partial r}\frac{\partial Z_{ij}}{\partial s},$$

where r and s are ζ or $\bar{\zeta}$.

Let us first discuss the meaning of the Weierstrass representations given by (2.13). We find that

(3.11) $$g_{\zeta\zeta} = (f^\dagger\varphi^2)^2 = 0, \quad g_{\bar{\zeta}\bar{\zeta}} = \left((\varphi^2)^\dagger f\right)^2 = 0,$$

which coincide with the algebraic constraints (2.6). The only nonzero term of the induced metric (3.10) is [**9**]

(3.12) $$g_{\zeta\bar{\zeta}} = A\left[\left|\sum_{k=2}^N \bar{f}_k\varphi_k^2\right|^2 + |\varphi_2|^2 + \cdots + |\varphi_N|^2\right], \quad A = 1 + |f_2|^2 + \cdots + |f_N|^2.$$

Expression (3.10) in terms of the Weierstrass variables φ_i and ψ_i appears to have a complicated form. However, if we write equation (3.10) in terms of f_i and ∂f_i through (2.8), and next use the homogeneous coordinates z, our expression simplifies considerably and we obtain

(3.13) $$g_{\zeta\bar{\zeta}} = |Dz|^2,$$

where

(3.14) $$Dz = \tfrac{1}{2}(D_1 - iD_2)z$$

and D_1 and D_2 are the covariant derivatives (1.2) involving ∂_μ, where $\mu = 1, 2$ denotes ζ and $\bar{\zeta}$. Note that in the special case of CP^1 maps for $N = 2$ the component of the induced metric $g_{\zeta\bar{\zeta}}$ takes a particularly simple form. It is given by

(3.15) $$g_{\zeta\bar{\zeta}} = \left(1 + \frac{|\psi_2|^2}{|\varphi_2|^2}\right)(|\varphi_2|^4 + |\psi_2|^2|\varphi_2|^2) = (|\psi_2|^2 + |\varphi_2|^2)^2 = \frac{|\partial f_2|^2}{(1+|f_2|^2)^2}$$

which is exactly $|Dz|^2$. For CP^2 maps (i.e., when $N = 3$) we get

(3.16) $$g_{\zeta\bar{\zeta}} = \frac{|\partial f_2|^2 + |\partial f_3|^2 + |f_2\partial f_3 - f_3\partial f_2|^2}{(1+|f_2|^2+|f_3|^2)^2}$$

which also coincides with the expression $|Dz|^2$. Moreover, it is easy to show that the representation given by (3.4) and (3.5) satisfies

$$(\partial\bar{\partial}Z, \partial\bar{\partial}Z) = (\partial Z, \bar{\partial}Z)^2.$$

This implies that the norm of the mean curvature vector $\vec{H} = (g_{\zeta\bar\zeta})^{-1}\partial\bar\partial Z$ is equal to one, $|\vec{H}| = 1$. At this point let us summarize our approach to constructing surfaces immersed into multi-dimensional Euclidean space.

PROPOSITION 2. *Assume that the set of $N^2 - 1$ real-valued functions $Z_l(z,\bar z)$ and $2N$ complex-valued functions (ψ_i, φ_i) satisfies the Weierstrass representation (3.4), (3.5) and the first order system (2.13). Then $Z\colon D^2 \to \mathbb{R}^{N^2-1}$ is a conformal immersion of a CMC-surface in $\mathbb{R}^{N^2-1}$. The induced metric on Σ is given by*

(3.17) $$ds^2 = 2|Dz|^2\, dz\, d\bar z,$$

where ζ and $\bar\zeta$ are the isothermal coordinates.

4. Final remarks

In this paper we have discussed links between the CP^{N-1} sigma model and our form of the Weierstrass representation for two-dimensional surfaces immersed in multi-dimensional Euclidean spaces. These links enabled us to present a new analytic method for the construction of CMC-surfaces immersed in $\mathbb{R}^{N^2-1}$ space. This method has been limited, however, by having to rely on several assumptions. For example, we have studied only the CP^{N-1} models defined over S^2 (i.e., when $J = 0$). The natural question arises as to whether our approach can be extended to Weierstrass systems describing surfaces immersed in multi-dimensional Riemannian or pseudo-Riemannian spaces. If this is the case, then can our method provide new classes of solutions which describe surfaces more diverse than those discussed in the multi-dimensional Euclidean case? Another situation worth investigating is the CP^{N-1} model involving maps from $\mathbb{R}^2$ (not necessarily S^2). We can expect that by taking $J \neq 0$ the applicability of our approach can be broadened.

In this paper we could not, due to the lack of space, include any examples of physical applications. It is worth mentioning, however, that the proposed method can be applied to several dynamical problems modeled by nonlinear PDEs and involving phenomena occurring on CMC-surfaces (e.g., propagation of flame fronts, growth of crystals, deformations of fluid membranes to world-sheets and their dynamics in the string theory [14, 16]). In particular, it is worth noting that there are cases involving known descriptions of CMC-surfaces in physical systems for which analytic methods have not yet been fully developed. However, using our approach we can select an appropriate CP^{N-1} sigma model corresponding to the given Weierstrass representation and characterize the class of equations describing the physical phenomena under investigation. This procedure has been successfully applied for the Weierstrass representation associated with the CP^1 sigma model for CMC-surfaces in three-dimensional Euclidean space [7] but, so far as we know, not for the multi-dimensional spaces. These and other questions of applications will be addressed in our future work.

References

1. D. J. Amit, *Field theory, the renormalization group and critical phenomena*, Internat. Ser. Pure Appl. Phys., McGraw-Hill, New York, 1978.
2. P. Bracken and A. M. Grundland, *Symmetry properties and explicit solutions of generalized Weierstrass system*, J. Math. Phys. **42** (2001), no. 3, 1250–1282.
3. P. Bracken, A. M. Grundland, and L. Martina, *The Weierstrass–Enneper system for CMC-surfaces and completely integrable sigma model*, J. Math. Phys. **40**, (1999), no. 7, 3379–3403.

4. A. M. Din and W. J. Zakrzewski, *General classical solutions in the CP^{N-1} model*, Nuclear Phys. B **174** (1980), no.2-3, 397-406.
5. A. Enneper, *Analytisch-geometrische Untersuchungen*, Nachr. Königl. Gesell. Wissensch. Georg-Augustus-Univ. Göttingen **12** (1868), 258–277.
6. D. J. Gross, T. Piran, and S. Weinberg (eds.), *Two-dimensional quantum gravity and random surfaces*, Jerusalem Winter School Theoret. Phys., vol. 8, World Scientific, Singapore, 1992.
7. K. Grosse-Brauckman and K. Polthier, *Constant mean curvature surfaces derived from Delauney's and Wente's examples*, Visualization and Mathematics: Experiments, Simulations and Environments (H. C. Hege and K. Polthier, eds.), Springer-Verlag, Berlin, 1997, pp. 96–149.
8. A. M. Grundland and W. J. Zakrzewski, *The Weierstrass representation for surfaces immersed into $\mathbb{R}^8$ and CP^2 maps*, J. Math. Phys. **43** (2002), no. 6, 3352–3362.
9. _____, *On certain geometrical aspects of CP^N harmonic maps*, J. Math. Phys. **44** (2003), no. 1, 328–338.
10. _____, *On CP^1 and CP^2 maps and Weierstrass representations for surfaces immersed into multi-dimensional Euclidean spaces*, J. Nonlinear. Math. Phys. **10** (2003), 1-2-6.
11. F. Hélein, *Constant mean curvature surfaces, harmonic maps and integrable systems*, Lectures Math. ETH Zürich, Birkhäuser Verlag, Basel, 2001.
12. B. G. Konopelchenko and G. Landolfi,*Induced surfaces and their integrable dynamics. II*, Stud. Appl. Math. **104** (2000), 129–169.
13. A. V. Mikhailov, *Integrable magnetic models*, Solitons, (S. E. Trullinger, V. E. Zakharov, and V. Pokrovski, eds.), Modern Problem in Condensed matter, vol. 17, North-Holland, Amsterdam, 1986, pp. 623–690.
14. D. Nelson, T. Piran and S. Weinberg (eds.), *Statistical mechanics of membranes and surfaces*, Jerusalem Winter School Theoret. Phys., vol. 5, World Scientific, Singapore, 1989.
15. R. Osserman, *A survey of minimal surfaces*, 2nd edition, Dover, New York, 1986.
16. Z.-C. Ou-Yang, J.-X. Liu and Y.-Z. Xie, *Geometric methods in the elastic theory of membranes in liquid crystal phases*, Adv. Ser. Theoret. Phys. Sci., vol. 2, World Scientific, Singapore, 1999.
17. B. L. Rozdestvenskiĭ and N. N. Janenko, *Systems of quasilinear equations and their applications to gas dynamics*, Transl. Math. Monogr., vol. 55, Amer. Math. Soc., Providence RI, 1983.
18. K. Weierstrass, *Fortsetzung Untersuchung über die Minimalflächen Mathematische Werke*, Vol. 3, Verlagsbuchhandlung, Hillesheim, 1866, 219–248.
19. W. J. Zakrzewski, *Low-dimensional sigma models*, Adam Hilger, Bristol 1989.

CENTRE DE RECHERCHES MATHÉMATIQUES, UNIVERSITÉ DE MONTRÉAL, C. P. 6128, SUCC. CENTRE-VILLE, MONTRÉAL, QC H3C 3J7, CANADA
E-mail address: `grundlan@crm.umontreal.ca`

DEPARTMENT OF MATHEMATICAL SCIENCES, UNIVERSITY OF DURHAM, DURHAM, DH1 3LE, UK
E-mail address: `W.J.Zakrzewski@durham.ac.uk`

Boson Realizations of Semi-Simple Lie Algebras

Čestmír Burdík and *Miloslav Havlíček*

ABSTRACT. A short review about the construction of the boson realizations for the semi-simple Lie algebras is given. The method is illustrated on the examples. Applications for the construction of the matrix differential realizations are mentioned and their properties are studied. It is shown how the concept of Lie field can be alternatively used.

1. Introduction

The purpose of this note is to refer on a special type of representations of semi-simple Lie algebras which could be interesting from the point of view quasi-exactly solvable [22, 24, 25] quantum problems or supersymmetric Hamiltonians [1, 23].

These representations were published in series of papers [2–5, 10–12, 14–18] during the seventies and we will try to describe roughly their construction and by several examples we will illustrate their main properties.

Initially, these representations were constructed,in a weel-defined sense, as skew-symmetric and Schurean (i.e., all Casimir operators were multiples of the identity operator) for the noncompact real simple Lie algebras. If such representations satisfy certain integrability condition, they can give unitary and irreducible representations of corresponding groups.

The skew-symmetricity of the representations does not seem important for the above mentioned problems but two further properties could be interesting.

First, generators in the considered representations have the form

$$(1.1) \qquad G_l = \sum_{i=1}^{N} A_i(x_1, x_2, \ldots, x_N) \frac{\partial}{\partial x_i} + B_i(x_1, x_2, \ldots, x_N)$$

with specified numbers N for a given algebra and with matrix coefficients $A_i(x_1, x_2, \ldots, x_N)$, $B_i(x_1, x_2, \ldots, x_N)$. It seems that this "matrix" representation is nontrivial in the sense that it cannot be reduced to the direct sum of representations of the type (1.1) but with matrix coefficients of smaller dimensions.

Second, generators depend on certain a number of parameters. By their suitable choice and choosing also appropriate representation spaces, such representation could have finite-dimensional invariant subspaces.

2000 *Mathematics Subject Classification.* Primary 17B45; Secondary 17B10, 17B81.
The authors were supported in part by GAČR Grant #1301013.
This is the final form of the paper.

As examples we use rank two Lie algebras sl(3, $\mathbb{C}$) and the exceptional Lie algebra g_2. In the concluding section we introduce very roughly the notion of the Lie field of the algebra gl(3, $\mathbb{C}$) and its simple algebraical extension by commutative elements. We show, in particular, that there exist rational functions in generators of gl(3, $\mathbb{C}$) commuting as canonical pairs P_i, Q_j, i.e., $[P_i, Q_j] = \delta_{ij}$. We include these considerations in our contribution as inspiration for possible applications.

2. The simplest example sl(2, $\mathbb{C}$) $\sim$ A_1

Recall some well-known results about representations of the Lie algebra sl(2, $\mathbb{C}$). Let

(2.1) $$e = \begin{pmatrix} 0 & 1 \\ 0 & 0 \end{pmatrix}, \quad h = \begin{pmatrix} 1 & 0 \\ 0 & -1 \end{pmatrix}, \quad f = \begin{pmatrix} 0 & 0 \\ 1 & 0 \end{pmatrix}$$

be the standard basis of sl(2, $\mathbb{C}$). Then

(2.2) $$[e, f] = h, \quad [h, e] = 2e, \quad [h, f] = -2f.$$

By an easy induction on n we deduce the following relations in the universal enveloping algebra of sl(2, $\mathbb{C}$):

(2.3) $$[h, f^n] = -2nf^n, \quad [h, e^n] = 2ne^n, \quad [e, f^n] = -n(n-1)f^{n-1} + nf^{n-1}h.$$

Now we define Verma modules by putting

(2.4) $$e|0\rangle = 0, \quad h|0\rangle = \lambda|0\rangle, \quad \text{and} \quad f|n\rangle = |n+1\rangle,$$

and by using the above commutation relations we obtain

(2.5) $$h|n\rangle = (\lambda - 2n)|n\rangle, \quad \text{and} \quad e|n\rangle = [\lambda n - n(n-1)]|n-1\rangle.$$

The representation has these properties: (a) It is Schurean (the Casimir operator C_2 is a multiple of identity $C_2 = \alpha I$). (b) For any $\lambda \notin N_0 = \{0, 1, 2, \ldots\}$ it is irreducible. (c) For $\lambda = m \in N_0$ it is reducible and the space $W = C\{|m+1\rangle, \ldots\}_{\text{lin}}$ is an invariant subspace and $\dim\{C\{|n\rangle, n = 0, 1 \ldots\}_{\text{lin}}/W\} = m < \infty$.

This representation is a starting point of our construction of boson realizations. We define

(2.6) $$\bar{a}|n\rangle = |n+1\rangle \quad \text{and} \quad a|n\rangle = n|n-1\rangle,$$

and the Verma modules we can rewrite in the operator form

(2.7) $$h = \lambda + 2 - 2a\bar{a}, \quad f = \bar{a}, \quad e = (\lambda + 2)a - a^2\bar{a};$$

the commutation relations (2.2) are simple consequence of usual relations $[a, \bar{a}] = 1$. Substituting

(2.8) $$a \to x, \quad \bar{a} \to -\partial_x \quad \text{and} \quad \lambda_1 \to (\lambda + 2)$$

the equations (2.7) reduce to

(2.9) $$-h = \lambda_1 - 2x\partial_x, \quad f = -\partial_x, \quad e = -\lambda_1 x + x^2\partial_x,$$

and it can be considered as representations on space of polynomials. Zero-order monomial is highest weight vector (with respect to $-h$) with weight λ_1 and if $\lambda_1 = m \in N_0$ the subspace $C\{1, x, x^2, \ldots, x^m\}_{\text{lin}}$ is invariant.

3. The general construction

The general construction is described in the paper [2]. We sum up this construction shortly. Let g be a simple Lie algebra and $\{\alpha_1, \ldots, \alpha_k\}$ a system of positive simple roots. There exists a Cartan–Weyl decomposition

$$g = n^+ \oplus h \oplus n^-,$$

where n^+ are the positive root vectors, n^- are the negative root vectors and h the Cartan subalgebra. Let α_i be one of the simple roots, e_{α_i} the appropriate positive root vector and f_{α_i} the appropriate negative root vector. Now we define

$$n_i^+ = \mathbb{C}\{e_{\alpha_i}, [e_{\alpha_i}, e_{\alpha_j}], \ldots\}_{\text{lin}} \subset n^+,$$
$$n_i^- = \mathbb{C}\{f_{\alpha_i}, [f_{\alpha_i}, f_{\alpha_j}], \ldots\}_{\text{lin}} \subset n^-,$$

and g_i as the subalgebra generated by remaining $e_\alpha \in n^+ - n_i^+$, $f_\alpha \in n^- - n_i^-$ and $h = \{h_1, \ldots, h_r\}$; evidently by this way we obtain a decomposition

$$g = n_i^+ \oplus g_i \oplus n_i^-.$$

For g_i we have the decomposition

$$g_i = g^1 \oplus g^2 \oplus Z$$

where the g^1, (g^2) is given by the root system $\{\alpha_1, \ldots, \alpha_{i-1}\}$, ($\{\alpha_{i+1}, \ldots, \alpha_k\}$) and Z is its one-dimensional center. For the construction we take the representation φ of the algebra $g^1 \oplus g^2$ on the representation space V. It is possible to extend the representation φ to the algebra $g_i \oplus n_i^-$ if we put

$$\varphi(f_\beta) v = 0 \quad \text{for all } \beta \in n_i^-, v \in V$$
$$\varphi(Z) v = \lambda v \quad \text{for } v \in V.$$

Let $U(g)$ be the enveloping algebra of g and consider the subspace $W \subset U(g) \otimes V$ generated by relations

$$xz \otimes v - x \otimes \varphi(z)v$$

where $x \in U(g)$, $z \in U(g_i \oplus n_i^-)$ and $v \in V$. The subspace W is invariant for the left regular representation ρ of g on $U(g) \otimes V$ and we can define the factor-representation of g

$$\rho_{i, \lambda, \varphi} = \rho/W.$$

It is possible to show that by suitable choice of basis in $U(n^+)$ we are able to rewrite this representation by means of $p = \dim(n_i^+)$ boson operators $\bar{a}_k$, a_k, the representation φ and one parameter $\lambda \in \mathbb{C}$.

In the next sections we will demonstrate this construction on the examples of algebra $\mathrm{sl}(3, \mathbb{C})$ and the exceptional Lie algebra g_2.

4. Examples

4.1. Lie algebra $\mathrm{sl}(3, \mathbb{C})$. The algebra $\mathrm{gl}(3, \mathbb{C})$ is the 9-dimensional complex Lie algebra with a standard basis $\{e_{ij} : i, j = 1, 2, 3\}$ the elements of which obey

(4.1) $$[e_{ij}, e_{kl}] = \delta_{jk} e_{il} - \delta_{il} e_{kj}.$$

This algebra is a direct sum if its one-dimensional centre (generated by the element $e = e_{11} + e_{22} + e_{33}$) and the simple subalgebra $\mathrm{sl}(3, \mathbb{C})$ whose generators are e_{ij}, $i \neq j$ and $h_i = e_{i+1, i+1} - e_{i, i}$, $i = 1, 2$. The simple roots are $e_{\alpha_1} = e_{21}$ and $e_{\alpha_2} = e_{32}$.

If we take $e_{\alpha_1} = e_{21}$ we obtain $n_1^+ \equiv \mathbb{C}\{e_{31}, e_{32}\}_{\text{lin}}$, $n_1^- \equiv \mathbb{C}\{e_{13}, e_{23}\}_{\text{lin}}$ and $g_1 \equiv \mathbb{C}\{h_i; i = 1, 2, e_{12}, e_{21}\}_{\text{lin}}$. Then

$$\text{(4.2)} \qquad g = n_1^+ \oplus g_1 \oplus n_1^-,$$

which is the starting decomposition of sl(3, $\mathbb{C}$).

For obtaining the explicit form of induced representation we need now these formulae

$$\text{(4.3)} \begin{aligned}
e_{31} e_{31}^{k_1} e_{32}^{k_2} &= e_{31}^{k_1+1} e_{32}^{k_2} \\
e_{32} e_{31}^{k_1} e_{32}^{k_2} &= e_{31}^{k_1} e_{32}^{k_2+1} \\
e_{21} e_{31}^{k_1} e_{32}^{k_2} &= -k_2 e_{31}^{k_1+1} e_{32}^{k_2-1} + e_{31}^{k_1} e_{32}^{k_2} e_{21} \\
e_{12} e_{31}^{k_1} e_{32}^{k_2} &= -k_1 e_{31}^{k_1-1} e_{32}^{k_2+1} + e_{31}^{k_1} e_{32}^{k_2} e_{12} \\
e_{13} e_{31}^{k_1} e_{32}^{k_2} &= k_2 e_{31}^{k_1+1} e_{32}^{k_2-1} e_{12} + e_{31}^{k_1} e_{32}^{k_2} e_{13} - k_1 e_{31}^{k_1-1} e_{32}^{k_2}(h_2 + h_1) \\
&\quad - k_1 k_2 e_{31}^{k_1-1} e_{32}^{k_2} - k_1(k_1 - 1) e_{31}^{k_1-1} e_{32}^{k_2} \\
e_{23} e_{31}^{k_1} e_{32}^{k_2} &= k_1 e_{31}^{k_1-1} e_{32}^{k_2} e_{21} + e_{31}^{k_1} e_{32}^{k_2} e_{23} - k_2 e_{31}^{k_1} e_{32}^{k_2-1} h_2 \\
&\quad - k_1 k_2 e_{31}^{k_1} e_{32}^{k_2-1} - k_2(k_2-1) e_{31}^{k_1} e_{32}^{k_2-1} \\
h_1 e_{31}^{k_1} e_{32}^{k_2} &= k_1 e_{31}^{k_1} e_{32}^{k_2} - k_2 e_{31}^{k_1} e_{32}^{k_2} + e_{31}^{k_1} e_{32}^{k_2} h_1 \\
h_2 e_{31}^{k_1} e_{32}^{k_2} &= k_1 e_{31}^{k_1} e_{32}^{k_2} + 2k_2 e_{31}^{k_1} e_{32}^{k_2} + e_{31}^{k_1} e_{32}^{k_2} h_2
\end{aligned}$$

Evidently

$$\text{(4.4)} \qquad Z = 2h_2 + h_1.$$

Now we obtain for boson realization formulae

$$\text{(4.5)} \begin{aligned}
\rho_{1,\lambda,\varphi}(h_1) &= +\bar{a}_1 a_1 - \bar{a}_2 a_2 + \varphi(h_1) \\
\rho_{1,\lambda,\varphi}(h_2) &= +\bar{a}_1 a_1 + 2\bar{a}_2 a_2 + \bigl(\lambda - \varphi(h_1)\bigr)/2 \\
\rho_{1,\lambda,\varphi}(e_{21}) &= -a_2 \bar{a}_1 + \varphi(e_{21}) \\
\rho_{1,\lambda,\varphi}(e_{31}) &= \bar{a}_1 \\
\rho_{1,\lambda,\varphi}(e_{32}) &= \bar{a}_2 \\
\rho_{1,\lambda,\varphi}(e_{12}) &= -a_1 \bar{a}_2 + \varphi(e_{12}) \\
\rho_{1,\lambda,\varphi}(e_{13}) &= -\bar{a}_1 a_1^2 - \bar{a}_2 a_2 a_1 - \bigl(\lambda + \varphi(h_1)\bigr)a_1/2 + a_2 \varphi(e_{12}) \\
\rho_{1,\lambda,\varphi}(e_{23}) &= -\bar{a}_1 a_1 a_2 - \bar{a}_2 a_2^2 - \bigl(\lambda - \varphi(h_1)\bigr)a_2/2 + a_1 \varphi(e_{21})
\end{aligned}$$

4.2. Exceptional Lie algebra g_2. The same construction for g_2 [8] gives

$$\text{(4.6)} \begin{aligned}
\rho(E_{\alpha_1}) &= a_5 \bar{a}_6 + a_2 \bar{a}_3 - a_2^2 \bar{a}_4 - 3a_2^2 a_3 \bar{a}_6 - a_2^3 \bar{a}_5 + \varphi(\widetilde{E}_{\alpha_1}) \\
\rho(E_{\alpha_2}) &= \bar{a}_2 + 3a_4 \bar{a}_5 - 2a_3 \bar{a}_4 - 3a_3^2 \bar{a}_6 \\
\rho(E_{-\alpha_1}) &= a_6 \bar{a}_5 + a_3 \bar{a}_2 - a_3^2 \bar{a}_4 - 2a_3^3 \bar{a}_6 + \varphi(\widetilde{E}_{-\alpha_1}) \\
\rho(E_{-\alpha_2}) &= a_5 \bar{a}_4 - 2a_4 \bar{a}_3 - 3a_4^2 \bar{a}_6 + 3a_2^2 a_3 \bar{a}_4 + 3a_2^3 a_3 \bar{a}_5 + 9a_2^2 a_3^2 \bar{a}_6 \\
&\quad - 3a_3 \varphi(\widetilde{E}_{\alpha_1}) - a_2[3a_3 \bar{a}_3 + a_2 \bar{a}_2 + \lambda - \tfrac{3}{2}\varphi(\widetilde{H}_1)]
\end{aligned}$$

Remaining 8 generators of g_2 we obtain by commutation of these four ones.

If representation φ in above examples will be Schurean, representations of sl(3, $\mathbb{C}$) and g_2 will be Schurean too. Note that finite-dimensional representation for

φ can be used and in this way, we obtain representations with matrix coefficients mentioned in the Introduction.

Formulas of the same type were derived in the series of papers [2, 10–12, 14–18] for all classical simple Lie algebras A_n, B_n, C_n and D_n. In the mentioned papers the authors were interested in the theory of unitary irreducible representations of real forms of simple classical Lie groups. Expressions were constructed for all corresponding Lie algebras with exceptions of so$(2n)^*$ and su$(2n)^*$.

5. Properties of the representation of sl(3, $\mathbb{C}$) and g_2 by differential operators

5.1. Lie algebra sl(3, $\mathbb{C}$).

First we will substitute to (4.5) $\varphi(h_1) = \Lambda_1 - 2x_3\partial_3$, $\varphi(e_{21}) = -\partial_3$, $\varphi(e_{12}) = -\Lambda_1 x_3 + x_3^2 \partial_3$, $\lambda = 2\Lambda_2 + \Lambda_1$ and further $\bar{a} \to -\partial_i$ $a_i \to x_i$ for $i = 1, 2$. We obtain eight differential operators

$$\rho^\Lambda(h_1) = -x_1\partial_1 + x_2\partial_2 - 2x_3\partial_3 + \Lambda_1$$
$$\rho^\Lambda(h_2) = -x_1\partial_1 - 2x_2\partial_2 + x_3\partial_3 + \Lambda_2$$
$$\rho^\Lambda(e_{21}) = +x_2\partial_1 - \partial_3$$
$$\rho^\Lambda(e_{31}) = -\partial_1$$
$$\rho^\Lambda(e_{32}) = -\partial_2$$
$$\rho^\Lambda(e_{12}) = +x_1\partial_2 + x_3(x_3\partial_3 - \Lambda_1)$$
$$\rho^\Lambda(e_{13}) = +x_1(x_1\partial_1 + x_2\partial_2 + x_3\partial_3 - \Lambda_1 - \Lambda_2) + x_2 x_3(x_3\partial_3 - \Lambda_1)$$
$$\rho^\Lambda(e_{23}) = +x_2(x_1\partial_1 + x_2\partial_2 - x_3\partial_3 - \Lambda_2) - x_1\partial_3$$

with $\Lambda = (\Lambda_2, \Lambda_1)$. Consider now these operators on the space P of polynomials in three variables x_1, x_2, x_3, which due to their special form, conserve P. As ∂_i, x_i fulfill the same commutation relations as a_i and $\bar{a}_j$, we have a representation of sl(3, $\mathbb{C}$) on the space P.

It is easy to prove that the restriction $\tilde{\rho}^\Lambda$ of ρ^Λ to the subspace $V^\Lambda \equiv \rho^\Lambda(U(g))x_0$, $x_0 \equiv 1 \in P$, is a representation of $g \equiv$ sl(3, $\mathbb{C}$) with the highest weight Λ and the vector x_0 as its highest-weight vector. Thus we have constructed for any $\Lambda = (\Lambda_1, \Lambda_2)$ the highest-weight representations $\tilde{\rho}^\Lambda$ of sl(3, $\mathbb{C}$). These representations have the following properties:

THEOREM 5.1. (a) $\tilde{\rho}^\Lambda$ is irreducible for any $\Lambda = (\Lambda_1, \Lambda_2)$.
(b) if $\Lambda \in N_0 \times N_0$ then $\dim\{V^\Lambda\} < \infty$.

PROOF. See [6, 7]. □

So, we have an irreducible finite-dimensional representation of sl(3, $\mathbb{C}$) in the form of a restriction of a representation by means of differential operators to an invariant finite-dimensional subspace of polynomials.

We will describe now one finite total set inside of V^Λ for a considered finite-dimensional case. Consider the representations for the weight $(1, 0)$ and $(0, 1)$. The spaces $V^{(1,0)}$ and $V^{(0,1)}$ we can construct explicitly by a very simple calculation. We obtain

$$V^{(1,0)} = \mathbb{C}\{1, x_3, x_1 + x_2 x_3\}_{\text{lin}}$$
$$V^{(0,1)} = \mathbb{C}\{1, x_2, x_1\}_{\text{lin}}.$$

The vectors $(x_1 + x_2 x_3)$ and x_1 in these representations are the vectors with the lowest weight.

Now we define the subspace $\widetilde{V}^\Lambda$ of P as

$$\widetilde{V}^\Lambda = \mathbb{C}\{x_3^{n_1}(x_1 + x_2 x_3)^{n_2} x_1^{m_1} x_2^{m_2}, \ n_1 + n_2 \leq \Lambda_1, \ m_1 + m_2 \leq \Lambda_2\}_{\text{lin}}.$$

THEOREM 5.2. (a) *The space $\widetilde{V}^\Lambda$ (for $\Lambda \in N_0 \times N_0$) is an invariant subspace of the representation ρ^Λ.*
(b) *The space $\widetilde{V}^\Lambda = V^\Lambda$ is irreducible.*

PROOF. (a) Due to commutation relations in the algebra $\mathrm{sl}(3, \mathbb{C})$ it is sufficient to show the invariance of $\widetilde{V}^\Lambda$ only with respect to the operators $\rho^\Lambda(e_{12})$, $\rho^\Lambda(e_{23})$, $\rho^\Lambda(e_{21})$ and $\rho^\Lambda(e_{32})$ and it is easy. For example

$$\rho^\Lambda(e_{12})(x_1^{m_1} x_2^{m_2} x_3^{n_1}(x_1 + x_2 x_3)^{n_2})$$
$$= m_2 x_1^{m_1+1} x_2^{m_2-1} x_3^{n_1}(x_1 + x_2 x_3)^{n_2} + (n_1 + n_2 - \Lambda_1) x_1^{m_1} x_2^{m_2} x_3^{n_1+1}(x_1 + x_2 x_3)^{n_2}$$

which is, for every $n_1, n_2, n_1 + n_2 \leq \Lambda_1$, a linear combination of monomials from our set. For other operators there are results similar to this.

(b) Irreducibility follows from the fact that ρ^Λ is Schurean and uniqueness of highest weight vector $x_0 = 1$. □

The last step is the construction of a basis in the space V^Λ. Evidently the set

$$\{x_3^{n_1}(x_1 + x_2 x_3)^{n_2} x_1^{m_1} x_2^{m_2}, \ n_1 + n_2 \leq \Lambda_1, m_1 + m_2 \leq \Lambda_2\}$$

is the basis of the V^Λ only for the weights $(0, \Lambda_2)$ and $(\Lambda_1, 0)$. Already for the weight $(1, 1)$ the set of these monomials is over complete. In the general case we can see this using the following consideration. The dimension of a finite-dimensional representation is very well known [**21**]. It is given by the formula

$$\dim(\Lambda) = (\Lambda_1 + 1)(\Lambda_2 + 1)\left(\frac{\Lambda_1 + \Lambda_2 + 2}{2}\right)$$

but the number of elements of the set equals

$$\frac{(\Lambda_1 + 1)(\Lambda_1 + 2)(\Lambda_2 + 1)(\Lambda_2 + 2)}{2}.$$

An example of the basis in V^Λ is given in the following theorem.

THEOREM 5.3. *Let*

$$N_1 \equiv \{(m_1, m_2, n_1) \in N_0 \times N_0 \times N_0; 0 \leq m_1 + m_2 \leq \Lambda_2, 0 \leq n_1 \leq \Lambda_1\},$$
$$N_2 \equiv \{(m_1, n_1, n_2) \in N_0 \times N_0 \times N_0; 0 \leq n_1 + n_2 \leq \Lambda_1, 0 \leq m_1 \leq \Lambda_2, n_2 \neq 0\}.$$

The two following sets of vectors

$$\{x_1^{m_1} x_2^{m_2} x_3^{n_1}; (m_1, m_2, n_1) \in N_1\}$$
$$\{x_1^{m_1} x_2^{\Lambda_2 - m_1} x_3^{n_1}(x_1 + x_2 x_3)^{n_2}, (m_1, n_1, n_2) \in N_2\}$$

constitute the basis in V^Λ.

PROOF. See [**6, 7**]. □

5.2. The case of the exceptional Lie algebra g_2. Here we will substitute to (4.6) $\varphi(\widetilde{H}_1) = \Lambda_1 - 2x_1\partial_1$ $\varphi(\widetilde{E}_{\alpha_1}) = -\partial_1$, $\varphi(\widetilde{E}_{-\alpha_1}) = -\Lambda_1 x_1 + x_1^2 \partial_1$, $\lambda = \frac{3}{2}\Lambda_1 + \Lambda_2$ and further $\bar{a} \to -\partial_i$ $a_i \to x_i$ for $i = 2, 3, 4, 5, 6$.

$$\rho^\Lambda(E_{\alpha_1}) = -x_5\partial_6 - x_2\partial_3 + x_2{}^2\partial_4 + 3x_2{}^2 x_3 \partial_6 + x_2{}^3\partial_5 - \partial_1$$

$$\rho^\Lambda(E_{\alpha_2}) = -\partial_2 - 3x_4\partial_5 + 2\,x_3\partial_4 + 3x_3{}^2\partial_6$$

$$\rho^\Lambda(E_{-\alpha_1}) = -x_6\partial_5 - x_3\partial_2 + x_3{}^2\partial_4 + 2x_3{}^3\partial_6 - \Lambda_1 x_1 + x_1{}^2\partial_1$$

$$\rho^\Lambda(E_{-\alpha_2}) = -x_5\partial_4 + 2\,x_4\partial_3 - 3x_2{}^2 x_3\partial_4 + 3x_4{}^2\partial_6 - 3x_2{}^3 x_3\partial_5 - 9x_2{}^2 x_3{}^2\partial_6$$
$$- x_2(-3x_3\partial_3 - x_2\partial_2 + \Lambda_2 + 3x_1\partial_1) + 3x_3\partial_1$$

where $\Lambda = (\Lambda_1, \Lambda_2)$. It is evident that the space P of polynomials in six variables is invariant under action of these operators, moreover it can be easily proved that $e_\alpha \to \rho^\Lambda(e_\alpha)$, generates the representation of g_2. Further the restriction $\tilde{\rho}^\Lambda$ of ρ^Λ to the subspace $V^\Lambda = \rho^\Lambda(U(g_2))1$ is the representation of g_2 with highest weight $\Lambda = (\Lambda_1, \Lambda_2)$ and the polynomial 1 is the highest weight vector. A little less trivial is the proof that this subspace is an irreducible one for any weight Λ.

In the special case if $\Lambda \in N_0 \times N_0$ we have [21]

$$(5.1) \quad \dim\bigl(\rho^\Lambda(U(g_2))1\bigr) = \frac{1}{5!}(1+\Lambda_2)(1+\Lambda_1)(\Lambda_1+\Lambda_2+2)$$
$$\times (2\Lambda_1+\Lambda_2+3)(3\Lambda_1+\Lambda_2+4)(3\Lambda_1+2\Lambda_2+5).$$

Consider representation $\tilde{\rho}^{(1,0)}$ and $\tilde{\rho}^{(0,1)}$. It is again easy to show that

$$V^{(1,0)} = \mathbb{C}\{z_{1,1}, \ldots, z_{1,14}\}_{\text{lin}}$$
$$V^{(0,1)} = \mathbb{C}\{z_{2,1}, \ldots, z_{2,7}\}_{\text{lin}}$$

where $z_{1,i} = z_{1,i}(x_1, x_2, \ldots, x_6)$ for $i = 1, 2, \ldots, 14$ and $z_{2,i} = z_{2,i}(x_2, \ldots, x_6)$ for $i = 1, 2, \ldots, 7$ are polynomials the explicit form of which is given in Appendix.

The polynomials $z_{1,14}$ and $z_{2,7}$ are the lowest weight vectors of $\tilde{\rho}^{(1,0)}$ or $\tilde{\rho}^{(0,1)}$ respectively. Now we define the subspace $\widetilde{V}^\Lambda$ of P as

$$(5.2) \quad \widetilde{V}^\Lambda = \mathbb{C}\left\{z_{1,1}^{n_{1,1}} \cdot z_{1,2}^{n_{1,2}} \cdots z_{1,14}^{n_{1,14}} \cdot z_{2,1}^{n_{2,1}} \cdot z_{2,2}^{n_{2,2}} \cdots z_{2,7}^{n_{2,7}}; \sum_{i=1}^{14} n_{1,i} \le \Lambda_1 \text{ and } \sum_{i=1}^{7} n_{2,i} \le \Lambda_2\right\}_{\text{lin}}$$

THEOREM 5.4. *For all $\Lambda \in N_0 \times N_0$, $\widetilde{V}^\Lambda = V^\Lambda = \rho^\Lambda(U(g_2))1$.*

PROOF. It can be easily verified by the same arguments as for case $sl(3, \mathbb{C})$. □

In the general case the set of polynomials entering in definition (5.2) of the space $\widetilde{V}^\Lambda$ forms a total set of $\widetilde{V}^\Lambda$ only and does not constitute a basis. The dimension of finite-dimensional representation is given by the formula (5.1) while the number of elements of our total set equals

$$\frac{(\Lambda_1+13)\cdots(\Lambda_1+1)(\Lambda_2+6)\cdots(\Lambda_2+1)}{13!6!}$$

The polynomial $z_{1,14}^{\Lambda_1} z_{2,7}^{\Lambda_2}$ is a lowest-weight vector of the irreducible representation of $\tilde{\rho}^{(\Lambda)}$ and following [20] polynomials

$$\rho^\Lambda(E_{\alpha_1})^{a_6} \rho^\Lambda(E_{\alpha_2})^{a_5} \rho^\Lambda(E_{\alpha_1})^{a_4} \rho^\Lambda(E_{\alpha_2})^{a_3} \rho^\Lambda(E_{\alpha_1})^{a_2} \rho^\Lambda(E_{\alpha_2})^{a_1} z_{1,14}^{\Lambda_1} z_{2,7}^{\Lambda_2}$$

where exponents $a_1, a_2, \ldots, a_6$ are given by inequalities from [**20**, Table XIV] form a basis in V^Λ.

6. The Lie field

In this chapter we will illustrate the usefulness of the concept of a Lie field [**9**], rederiving above presented representations of the Lie algebra $\mathrm{gl}(3,\mathbb{C})$.

Lie field $D[\mathrm{gl}(3,\mathbb{C})]$ of Lie algebra $\mathrm{gl}(3,\mathbb{C})$ consists of fractions $A \cdot B^{-1}$ where elements A, B belong to enveloping algebra $U[\mathrm{gl}(3,\mathbb{C})]$. Define two such fractions $Q_1, Q_2 \in D[\mathrm{gl}(3,\mathbb{C})]$:

(6.1) $$Q_1 = A_1 B^{-1}, \quad Q_2 = A_2 B^{-1}$$

(6.2) $$A_1 = E_{13}(E_{22} + X - 1) - E_{23}E_{12}$$

(6.3) $$A_2 = -E_{13}E_{21} + E_{23}(E_{11} + X - 1)$$

(6.4) $$B = -\tfrac{1}{2}C_2^{(2)} + \tfrac{1}{2}(C_2^{(1)})^2 - \tfrac{1}{2}(C_2^{(1)}) + X(C_2^{(1)} - 1) + X^2.$$

Here $C_n^{(p)} = E_{i_1 i_2} \cdots E_{i_p i_1}$ $1 \leq p$, $i_1, \ldots, i_p \leq n$ are Casimir operators of $\mathrm{gl}(n,\mathbb{C})$ and X commutes with generators of $\mathrm{gl}(3,\mathbb{C})$, and it will be specified later.

LEMMA 6.1.

(a) $\quad [E_{ij}, Q_k] = \delta_{jk} Q_i, \quad$ where $i, j, k = 1, 2,$

(b) $\quad [Q_i, Q_j] = 0.$

PROOF. By direct verification. For the second part, use identities

(6.5) $$(B - C_2^{(1)} - 2(X - 1))A_1 = (A_1 - E_{13})B,$$

(6.6) $$(B - C_2^{(1)} - 2(X - 1))A_2 = (A_2 - E_{23})B,$$

what makes possible to express the right fractions Q_1 and Q_2 as left fractions:

(6.7) $$Q_1 = (B - C_2^{(1)} - 2(X-1))^{-1}(A_1 - E_{13}),$$
$$Q_2 = (B - C_2^{(1)} - 2(X-1))^{-1}(A_2 - E_{23}).$$

Choose two commuting elements and denote

(6.8) $$P_1 = -E_{31}, \quad P_2 = -E_{32}.$$

Evidently

(6.9) $$[E_{ij}, P_k] = -\delta_{ik} P_j, \quad \text{where } i, j, k = 1, 2$$

and

(6.10) $$[P_i, P_j] = 0. \qquad \square$$

LEMMA 6.2. *Let X be a generating element of a commutative extension of the center of the Lie algebra* $\mathrm{gl}(3,\mathbb{C})$ *satisfying equation*

(6.11) $$X^3 + X^2(C_3^{(2)} - 3) + \tfrac{1}{2}X[(C_3^{(1)} - 2)^2 - C_3^{(2)}] + a(C_3^{(1)}, C_3^{(2)}, C_3^{(3)}) = 0$$

where $a(C_3^{(1)}, C_3^{(2)}, C_3^{(3)}) = -\tfrac{1}{3}[C_3^{(3)} - \tfrac{3}{2}C_3^{(2)}C_3^{(1)} + \tfrac{1}{2}(C_3^{(1)})^3] + \tfrac{1}{3}(C_3^{(1)})^2 - \tfrac{2}{3}C_3^{(1)}.$
Then

(6.12) $$[P_j, Q_k] = \delta_{jk}.$$

PROOF. Expressing Q_j as right fraction in product $P_i Q_j$ and as left fraction in the product $Q_j P_i$, the proof consist in verifying the "enveloping algebra identity"

$$(B - C_2^{(1)} - 2(X-1))E_{3i}A_j - (A_j - E_{j3})E_{3i}B = (B - C_2^{(1)} - 2(X-1))B\delta_{ij},$$

what is a consequence of equation (6.11). □

LEMMA 6.3. *Let*

(6.13) $$F_{ij} = E_{ij} - Q_i P_j \in D[\mathrm{gl}(3,\mathbb{C})]$$

then:

(a) $\qquad [F_{ij}, Q_k] = [F_{ij}, P_k] = 0,$

(b) $\qquad [F_{ij}, F_{kl}] = \delta_{jk} F_{il} - \delta_{kj} F_{li}.$

PROOF. It is a simple consequence of the derived commutation relations. □

Now we are in position to reconstruct realizations mentioned above. This means to express eight generators $E_{\mu\nu}$ of $\mathrm{gl}(3,\mathbb{C})$ by means of nine "letters" F_{ij}, Q_i, P_i and X. Note that X is in fact a generator depending on generators of the center of $U[\mathrm{gl}(3,\mathbb{C})]$. So, it is not a new ninth independent generator. Equation (6.13) gives

(6.14) $$E_{ij} = F_{ij} + Q_i P_j.$$

Further, equation (6.8) expres generators E_{3j} as functions of P_i. Remaining generators E_{i3} we can calculate from equations (6.1) using identities $E_{12}(E_{22} + X - 1) = (E_{22} + X)E_{12}$, $(E_{11} + X)E_{21} = E_{21}(E_{11} + X - 1)$ and we have:

(6.15) $$E_{13} = Q_1(E_{11} + X) + Q_2 E_{12},$$

(6.16) $$E_{23} = Q_1 E_{21} + Q_2(E_{22} + X).$$

Substituting from equation (6.14) we obtain the final result. For generator E_{33} we use for example $E_{33} = [E_{31}, E_{13}] + E_{11}$.

REMARK 6.4. Exactly speaking, we did not work with the Lie field of $\mathrm{gl}(3,\mathbb{C})$ but with the simple algebraical extension $\widetilde{D}[\mathrm{gl}(3,\mathbb{C})] = \{a_0 + a_1 X + a_2 X^2;\ a_i \in D[\mathrm{gl}(3,\mathbb{C})]\}$ of $D[\mathrm{gl}(3,\mathbb{C})]$ by commutative element X fulfilling equation (6.11). (our elements B^{-1} and $(B - C_2^{(1)} - 2(X-1))^{-1}$ are of that form). If polynomial (6.11) is irreducible in $D[\mathrm{gl}(3,\mathbb{C})]$, an extension $\widetilde{D}[\mathrm{gl}(3,\mathbb{C})]$ is a field.

REMARK 6.5. Consider representation of $\mathrm{gl}(2,\mathbb{C})$,

(6.17) $$F_{11} = Q_3 P_3 + Y + C_3^{(1)},$$

(6.18) $$F_{21} = -P_3,$$

(6.19) $$F_{12} = Q_3(Q_3 P_3 + 2Y + C_3^{(1)}),$$

(6.20) $$F_{22} = -Q_3 P_3 - Y,$$

where $C_3^{(1)}$ and Y are assumed to commute with P_3 and Q_3 which satisfy

(6.21) $$[P_3, Q_3] = 1.$$

We can express Q_3 as a function of generators F_{ij} and Y

(6.22) $$Q_3 = F_{12}(F_{11} + Y)^{-1} = (F_{11} + Y - 1)^{-1} F_{12}.$$

Calculating the Casimir operators $\widetilde{C}_2^{(1)} = F_{11} + F_{22}$ and $\widetilde{C}_2^{(2)} = F_{11}^2 + F_{12}F_{21} + F_{21}F_{12} + F_{22}^2$ we obtain

(6.23) $$\widetilde{C}_2^{(1)} = C_3^{(1)}$$

and

(6.24) $$\widetilde{C}_2^{(2)} = 2Y^2 + 2Y(\widetilde{C}_2^{(1)} - 1) + [\widetilde{C}_2^{(1)}]^2 - \widetilde{C}_2^{(1)}.$$

Conversely, we can easily show that expressions P_3 and Q_3 defined by equations (6.18) and (6.22) fulfill relation (6.12) using commutation relations (6.13) and equation (6.24) only.

Let $D_{3,3}$ denotes Lie field generated by commutation relations

(6.25) $$[p_\mu, q_\nu] = \delta_{\mu\nu},$$
(6.26) $$[p_\mu, x] = [p_\mu, y] = [p_\mu, c] = 0,$$
(6.27) $$[q_\mu, x] = [q_\mu, y] = [q_\mu, c] = 0,$$
(6.28) $$[x, y] = [x, c] = [y, c] = 0,$$

where $\mu, \nu = 1, 2, 3$.

Denote by $\widetilde{\widetilde{D}}[\mathrm{gl}(3,\mathbb{C})]$ the extension of $\widetilde{D}[\mathrm{gl}(3,\mathbb{C})]$ by further commuting Y fulfilling equation (6.24). (Note that $\widetilde{C}_2^{(1)}$ and $\widetilde{C}_2^{(2)}$ are simple functions of $C_3^{(i)}$ (see equation (6.23)) and for $\widetilde{C}_2^{(2)}$ we easily derive $\widetilde{C}_2^{(2)} = C_3^{(2)} + C_3^{(1)} - X^2 - 2X$) and define the mapping $\phi \colon D_{3,3} \to \widetilde{\widetilde{D}}[\mathrm{gl}(3,\mathbb{C})]$ by

$$\phi(p_\mu) = P_\mu, \quad \phi(q_\mu) = Q_\mu,$$
$$\phi(x) = X, \quad \phi(y) = Y, \quad \phi(c) = C_3^{(1)}.$$

If $\widetilde{\widetilde{D}}[\mathrm{gl}(3,\mathbb{C})]$ is a field (i.e., if considered polynomials (6.11) and (6.24) are irreducible) ϕ is an explicit example of the Gelfand–Kirrilov isomorphism [**19**] of two considered fields.

REMARK 6.6. Some physically interesting Hamiltonians [**22, 24, 25**] can be rewritten as elements of enveloping algebras of some Lie algebra in appropriate representation. Commutation relations of Lie algebra then can help to solve, at least partially, some problems as eigenvalue problems. Considering the Lie field instead of enveloping algebra, wider class of Hamiltonians could be represented by function of generators, because Lie field contains elements such as P_i and Q_i. Hamiltonians expressed as "right" rational function of generators can be, however, considered only in representation in which the denominator is an invertible operator.

Acknowledgments

The authors are grateful to Dr. O. Navrátil for stimulating discussions. The work was partially supported by the Committee for collaboration of the Czech Republic with CERN. One of us (M.H.) thanks tthe organizers of the of Workshop on Symmetry in Physics in Memory of R. T. Sharp.

Appendix

In this appendix we present the explicit formulas for the polynomials of the exceptional Lie algebra g_2.

$$z_{2,1} = 1, \quad z_{2,2} = x_2, \quad z_{2,3} = x_3, \quad z_{2,4} = x_4 + x_2x_3, \quad z_{2,5} = x_5 - x_2x_4$$
$$z_{2,6} = x_6 - x_3x_4 + x_2x_3^2, \quad z_{1,7} = x_4^2 - x_2x_6 + x - 3x_5 + 2x_2x_3x_4$$
$$z_{1,1} = 1, \quad z_{1,2} = x_1, \quad z_{1,3} = x_3 - x_1x_2, \quad z_{1,4} = x_4 + x_1x_2^2,$$
$$z_{1,5} = x_3^2 - x_1x_4 - 2x - 1x_2x_3, \quad z_{1,6} = x_5 + x_1x_2^3 \quad z_{1,7} = x_6 + x_1x_5 + 3x_1x_2^2x_3,$$
$$z_{1,8} = x_3x_4 - x_1x_5 - x_1x_2x_4 + x_1x_2^2x_3, \quad z_{1,9} = x_3^3 - 3x_1x_2x_3^2$$
$$z_{1.10} = x_4^2 + x_3x_5 - x_1x_2^2x_5 + 2x_1 + 2x_1x_2,$$
$$z_{1,11} = 4x_4x_3x_2x_1 - 2\,x_4x_3^2 + x_4{}^2x_1 + x_2{}^2x_3{}^2x_1 - x_2x_6x_1 + x_3x_6$$
$$z_{1,12} = -2x_5x_3x_2x_1 + x_5x_3{}^2 - x_5x_1x_4 + x_4{}^2x_2x_1 - x_4{}^2x_3 + x_4x_2{}^2x_3x_1$$
$$+ x_4x_6 + x_2{}^3x_3{}^2x_1 + x_2{}^2x_6x_1$$
$$z_{1,13} = -x_2{}^3x_6x_1 + x_4{}^3 + 3x_5x_4x_3 + 3x_4{}^2x_2{}^2x_1$$
$$+ 3x_4x_2{}^3x_3x_1 - x_5x_6 + x_5{}^2x_1 - 3x_5x_4x_2x_1$$
$$z_{1,14} = -3x_3{}^2x_5x_2x_1 - 6x_3x_4{}^2x_2x_1 + 3x_3x_2{}^2x_6x_1 - 3x_4x_2{}^2x_3{}^2x_1 + x_6{}^2 - 3x_6x_4x_3$$
$$+ 3x_3{}^2x_4{}^2 + x_3{}^3x_5 - x_6x_5x_1 + x_3{}^3x_2{}^3x_1 + 3x_6x_4x_2x_1 - x_1x_4{}^3$$

References

1. M. A. Ayari, V. Hussin, and P. Winternitz, *Group invariant solutions for the $N = 2$ super Korteweg–de Vries equation*, J. Math. Phys. **40** (1999), no. 4, 1951–1965.
2. Č. Burdík, *Realisations of the real semisimple Lie algebras: a method of construction*, J. Phys. A **18** (1985), no. 16, 3101–3111.
3. _____, *A new class of realizations of the Lie algebra* $\mathrm{gl}(n+1,\mathbb{R})$, Czechoslovak J. Phys. B **36** (1986), no. 11, 1235–1241.
4. _____, *A new class of realizations of the Lie algebra* $\mathrm{sp}(n,\mathbb{R})$, J. Phys. A **19** (1986), no. 13, 2465–2471.
5. _____, *A new class of realizations of the Lie algebra* $\mathrm{so}(q, 2n - q)$, J. Phys. A **21** (1988), no. 2, 289–295.
6. Č. Burdík, P. Exner, and M. Havlíček, *Highest-weight representation of the* $\mathrm{sl}(2,\mathbb{C})$ *and* $\mathrm{sl}(3,\mathbb{C})$ *algebras: via canonical realizations*, Czechoslovak J. Phys. B **31** (1981), no. 5.459–469.
7. Č. Burdík, M. Havlíček, and P. Exner, *Highest-weight representations of the* $\mathrm{sl}(n+1,\mathbb{C})$ *algebras: maximal representations*, J. Phys. A **14** (1981), no. 5, 1039–1054.
8. Č. Burdík, O. Navrátil, and M. Thoma, *A new class of realisations of Lie algebra* G_2, Czechoslovak J. Phys. B **43** (1993), no. 7, 697–776.
9. J. Dixmier *Algèbres enveloppantes*, Cahiers Sci., vol. 37, Gauthier-Villars, Paris, 1974.
10. P. Exner and M. Havlíček, *On the minimal canonical realizations of the Lie algebra* $O_C(n)$, Ann. Inst. H. Poincaré Sect. A (N.S.) **23** (1975), no. 4, 313–333.
11. _____, *Matrix canonical realizations of the Lie algebra* $\mathrm{o}(m,n)$. *I. Basic formulae and classification*, Ann. Inst. H. Poincaré Sect. A (N.S.) **23** (1975), no. 4, 335–347.
12. _____, *Matrix canonical realizations of the Lie algebra* $\mathrm{o}(m,n)$. *II. Casimir operators*, Czech. J. Phys. B **28** (1978), no. 9, 949–962.
13. F. Finkel, D. Gomez-Ullate, A. González-López, M. A. Rodríguez and R. Zhdanov, A_N-*type Dunkl operators and new spin Calogero–Sutherland models*, Comm. Math. Phys. **221** (2001), no. 3, 477–497.
14. M. Havlíček and W. Lassner, *Canonical realizations of the Lie algebras* $\mathrm{gl}(n,\mathbb{R})$ *and* $\mathrm{sl}(n,\mathbb{R})$. *I. Formulae and classification*, Rep. Mathematical Phys. **8** (1975), no. 3, 391–399.
15. _____, *Canonical realizations of the Lie algebras* $\mathrm{gl}(n,\mathbb{R})$ *and* $\mathrm{sl}(n,\mathbb{R})$. *II. Casimir operators*, Rep. Mathematical Phys. **9** (1976), no. 2, 177–185.
16. _____, *Canonical realizations of the Lie algebra* $\mathrm{sp}(2n,\mathbb{R})$, Internat. J. Theoret. Phys. **15** (1976), no. 11, 867–876.
17. _____, *On the "near to minimal" canonical realizations of the Lie algebra* C_n, Internat. J. Theoret. Phys. **15** (1976), no. 11, 877–884.

18. _____, *Matrix canonical realizations of the Lie algebra* $u(p,q)$, Rep. Mathematical Phys. **12** (1977), no. 1, 1–8.
19. I. M. Gel'fand and A. A. Kirillov, *Structure of the Lie field connected with a semisimple decomposable Lie algebra* Funkcional. Anal. i Priložen. **3** (1969), no. 1, 7–26 (Russian).
20. S. P. Li, R. V. Moody, M. Nicolescu and J. Patera, *Verma bases for representations of classical simple Lie algebras*, J. Math. Phys. **27** (1986), no. 3, 668–677.
21. M. A. Naimark, *Representation theory of groups*, Izdat. "Nauka", Moscow, 1976 (Russian).
22. W. Rühl and A. Turbiner, *Exact solvability of the Calogero and Sutherland models*, Modern Phys. Lett. A **10** (1995), no. 29, 2213–2221.
23. P. Tempesta, A. Turbiner, and P. Winternitz, *Exact solvability of superintegrable systems*, J. Math. Phys. **42** (2001), no. 9, 4248–4257.
24. A. Turbiner, *Quasi-exactly-solvable problems and* $sl(2)$ *algebra*, Comm. Math. Phys. **118** (1988), no. 3, 467–474.
25. A. G. Ushveridze, *Quasi-exactly solvable models in quantum mechanics*, Institute of Physics Publishing, Bristol, 1994.

DEPARTMENT OF MATHEMATICS AND DOPPLER INSTITUTE, FACULTY OF NUCLEAR SCIENCES AND PHYSICAL ENGINEERING, CZECH TECHNICAL UNIVERSITY, TROJANOVA 13, 120 00 PRAGUE 2, CZECH REPUBLIC

E-mail address, Č. Burdík: `burdik@km1.fjfi.cvut.cz`

E-mail address, M. Havlíček: `havlicek@km1.fjfi.cvut.cz`

Stretched Littlewood–Richardson and Kostka Coefficients

R. C. King, C. Tollu, and F. Toumazet

This paper is dedicated to the memory of Professor R. T. Sharp, a great colleague and friend.

ABSTRACT. Littlewood–Richardson and Kostka coefficients are specified by means of partitions. If the parts of the partitions are all multiplied by some fixed positive integer N, then the values of the coefficients will change as a function of the stretching parameter N. It is conjectured that this behaviour is always polynomial in N. The study makes use of a hive model for evaluating the coefficients, thereby making a connection with integer points of certain rational convex polytopes. Many examples of the relevant polynomials are given, along with further conjectures regarding their general form and that of their generating functions.

1. Introduction

The objective here is to describe some speculative work, still in progress, on the behaviour of Littlewood–Richardson coefficients, $c_{\lambda\mu}^{\nu}$, and Kostka coefficients, $K_{\lambda\mu}$, under the simultaneous stretching of the partitions λ, μ and ν by some scale factor N, where N is a positive integer. The work was motivated by Fulton's Conjecture that $c_{N\lambda,N\mu}^{N\nu} = 1$ for all positive integers N if and only if $c_{\lambda\mu}^{\nu} = 1$, and by the observation that $c_{N\lambda,N\mu}^{N\nu}$ appears always to be $N+1$ if $c_{\lambda\mu}^{\nu} = 2$.

We offer here, along with a considerable amount of supporting evidence, a series of stronger conjectures. The most important of these is that the stretched Littlewood–Richardson coefficients are polynomial in the stretching parameter. A similar conjecture applies to stretched Kostka coefficients.

In order to define both Littlewood-Richardson coefficients and Kostka coefficients it is necessary to introduce some standard notation and terminology [8, 9]. Let $\lambda = (\lambda_1, \lambda_2, \ldots, \lambda_p)$ with $\lambda_1 \geq \lambda_2 \geq \cdots \geq \lambda_p > 0$ be a partition of weight

---

2000 *Mathematics Subject Classification.* Primary 05A15; Secondary 05E05, 52B11.

The first author is grateful for the hospitality he was afforded at the Laboratoire d'Informatique de Paris-Nord, where much of this work was carried out. The second and third authors are supported by a JemSTIC grant from the CNRS (France). Thanks are also due to Professor Wybourne, both for useful discussions and for the explicit evaluation of many stretched coefficients using SCHUR.

This is the final form of the paper.

$|\lambda| = \lambda_1 + \cdots + \lambda_p$ and length $\ell(\lambda) = p$. The corresponding Young diagram F^λ consists of $|\lambda|$ boxes arranged in $\ell(\lambda) = p$ left-adjusted rows of lengths $\lambda_1, \lambda_2, \ldots, \lambda_p$. For example, for $\lambda = (3, 2)$ we have

(1.1) $$F^{3,2} = \begin{array}{|c|c|c|}\hline & & \\\hline & & \multicolumn{1}{c}{}\\\cline{1-2}\end{array}$$

A semistandard Young tableau, T^λ, is a numbering of the boxes of F^λ with entries from $\{1, 2, \ldots, n\}$, weakly increasing across rows and strictly increasing down columns. Typically, for $\lambda = (3, 2)$ we have amongst others the following semistandard Young tableaux, T^{32} of shape F^{32} provided that $n \geq 3$:

(1.2) $\begin{array}{|c|c|c|}\hline 1&1&1\\\hline 2&2\\\cline{1-2}\end{array}$, $\begin{array}{|c|c|c|}\hline 1&1&1\\\hline 2&3\\\cline{1-2}\end{array}$, $\begin{array}{|c|c|c|}\hline 1&1&2\\\hline 2&3\\\cline{1-2}\end{array}$, $\begin{array}{|c|c|c|}\hline 1&1&3\\\hline 2&2\\\cline{1-2}\end{array}$, $\ldots$.

A basis for the ring of symmetric functions of the components of $\mathbf{x} = (x_1, x_2, \ldots, x_n)$ is provided by the Schur functions $s_\lambda(\mathbf{x})$, with λ ranging over the set of all partitions of length $\ell(\lambda) \leq n$. The Schur function $s_\lambda(\mathbf{x})$ may be defined by

(1.3) $$s_\lambda(\mathbf{x}) = \sum_{T^\lambda} x_1^{m_1} x_2^{m_2} \cdots x_n^{m_n},$$

where m_i is the number of entries i in T^λ for $i = 1, 2, \ldots, n$.

The Littlewood–Richardson coefficients, $c_{\lambda\mu}^\nu$, are then defined by the expansion of the product of two such Schur functions which necessarily takes the form:

(1.4) $$s_\lambda(\mathbf{x}) s_\mu(\mathbf{x}) = \sum_\nu c_{\lambda\mu}^\nu s_\nu(\mathbf{x}).$$

The coefficients may be evaluated by means of the famous Littlewood–Richardson rule (see [8, p. 94], [9, p. 142]). This states that $c_{\lambda\mu}^\nu$ is the number of distinctly labelled diagrams of shape F^ν obtained by adding to F^λ the boxes of F^μ labelled by their row number in such a way that: (i) the entries increase weakly across rows; (ii) the entries increase strictly down columns; (iii) any initial segment of the sequence of entries read from right to left across rows taken in turn from top to bottom is such that $\#1's \geq \#2's \geq \cdots \geq \#n's$ (the lattice permutation rule). Although it is not obvious from this rule, the product is commutative so that $c_{\lambda\mu}^\nu = c_{\mu\lambda}^\nu$

For example, if $\lambda = (3, 2)$, $\mu = (2, 1)$ and $\nu = (4, 3, 1)$, then $c_{32,21}^{431} = 2$ since the Littlewood–Richardson rule gives rise to just to just two appropriately labelled diagrams:

$$\begin{array}{|c|c|c|}\hline&&\\\hline&&\multicolumn{1}{c}{}\\\cline{1-2}\end{array} \times \begin{array}{|c|c|}\hline 1&1\\\hline 2\\\cline{1-1}\end{array} \to \cdots + \begin{array}{|c|c|c|c|}\hline&&&1\\\hline&&&\multicolumn{1}{c}{}\\\cline{1-3}&\multicolumn{3}{c}{}\\\cline{1-1}\end{array}\;\;\text{(with 1, 2 placement)} + \begin{array}{|c|c|c|c|}\hline&&&1\\\hline&&&\multicolumn{1}{c}{}\\\cline{1-3}&\multicolumn{3}{c}{}\\\cline{1-1}\end{array}\;\;\text{(with 2, 1 placement)} + \cdots .$$

Similarly, if $\lambda = (2, 1)$, $\mu = (3, 2)$ and $\nu = (4, 3, 1)$, then $c_{21,32}^{431} = 2$, where the only relevant diagram are:

$$\begin{array}{|c|c|}\hline&\\\hline&\multicolumn{1}{c}{}\\\cline{1-1}\end{array} \times \begin{array}{|c|c|c|}\hline 1&1&1\\\hline 2&2\\\cline{1-2}\end{array} \to \cdots + \begin{array}{|c|c|c|c|}\hline&&1&1\\\hline&1&2\\\cline{1-3}&2\\\cline{1-1}\end{array} + \begin{array}{|c|c|c|c|}\hline&&1&1\\\hline&2&2\\\cline{1-3}&1\\\cline{1-1}\end{array} + \cdots .$$

Kostka coefficients may also be defined combinatorially (see [8, p. 191], [9, p. 101]): $K_{\lambda\mu}$ is the number of distinctly labelled semistandard Young tableaux T^λ of shape F^λ and weight μ, that is to say with μ_i entries i for $i = 1, 2, \ldots, \ell(\mu)$.

For example, if $\lambda = (3,2)$ and $\mu = (2,2,1)$ then there are just two semistandard Young tableaux T^{32} of weight $(2,2,1)$:

$$\begin{array}{|c|c|c|} \hline 1 & 1 & 2 \\ \hline 2 & 3 \\ \cline{1-2} \end{array} \;,\quad \begin{array}{|c|c|c|} \hline 1 & 1 & 3 \\ \hline 2 & 2 \\ \cline{1-2} \end{array}.$$

This implies that $K_{32,221} = 2$.

Kostka coefficients can be viewed as special types of Littlewood–Richardson coefficients. To be precise $K_{\lambda\mu} = c^\tau_{\sigma\lambda}$ where

(1.5) $$\begin{cases} \tau_i = \mu_i + \mu_{i+1} + \ldots \\ \sigma_i = \mu_{i+1} + \mu_{i+2} + \cdots \end{cases} \quad \text{for } i = 1, 2, \ldots.$$

For example, if $\lambda = (3,2)$ and $\mu = (2,2,1,)$, then $\tau = (5,3,1)$ and $\sigma = (3,1)$. The calculation of $c^\tau_{\sigma\lambda}$ by means of the Littlewood–Richardson rule then leads to the diagrams:

$$\;\times\; \begin{array}{|c|c|c|} \hline 1 & 1 & 1 \\ \hline 2 & 2 \\ \cline{1-2} \end{array} \;\to\; \cdots + \;\cdots\; + \;\cdots\; + \;\cdots\; + .$$

It follows that $K_{32,221} = c^{531}_{31,32} = 2$, as obtained before.

Now we are in a position to define and evaluate stretched Littlewood–Richardson and Kostka coefficients. The partition obtained from $\lambda = (\lambda_1, \lambda_2, \ldots, \lambda_p)$ by multiplying all of its parts by the same positive integer, N, is denoted by $N\lambda = (N\lambda_1, N\lambda_2, \ldots, N\lambda_p)$. With this notation, we refer to $c^{N\nu}_{N\lambda,N\mu}$ and $K_{N\lambda,N\mu}$ as stretched Littlewood-Richardson and Kostka coefficients, respectively, where N is said to be the stretching parameter. It is not difficult to evaluate these stretched coefficients.

For example, if $\lambda = (2,1)$, $\mu = (3,2)$ and $\nu = (4,3,1)$ then the stretched Littlewood–Richardson coefficient $c^{N\nu}_{N\lambda,N\mu}$ may be evaluated for various N as follows:

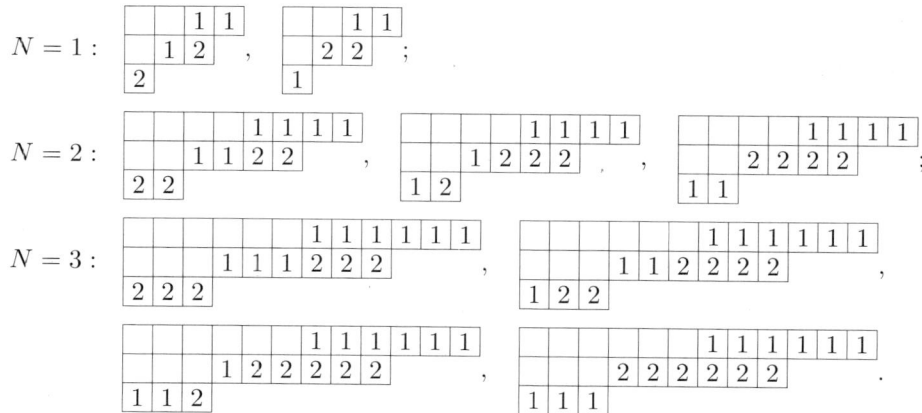

It follows that $c^\nu_{\lambda\mu} = 2$, $c^{2\nu}_{2\lambda,2\mu} = 3$, and $c^{3\nu}_{3\lambda,3\mu} = 4$. Moreover, by noting that the only freedom in these diagrams lies in the labelling of the boxes in the third row, it is easy to see that in this case $c^{N\nu}_{N\lambda,N\mu} = N + 1$.

Similarly, if $\lambda = (3,2)$ and $\mu = (2,1,1,1)$ then the stretched Kostka coefficients $K_{N\lambda,N\mu}$ are determined for various N by the enumeration of the following semistandard tableaux.

$N=1:$ $\begin{array}{|c|c|c|}\hline 1&1&2\\\hline 3&4\\\cline{1-2}\end{array}$, $\begin{array}{|c|c|c|}\hline 1&1&3\\\hline 2&4\\\cline{1-2}\end{array}$, $\begin{array}{|c|c|c|}\hline 1&1&4\\\hline 2&3\\\cline{1-2}\end{array}$;

$N=2:$ $\begin{array}{|c|c|c|c|c|c|}\hline 1&1&1&1&2&2\\\hline 3&3&4&4\\\cline{1-4}\end{array}$, $\begin{array}{|c|c|c|c|c|c|}\hline 1&1&1&1&2&3\\\hline 2&3&4&4\\\cline{1-4}\end{array}$, $\begin{array}{|c|c|c|c|c|c|}\hline 1&1&1&1&3&3\\\hline 2&2&4&4\\\cline{1-4}\end{array}$,

$\begin{array}{|c|c|c|c|c|c|}\hline 1&1&1&1&2&4\\\hline 2&3&3&4\\\cline{1-4}\end{array}$, $\begin{array}{|c|c|c|c|c|c|}\hline 1&1&1&1&3&4\\\hline 2&2&3&4\\\cline{1-4}\end{array}$, $\begin{array}{|c|c|c|c|c|c|}\hline 1&1&1&1&4&4\\\hline 2&2&3&4\\\cline{1-4}\end{array}$

$N=3:$ $\begin{array}{|c|c|c|c|c|c|c|c|c|}\hline 1&1&1&1&1&1&2&2&2\\\hline 3&3&3&4&4&4\\\cline{1-6}\end{array}$, $\begin{array}{|c|c|c|c|c|c|c|c|c|}\hline 1&1&1&1&1&1&2&2&3\\\hline 2&3&3&4&4&4\\\cline{1-6}\end{array}$,

$\begin{array}{|c|c|c|c|c|c|c|c|c|}\hline 1&1&1&1&1&1&2&3&3\\\hline 2&2&3&4&4&4\\\cline{1-6}\end{array}$, $\begin{array}{|c|c|c|c|c|c|c|c|c|}\hline 1&1&1&1&1&1&3&3&3\\\hline 2&2&2&4&4&4\\\cline{1-6}\end{array}$,

$\begin{array}{|c|c|c|c|c|c|c|c|c|}\hline 1&1&1&1&1&1&2&2&4\\\hline 2&3&3&3&4&4\\\cline{1-6}\end{array}$, $\begin{array}{|c|c|c|c|c|c|c|c|c|}\hline 1&1&1&1&1&1&2&3&4\\\hline 2&2&3&3&4&4\\\cline{1-6}\end{array}$,

$\begin{array}{|c|c|c|c|c|c|c|c|c|}\hline 1&1&1&1&1&1&3&3&4\\\hline 2&2&2&3&4&4\\\cline{1-6}\end{array}$, $\begin{array}{|c|c|c|c|c|c|c|c|c|}\hline 1&1&1&1&1&1&2&4&4\\\hline 2&2&3&3&3&4\\\cline{1-6}\end{array}$,

$\begin{array}{|c|c|c|c|c|c|c|c|c|}\hline 1&1&1&1&1&1&3&4&4\\\hline 2&2&2&3&3&4\\\cline{1-6}\end{array}$, $\begin{array}{|c|c|c|c|c|c|c|c|c|}\hline 1&1&1&1&1&1&4&4&4\\\hline 2&2&2&3&3&3\\\cline{1-6}\end{array}$

It follows that in this case $K_{\lambda\mu} = 3$, $K_{2\lambda,2\mu} = 6$ and $K_{3\lambda,3\mu} = 10$. Quite generally, the right-most N boxes in the first row can only be filled with entries 2, 3 or 4, in standard order. Once these are chosen the remaining entries are all fixed. It follows that

(1.6) $$K_{N\lambda,N\mu} = \tfrac{1}{2}(N+1)(N+2).$$

2. Known results

Having exposed the nature of the problem, it is worth listing some known results.

THEOREM 2.1 (Saturation Conjecture [**3, 6**]).

(2.1) $$c^{N\nu}_{N\lambda,N\mu} > 0 \iff c^{\nu}_{\lambda\mu} > 0.$$

COROLLARY 2.2 (K-Saturation Conjecture).

(2.2) $$K_{N\lambda,N\mu} > 0 \iff K_{\lambda\mu} > 0.$$

THEOREM 2.3 (Fulton's Conjecture [**5, 7**]).

(2.3) $$c^{N\nu}_{N\lambda,N\mu} = 1 \iff c^{\nu}_{\lambda\mu} = 1.$$

COROLLARY 2.4 (K-Fulton's Conjecture).

(2.4) $$K_{N\lambda,N\mu} = 1 \iff K_{\lambda\mu} = 1.$$

The dominance ordering of partitions is such that λ is said to dominate μ, and we write $\lambda \geq \mu$ if and only if

(2.5) $$\lambda_1 + \lambda_2 + \cdots + \lambda_i \geq \mu_1 + \mu_2 + \cdots + \mu_i$$

for all $i = 1, 2, \ldots$. With this notation it is possible to establish the following:

LEMMA 2.5.
$$K_{\lambda\mu} > 0 \iff |\lambda| = |\mu| \text{ and } \lambda \geq \mu. \tag{2.6}$$

Thanks to this lemma one can write down a rather simple direct proof of Corollary 2.2, namely the argument

$$\begin{aligned}K_{\lambda\mu} > 0 &\iff |\lambda| = |\mu| \text{ and } \lambda \geq \mu \\ &\iff |N\lambda| = |N\mu| \text{ and } N\lambda \geq N\mu \iff K_{N\lambda,N\mu} > 0,\end{aligned} \tag{2.7}$$

where use has been made of the rather obvious fact that the simultaneous scaling of all the parts of the partitions λ and μ by N preserves both the equality of weights of these partitions and their dominance order.

Results of a quite different nature may be found by noting the connection between Littlewood–Richardson coefficients and tensor products, and between Kostka coefficients and weight multiplicities, all within the context of the representation theory of the general linear group, $GL(n)$, over $\mathbb{C}$.

To be precise, for each partition λ with $\ell(\lambda) \leq n$ there exists an irreducible representation V^λ of $GL(n)$ of highest weight λ and character $\operatorname{ch} V^\lambda$. With a suitable identification of the indeterminates $x_1, x_2, \ldots, x_n$, the character $\operatorname{ch} V^\lambda$ is nothing other than the Schur-function $s_\lambda(\mathbf{x})$. It follows that the Littlewood–Richardson coefficients govern the decomposition of tensor products of irreducible representations of $GL(n)$ in accordance with the formula

$$V^\lambda \otimes V^\mu = \sum_\nu c_{\lambda\mu}^\nu V^\nu. \tag{2.8}$$

Moreover, the weight space decomposition of the irreducible representation V^λ takes the form $V^\lambda = \bigoplus_\mathbf{m} V_\mathbf{m}^\lambda$, where the sum is taken over all weights $\mathbf{m} = (m_1, m_2, \ldots, m_n)$. This decomposition is such that the multiplicity of the weight specified by $\mathbf{m} = \mu$ in the irreducible representation V^λ is given by $\dim V_\mu^\lambda = K_{\lambda\mu}$.

This allows us to lean on some well known results in the physics literature. First, the quantum theory of angular momentum is such that the coupling of two angular momenta is multiplicity free. This implies that all tensor product and weight decompositions are multiplicity free for not only the rotation group $SO(3)$, but also for $SU(2)$ and even $GL(2)$. This leads to the following:

OBSERVATION 2.6. For all λ, μ, ν with $\ell(\lambda)$, $\ell(\mu)$, $\ell(\nu) \leq 2$

$$c_{N\lambda,N\mu}^{N\nu} = c_{\lambda\mu}^\nu = 1 \quad \text{and} \quad K_{N\lambda,N\mu} = K_{\lambda\mu} = 1. \tag{2.9}$$

In connection with the special unitary group $SU(3)$ Wigner [11] established a formula for weight multiplicities which applies equally well to $GL(3)$, namely

$$K_{\lambda\mu} = 1 + \min\{\lambda_1 - \lambda_2, \lambda_2 - \lambda_3, \lambda_1 - \mu_1, \mu_3 - \lambda_3\}. \tag{2.10}$$

This implies that for any positive integer N

$$\begin{aligned}K_{N\lambda,N\mu} - 1 &= \min\{N\lambda_1 - N\lambda_2, N\lambda_2 - N\lambda_3, N\lambda_1 - N\mu_1, N\mu_3 - N\lambda_3\} \\ &= N\min\{\lambda_1 - \lambda_2, \lambda_2 - \lambda_3, \lambda_1 - \mu_1, \mu_3 - \lambda_3\} \\ &= N(K_{\lambda\mu} - 1).\end{aligned} \tag{2.11}$$

Hence we have

OBSERVATION 2.7. If $K_{\lambda\mu} = k > 0$ with $\ell(\lambda)$, $\ell(\mu) \leq 3$ then

$$K_{N\lambda,N\mu} = 1 + N(k-1). \tag{2.12}$$

3. Polynomial conjectures

We are now in a position to state our main conjecture, namely

CONJECTURE 3.1. *For all partitions λ, μ, ν such that $c_{\lambda\mu}^\nu > 0$ there exists a polynomial $P_{\lambda\mu}^\nu(N)$ in N with non-negative rational coefficients such that $P_{\lambda\mu}^\nu(0) = 1$ and $P_{\lambda\mu}^\nu(N) = c_{N\lambda,N\mu}^{N\nu}$ for all positive integers N.*

An immediate corollary of this conjecture, if proved true, but one that also stands in its own right as a new conjecture, is the following:

CONJECTURE 3.2. *For all partitions λ, μ such that $K_{\lambda\mu} > 0$ there exists a polynomial $P_{\lambda\mu}(N)$ in N with non-negative rational coefficients such that $P_{\lambda\mu}(0) = 1$ and $P_{\lambda\mu}(N) = K_{N\lambda,N\mu}$ for all positive integers N.*

The nature of the polynomials encountered in these two conjectures is certainly constrained in some way, and as preliminary comment we offer the following:

CONJECTURE 3.3. *Given k such that $c_{\lambda\mu}^\nu = k$ or $K_{\lambda\mu} = k$ there exists only a restricted number of distinct polynomials $P(N)$ such that $P(N) = P_{\lambda\mu}^\nu(N)$ or $P(N) = P_{\lambda\mu}(N)$. In particular:*

(3.1)
$$\begin{aligned}
&k = 0 \implies P(N) = 0; &&\text{(Saturation Conjecture)}\\
&k = 1 \implies P(N) = 1; &&\text{(Fulton's Conjecture)}\\
&k = 2 \implies P(N) = N+1;\\
&k = 3 \implies P(N) = 2N+1 \text{ or } P(N) = \tfrac{1}{2}(N+1)(N+2)
\end{aligned}$$

Evidence in support of our polynomial conjectures comes to a large extent from fitting data on Littlewood-Richardson and Kostka coefficients calculated for many partitions λ, μ and ν for a wide range of values of N. For example, for $\lambda = (4,2,1)$ and $\mu = (5,3,2)$ the unstretched Littlewood–Richardson coefficient $c_{421,532}^\nu = k$ takes values ranging from $k = 1$ to $k = 7$ for various ν. The data on the corresponding stretched Littlewood–Richardson coefficients for representative partitions ν can all be fitted by the polynomials $P(N) = c_{N(421),N(532)}^{N\nu}$ given below.

(3.2)
$$\begin{aligned}
&k = 1 &&\nu = (953) &&P(N) = 1\\
&k = 2 &&\nu = (9431) &&P(N) = (N+1)\\
&k = 3 &&\nu = (8441) &&P(N) = (N+1)(N+2)/2\\
&k = 4 &&\nu = (8531) &&P(N) = (N+1)(N+2)(N+3)/6\\
&k = 4 &&\nu = (7442) &&P(N) = (N+1)^2\\
&k = 5 &&\nu = (7541) &&P(N) = (N+1)(N+2)(2N+3)/6\\
&k = 6 &&\nu = (7532) &&P(N) = (N+1)^2(N+2)/2\\
&k = 7 &&\nu = (74321) &&P(N) = (N+1)(N+2)(N^2+3N+6)/12
\end{aligned}$$

It is to be noted that in the case $k = 4$, not only do two different polynomials arise, but they are of different degrees.

Similarly, for various values, $k = K_{\lambda\mu}$, of the unstretched Kostka coefficients specified by some quite representative partitions λ and μ, we find that the data on the corresponding stretched Kostka coefficients can all be fitted by polynomials

$P(N) = K_{N\lambda, N\mu}$ as illustrated below.

(3.3)
$$\begin{array}{llll}
k=1 & \lambda=(32) & \mu=(31^2) & P(N)=1 \\
k=2 & \lambda=(32) & \mu=(2^21) & P(N)=(N+1) \\
k=3 & \lambda=(32) & \mu=(21^3) & P(N)=(N+1)(N+2)/2 \\
k=4 & \lambda=(41) & \mu=(1^5) & P(N)=(N+1)(N+2)(N+3)/6 \\
k=5 & \lambda=(32) & \mu=(1^5) & P(N)=(N+1)(N^2+2N+2))/2 \\
k=5 & \lambda=(221) & \mu=(1^5) & P(N)=(N+1)(N+2)(N^2+3N+6)/12 \\
k=6 & \lambda=(321) & \mu=(1^6) &
\end{array}$$
$$P(N) = (N+1)(N+2)(N+3)(17N^2 + 80N^3 + 187N^2 + 220N + 168)/1008.$$

In general, even for Kostka coefficients, let alone the more general Littlewood–Richardson coefficients, it appears to be rather difficult to predict the nature of the corresponding polynomial $P(N)$, or even its degree.

However, in the case of a hook shaped partition $\lambda = (a+1, 1^b)$ we do have the following:

PROPOSITION 3.4. *Let* $\lambda = (a+1, 1^b)$ *and* μ *be partitions of* m, *so that* $\ell(\lambda) = b+1$, *and let* $\ell(\mu) = c+1$. *Then*

(3.4) $$K_{\lambda\mu} = \binom{c}{b} \quad \text{and} \quad K_{N\lambda, N\mu} = \prod_{i=1}^{b} \prod_{j=1}^{c-b} \frac{N+i+j-1}{i+j-1}.$$

PROOF. Each hook-shaped semistandard tableau T^λ contains μ_i entries i for $i = 1, 2, \ldots, \ell(\mu)$. All the 1's must go in the top row, so that in particular the entry at the top row of the first column must be 1. The b entries β_j for $j = 1, \ldots, b$ in the leg are all distinct and must be chosen from $\{2, 3, \ldots, c+1\}$ without restriction since $\mu_i \geq 1$ for $i = 2, \ldots, c+1$. The remaining a entries α_i for $i = 1, \ldots, a$ are then all placed in the arm in weakly increasing order. This is illustrated below.

(3.5)
$$T^\lambda = \begin{array}{|c|c|c|c|c|}\hline 1 & \alpha_1 & \alpha_2 & \cdots & \alpha_a \\\hline \beta_1 \\\cline{1-1} \beta_2 \\\cline{1-1} \vdots \\\cline{1-1} \beta_b \\\cline{1-1}\end{array}$$

It follows by a simple counting argument that $K_{\lambda\mu} = \binom{c}{b}$.

If we now consider each fat hook-shaped semistandard tableau $T^{N\lambda}$ it must contains $N\mu_i$ entries i for $i = 1, 2, \ldots, \ell(\mu)$. There are at least N 1's, all to go in the top row, filling in particular all the boxes at the top of the fat column. The remaining Nb entries β_{ij} with $i = 1, \ldots, b$ and $j = 1, \ldots, N$ in the fat leg are chosen from $\{2, 3, \ldots, c+1\}$ with repetitions allowed across rows but not down columns, but without restriction of choice since $N\mu_i \geq N$ for $i = 2, \ldots, \ell(\mu)$. Finally the remaining Na entries α_{ij} for $i = 1, \ldots, N$ and $j = 1, 2, \ldots, a$ are placed in weakly increasing order in the stretched thin arm. Once again this is illustrated by:

(3.6) $\quad T^{N\lambda} =$

1	1	$\cdots$	1	α_{11}	α_{12}	$\cdots$	α_{Na}
β_{11}	β_{12}	$\cdots$	β_{1N}				
β_{21}	β_{22}	$\cdots$	β_{2N}				
$\vdots$	$\vdots$	$\cdots$	$\vdots$				
β_{b1}	β_{b2}	$\cdots$	β_{bN}				

It follows that $\#\{T^{N\lambda}\} = \#\{T^{N^b}\} = \dim_{\mathrm{GL}(c)} V^{N^b}$. The famous hook-length formula (see [**9**, p. 45]) then gives the required result, which is a polynomial in N. $\square$

4. The hive model and convex polytopes

To try to understand the nature of the various polynomials fits that we have found it is worth considering a model for all the inequalities underlying the calculation of Littlewood-Richardson and Kostka coefficients. This model is known as the hive model and was proposed by Knutson and Tao [**2, 6, 7**].

An n-hive is a triangular array of numbers a_{ij} with $0 \leq i, j \leq n$. Typically, for $n = 4$ their arrangement is as shown below

(4.1)
$$\begin{array}{ccccccccc}
& & & & a_{00} & & & & \\
& & & a_{10} & & a_{01} & & & \\
& & a_{20} & & a_{11} & & a_{02} & & \\
& a_{30} & & a_{21} & & a_{12} & & a_{03} & \\
a_{40} & & a_{31} & & a_{22} & & a_{13} & & a_{04}
\end{array}$$

Such an n-hive is said to be an integer hive if all of its entries are non-negative integers.

Neighbouring entries define three distinct types of rhombus, each with its own constraint condition.

(4.2) $\quad$ R1 : $\begin{array}{cc} a & b \\ c & d \end{array}$ $\qquad$ R2 : $\begin{array}{c} a \\ b \quad c \\ d \end{array}$ $\qquad$ R3 : $\begin{array}{cc} b & d \\ a & c \end{array}$

In each case, with the labelling as shown, the hive condition takes the form:

(HC) $\qquad\qquad b + c \geq a + d$

DEFINITION 4.1. An LR-hive is an integer hive satisfying the hive condition (HC) for all its constituent rhombi of type R1, R2 and R3, with border entries determined by partitions λ, μ, ν in such a way that $a_{00} = 0$, $a_{0j} = \lambda_1 + \lambda_2 + \cdots + \lambda_j$ for $j = 1, 2, \ldots, n$, $a_{i0} = \nu_1 + \nu_2 + \cdots + \nu_i$ for $i = 1, 2, \ldots, n$, $a_{k,n-k} = a_{0n} + \mu_1 + \mu_2 + \cdots + \mu_k$ for $k = 1, 2, \ldots, n$, with $\ell(\lambda), \ell(\mu), \ell(\nu) \leq n$ and $|\lambda| + |\mu| = |\nu|$.

Schematically, we have

(4.3)
$$\begin{array}{ccccccc} & & & 0 & & & \\ & & \nu_1 & & \lambda_1 & & \\ & \nu_1+\nu_2 & & a_{11} & & \lambda_1+\lambda_2 & \\ & \cdots & a_{21} & & a_{12} & & \cdots \\ |\nu| & |\nu|-\mu_n & & \cdots & & |\lambda|+\mu_1 & |\lambda| \end{array}$$

With the above definition

PROPOSITION 4.2. *The Littlewood–Richardson coefficient $c_{\lambda\mu}^{\nu}$ is the number of LR-hives with border labels determined as above by λ, μ, ν.*

By way of example if $\lambda = (3,2)$, $\mu = (2,1)$ and $\nu = (4,3,1)$ the corresponding LR-hives take the form

(4.4)
$$\begin{array}{ccccccc} & & & 0 & & & \\ & & 4 & & 3 & & \\ & 7 & & a & & 5 & \\ 8 & & 8 & & 7 & & 5 \end{array}$$

It follows that $c_{32,21}^{431} = 2$ since the only integer values of a satisfying the hive condition (HC) for all constituent rhombi of type R1, R2 and R3 are $a = 6$ and $a = 7$.

In the same way we introduce

DEFINITION 4.3. A K-hive is an integer hive satisfying the hive condition HC for all its constituent rhombi of type R1 and R2 (but not R3), with border labels determined by partitions λ, μ in such a way that $a_{i0} = 0$ for $i = 0, 1, \ldots, n$, $a_{0j} = \lambda_1 + \lambda_2 + \cdots + \lambda_j$ for $j = 1, 2, \ldots, n$, $a_{n-k,k} = \mu_1 + \mu_2 + \cdots + \mu_k$ for $k = 1, 2, \ldots, n$, with $\ell(\lambda), \ell(\mu) \leq n$ and $|\lambda| = |\mu|$.

Schematically, we have

(4.5)
$$\begin{array}{ccccccc} & & & 0 & & & \\ & & 0 & & \lambda_1 & & \\ & 0 & & a_{11} & & \lambda_1+\lambda_2 & \\ & \cdots & a_{21} & & a_{12} & & \cdots \\ 0 & & \mu_1 & & \mu_1+\mu_2 & \cdots & |\lambda| \end{array}$$

PROPOSITION 4.4. *The Kostka coefficient $K_{\lambda\mu}$ is the number of K-hives with border labels determined as above by λ, μ.*

For example, if $\lambda = (3,2)$ and $\mu = (2,2,1)$ the corresponding K-hives take the form:

(4.6)
$$\begin{array}{ccccccc} & & & 0 & & & \\ & & 0 & & 3 & & \\ & 0 & & a & & 5 & \\ 0 & & 2 & & 4 & & 5 \end{array}$$

Since the only integer values of a satisfying the hive condition (HC) for all the constituent rhombi of type R1 and R2 are $a = 2$ and $a = 3$, it follows that $K_{32,221} = 2$.

The adoption of the hive model, as a way of dealing with all the inequalities implicit in the definition of Littlewood-Richardson and Kostka coefficients, has the advantage that it is easy to determine individual coefficients rather than full tensor product or weight space decompositions. The problem is reduced to solving the rhombus constraints for $m = (n-1)(n-2)/2$ interior vertex labels a_{ij}. The corresponding set of linear constraints with integer coefficients defines a convex rational polytope, $\mathcal{P}$, in $\mathbb{R}^m$, known as the hive polytope. The unstretched coefficient k is just the number of points in the intersection of the hive polytope, $\mathcal{P}$, with the integer lattice, $\mathbb{Z}^m$, that is $k = \#\{\mathcal{P} \cap \mathbb{Z}^m\}$. Similarly, the stretched coefficient $P(N)$, which may or may not be polynomial, is the number of points in the intersection of the stretched convex rational polytope $N\mathcal{P} = \{Nv \mid v \in \mathcal{P}\}$ with the integer lattice, that is $P(N) = \#\{N\mathcal{P} \cap \mathbb{Z}^m\}$.

By way of a simple application of this approach we can derive the following:

PROPOSITION 4.5. *For partitions of lengths ≤ 3, if the unstretched Littlewood–Richardson or Kostka coefficient is k then the stretched coefficient is $1 + N(k-1)$.*

PROOF. For $n = 3$ there is just one interior vertex label $a = a_{11}$. All the unstretched constraints must reduce to a single constraint of the form $a_1 \leq a \leq a_2$ with $a_1, a_2 \in \mathbb{Z}$ both linear in the parts of the relevant partitions. With this notation, $k = 1 + a_2 - a_1$. Correspondingly, the stretched constraints must reduce to the single constraint $Na_1 \leq a \leq Na_2$ so that $k(N) = 1 + Na_2 - Na_1 = 1 + N(k-1)$, as required. $\square$

As a more complicated example, the LR-hives showing that $c^{9964}_{753,742} = 6$ take the form:

(4.7)
$$\begin{array}{ccccc} & & 0 & & \\ & 9 & & 7 & \\ & 18 & a & 12 & \\ 24 & & c & b & 15 \\ 28 & 28 & 26 & 22 & 15 \end{array}$$

where the hive conditions (HC) for all the constituent rhombi of type R1, R2 and R3 are such that $(a, b, c) = (16, 19, 23), (16, 20, 22), (16, 20, 23), (16, 20, 24), (16, 21, 23)$ or $(16, 21, 24)$. Plotting these points in $\mathbb{R}^3$ gives the intersection of the corresponding hive polytope $\mathcal{P}$ with $\mathbb{Z}^3$, as shown below. The plot is two-dimensional since $a = 16$. The b and c axes are horizontal and vertical, respectively.

(4.8)

Expanding $\mathcal{P}$ to give $N\mathcal{P}$ yields the following sets of integer points for $N = 2$ and $N = 3$.

(4.9)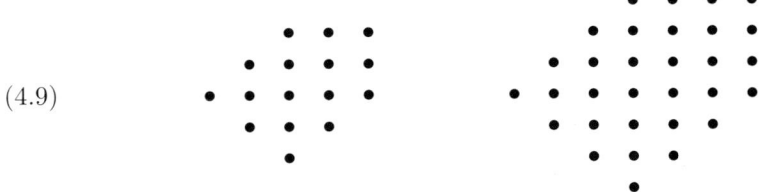

Extending these diagrams to higher values of N is straightforward and counting points leads to the polynomial formula $P(N) = \frac{1}{2}(5N^2 + 5N + 2)$.

More generally, the determination of Littlewood-Richardson and Kostka coefficients by the enumeration of integer hives, and hence by the enumeration of integer points in convex rational polytopes, can be exploited in a manner which at first sight comes close to proving our polynomial conjectures.

Let $\mathcal{P} \in \mathbb{R}^m$ be a convex rational polytope of dimension $d \leq m$, let $N\mathcal{P} = \{Nv \mid v \in \mathcal{P}\}$ and let $i(\mathcal{P}, N) = \#\{N\mathcal{P} \cap \mathbb{Z}^m\}$ be the number of integer points in $N\mathcal{P}$. A convex rational polytope $\mathcal{P} \in \mathbb{R}^m$ is an integer polytope if all of its vertices lie in $\mathbb{Z}^m$.

Then we have the following:

THEOREM 4.6 (Ehrhart [4]; see [10, p. 235]). *For any d-dimensional convex rational polytope $\mathcal{P} \in \mathbb{R}^m$, with $d \leq m$, $i(\mathcal{P}, N)$ is a quasi-polynomial of degree d in N, that is there exists $r \geq 1$ and polynomials $P_s(N)$ of degree d in N such that $i(P, N) = P_s(N)$ for $N \equiv s \bmod r$. If the convex rational polytope $\mathcal{P} \in \mathbb{R}^m$ is an integer polytope, then $i(\mathcal{P}, N)$ is a polynomial in N of degree d.*

COROLLARY 4.7. *If the convex rational polytope $\mathcal{P}$ associated with the enumeration of LR-hives or K-hives of given border is an integer polytope, then the corresponding stretched Littlewood–Richardson or Kostka coefficients are polynomial in the stretching parameter N.*

To exploit this corollary it is necessary to determine whether or not the integer hives define convex rational polytopes that are integer. We can prove the following:

THEOREM 4.8. *Let λ, μ, ν be partitions of lengths ≤ 4, then the hive polytope $\mathcal{P}$ over the border defined by λ, μ, ν is an integer polytope.*

METHOD OF PROOF. Since $\ell(\lambda)$, $\ell(\mu)$, $\ell(\nu) \leq 4$ we need only consider hives in which the elements to be enumerated constitute a vector $v = (a_{11}, a_{12}, a_{21})$. To prove the theorem it is shown that for any $v = (x, y, z) \in \mathcal{P}$ such that $(x, y, z) \notin \mathbb{Z}^3$ there exists some non-zero perturbation $(p, q, r) \in \mathbb{R}^3$ such that all $v' = (x', y', z') \in [(x-p, y-q, z-r), (x+p, y+q, z+r)]$ are in $\mathcal{P}$. Hence v is not a vertex of $\mathcal{P}$. Thus the vertices of $\mathcal{P}$ are all integer vertices. □

COROLLARY 4.9. *If $\ell(\lambda)$, $\ell(\mu)$, $\ell(\nu) \leq 4$ then both $c^{N\nu}_{N\lambda,N\mu}$ and $K_{N\lambda,N\mu}$ are polynomial in N.*

However, there do exist non-integer hive polytopes. Moreover, not all convex rational polytopes $\mathcal{P}$ associated with LR-hives or K-hives of given border are integer polytopes. For example, the K-hives used to evaluate $K_{32,1^5}$ take the form

(4.10)
$$\begin{array}{ccccccc} & & & & 0 & & \\ & & & 0 & & 3 & \\ & & 0 & & a & & 5 \\ & 0 & & b & & c & & 5 \\ 0 & & d & & e & & f & & 5 \\ 0 & 1 & 2 & 3 & 4 & 5 \end{array}$$

There exist precisely 5 distinct K-hives of this type, so that $K_{\lambda\mu} = 5$.

However, it can be shown from the hive inequalities that the vertices $v = (a, b, c, d, e, f)$ of the corresponding convex polytope $\mathcal{P}$ are given by

(4.11)
$$v_1 = (2, 4, 4, 2, 3, 1)$$
$$v_2 = (2, 4, 4, 2, 3, 2)$$
$$v_3 = (3, 4, 4, 2, 3, 1)$$
$$v_4 = (3, 4, 4, 2, 3, 2)$$
$$v_5 = (3, 4, 4, 3, 3, 2)$$
$$v_6 = (\tfrac{5}{2}, 4, 4, \tfrac{3}{2}, 3, \tfrac{3}{2})$$

Of these the first 5 are integral and specify the five K-hives to which we have referred. The sixth vertex v_6 is manifestly not integral. The corresponding hive is not a K-hive, it takes the form

(4.12)
$$\begin{array}{ccccccc} & & & & & & 0 \\ & & & & & 0 & 3 \\ & & & & 0 & \tfrac{5}{2} & 5 \\ & & & 0 & \tfrac{3}{2} & 4 & 5 \\ & & 0 & \tfrac{3}{2} & 3 & 4 & 5 \\ & 0 & 1 & 2 & 3 & 4 & 5 \end{array}$$

The convex polytope $\mathcal{P}$ is not integral, so that Ehrhart's Theorem cannot be used to infer that $P(N) = K_{N\lambda,N\mu}$ is a polynomial, only that it is a quasi-polynomial. However, the direct enumeration of K-hives in this case leads to the conclusion that $P(N)$ is in fact polynomial. We find that the data can be fitted by $P(N) = (N + 1)(N^2 + 2N + 2)/2$.

5. Generating functions

In the spirit of much of the work of Bob Sharp, we can try to encapsulate our results by means of generating functions. All our data on stretched Littlewood-Richardson and Kostka coefficients, $P(N) = c_{N\lambda,N\mu}^{N\nu}$ or $P(N) = K_{N\lambda,N\mu}$, can be fitted by generating functions of the form

(5.1) $$F(z) = \frac{G(z)}{(1-z)^{d+1}} = \sum_{N=0}^{\infty} P(N) z^N,$$

with $G(z)$ a polynomial of degree $\deg G \leq d$. This is of course a consequence of our finding that that in every case $P(N)$ is a polynomial.

Typically, we find the following generating functions for $K_{N\lambda,N1^6}$ for various λ.

(5.2)
$$\begin{array}{ll} (6): & 1/(1-z) \\ (51): & 1/(1-z)^5 \\ (42): & (1+4z+5z^2)/(1-z)^5 \\ (41^2): & (1+3z+z^2)/(1-z)^7 \\ (3^2): & (1+z+z^2)/(1-z)^4 \\ (321): & (1+8z+35z^2+32z^3+9z^4)/(1-z)^8 \\ (31^3): & (1+3z+z^2)/(1-z)^7 \\ (2^3): & (1+z^2)/(1-z)^5 \\ (2^21^2): & (1+z+6z^2+z^3+z^4)/(1-z)^8 \\ (21^4): & 1/(1-z)^5 \\ (1^6): & 1/(1-z) \end{array}$$

On the basis of a wealth of similar results we are led to:

CONJECTURE 5.1. *The generating functions for both $c_{N\lambda,N\mu}^{N\nu}$ and $K_{N\lambda,N\mu}$ take the form $F(z) = G(z)/(1-z)^{d+1}$, with $G(z)$ a polynomial of degree $\deg G \leq d$ having non-negative integer coefficients.*

If true, then it would immediately follow that both $c_{N\lambda,N\mu}^{N\nu}$ and $K_{N\lambda,N\mu}$ are polynomial in N. Of course if the corresponding unstretched K-hives define convex integer polytopes then the conjecture follows from the known properties of Ehrhart polynomials. These properties include the following (see [**10**, p. 208]):

THEOREM 5.2. *If $\mathcal{P}$ is a d-dimensional convex integer polytope the generating function for the Ehrhart polynomial $i(\mathcal{P}, N)$ takes the form $F(z) = G(z)/(1-z)^{d+1}$, with $G(z)$ a polynomial of degree $\deg G \leq d$ having non-negative integer coefficients.*

However, as we have seen not all polytopes $\mathcal{P}$ associated with LR-hives or K-hives of given border are integer. The challenge is therefore, either to prove the validity of our conjectures or provide counter examples.

In the first case the task is to prove that both $P(N) = c_{N\lambda,N\mu}^{N\nu}$ and $P(N) = K_{N\lambda,N\mu}$ are polynomial in N for all partitions λ, μ, ν. If this can be done, it would then be desirable to determine the degree d of $P(N)$ as a function of λ, μ, ν, and to show if possible that all its coefficients are non-negative rational numbers. If $F(z) = G(z)/(1-z)^{d+1}$ is the generating function of $P(N)$, it is then necesary to show that $G(z)$ is a polynomial whose coefficients are all non-negative integers, and to determine the degree, $\deg G$, of this polynomial.

In the case of counter examples, one must identify partitions λ, μ, ν such that $P(N)$ is quasi-polynomial but not polynomial. In the case of such a quasi-polynomial it would then be desirable to determine the form of the corresponding generating function $F(z)$ and its properties.

We close with an example that might either whet the appetite or dull the senses. In the case of the Kostka coefficients $K_{\lambda\mu}$ with $\lambda = (2^p)$ and $\mu = (1^{2p})$ we believe on the basis of extensive calculations that $P(N) = K_{N\lambda,N\mu}$ is a polynomial of degree $d = (p-1)^2$, and that the numerator of the generating function $G(z)$ is a symmetric polynomial of degree $(p-1)(p-2)$ with non-negative integer coefficients. To exemplify this, in the case $\lambda = (2^4)$ and $\mu = (1^8)$ we can fit the data with the following:

$$
\begin{aligned}
P(N) &= \frac{1}{3240}(N+1)(N+2)(N+3)(N^2+4N+9) \\
&\quad \cdot (N^4 + 8N^3 + 26N^2 + 40N + 60) \\
F(z) &= (1 + 4z + 31z^2 + 40z^3 + 31z^4 + 4z^5 + z^6)/(1-z)^{10}
\end{aligned}
\tag{5.3}
$$

Note added in proof

Since the submission of the present paper, the attention of the authors has been drawn to a preprint by H. Derkson and J. Weyman, *On the Littlewood–Richardson polynomials*, in which they use results on the semi-invariants of quivers to prove the key component of our Conjecture 3.1, namely that the stretched Littlewood–Richardson coefficients, $c_{N\mu,N\nu}^{N\lambda}$, are polynomials in the stretching parameter N.

References

1. A. D. Berenstein and A. V. Zelevinsky, *Triple multiplicities for* $\mathrm{sl}(r+1)$ *and the spectrum of the exterior algebra of the adjoint representation*, J. Algebraic Combin. **1** (1992), no. 1, 7–22.
2. A.S. Buch, *The saturation conjecture (after A Knutson and T Tao)*, Enseign. Math. (2) **46** (2000), no. 1-2, 43–60; `math.CO/9810180`.
3. H. Derksen and J. Weyman, *Semi-invariants of quivers and saturation for Littlewood–Richardson coefficients*, J. Amer. Math. Soc. **13** (2000), no. 3, 467–479 (electronic).
4. E. Ehrhart, *Polynômes arithmétiques et méthode des polyèdres en combinatoire*, Internat. Ser. Numer. Math., vol. 35, Birkhäuser Verlag, Basel, 1977.
5. W. Fulton, *Eigenvalues, invariant factors, highest weight, and Schubert calculus*, Bull. Amer. Math. Soc. **37** (2000), no. 3, 209–249.
6. A. Knutson and T. Tao, *The honeycomb model of* $\mathrm{GL}_n(\mathbb{C})$ *tensor products,* I. *Proof of the saturation conjecture*, J. Amer. Math. Soc. **12** (1999), no. 4, 1055–1090.
7. A. Knutson, T. Tao, and C. Woodward, *The honeycomb model of* $\mathrm{GL}_n(\mathbb{C})$ *tensor products.* II. *Puzzles determine facets of the Littlewood–Richardson cone*, `math.CO/0107011`.
8. D. E. Littlewood, *The theory of group characters* Oxford Univ. Press, New York, 1940.
9. I. G. Macdonald, *Symmetric functions and Hall polynomials*, 2nd edition, Oxford Math. Monogr., Clarendon Press, Oxford, 1995.
10. R. P. Stanley, *Enumerative combinatorics*, vol. 1, Cambridge Stud. Adv. Math., vol. 49, Cambridge Univ. Press, Cambridge, 1997.
11. E. Wigner, *On the consequences of the symmetry of the nuclear Hamiltonian on the spectroscopy of nuclei* Phys. Rev. **51** (1937), 106–119.

FACULTY OF MATHEMATICAL STUDIES, UNIVERSITY OF SOUTHAMPTON, SOUTHAMPTON SO17 1BJ, ENGLAND
E-mail address: `R.C.King@maths.soton.ac.uk`

LABORATOIRE D'INFORMATIQUE DE PARIS-NORD, CNRS UMR 7030, UNIVERSITÉ PARIS 13, 93430 VILLETANEUSE, FRANCE
E-mail address, C. Tollu: `ct@lipn.univ-paris13.fr`
E-mail address, F. Toumazet: `ft@lipn.univ-paris13.fr`

Group Actions on Compact Hyperbolic Manifolds and Closed Geodesics

Peter Kramer

This paper is dedicated to the memory of Robert T. Sharp.

ABSTRACT. Compact hyperbolic 3-manifolds are used in cosmological models. In all cases, the universal covering of the compact manifold M is the hyperbolic space H^3 or the hyperbolic ball B^3. The action on H^3 of a hyperbolic element g from the holonomy group, isomorphic to the discrete homotopy group $\pi_1(M)$ of M but acting on H^3, produces preimage and image points of geodesic sections which by self-intersection close on M. For any fixed hyperbolic $g \in \pi_1(M)$ we construct a continuous commutative two-parameter normalizer $N_g < \mathrm{Sl}(2, C)$ and its orbit surfaces on H^3, B^3. The orbit surfaces classify sets of closed geodesic sections of equal length. We give general expressions for the length of closed geodesic sections and for the defect angle at the self-intersection on M in terms of the parameters of g and of the orbit. Sets of closed geodesics with minimal length, given from the character $\chi(g)$, belong to a single orbit. These and only these minimal closed geodesics have vanishing defect angle and hence are smooth. The role of additional symmetries is illuminated by the example of the dodecahedral hyperbolic Weber-Seifert manifold.

1. Introduction

Models of a closed cosmos with nontrivial topology were reviewed by Lachieze and Luminet in [8] and by Levin in [9]. The predictions from the models may be compared directly with astronomical data, compare Fagundes [2, 3], or with the autocorrelation of the observed cosmic mass density [8, pp. 200–201]. Examples of compact hyperbolic manifolds are the well-known dodecahedral manifold of Weber and Seifert [12], see Thurston [11, pp. 36-37], and its companions described by Best [1]. Closed hyperbolic manifolds can be ordered by their volume. Those of small volume due to Thurston and to Weeks have found particular attention in cosmology [9, p. 265]. As pointed out in [9, p. 266], closed geodesics on the manifold M convey important information for the spacing of ghost images and for the autocorrelation of the mass density. In [4, pp. 340–369], the compact hyperbolic double torus manifold with the universal covering H^2 is reviewed. It is pointed out that the

---

2000 *Mathematics Subject Classification.* 37D05, 83F05.
This is the final form of the paper.

length spectrum of closed geodesics is related via Selberg's trace formula to the eigenmodes of the Laplacian.

We use continuous and discrete groups for the analysis of closed geodesics. We consider a compact hyperbolic manifold M whose universal covering is the hyperbolic space H^3. H^3 is equivalent to the coset space $SO_\uparrow^+(1,3,\mathbb{R})/SO(3,\mathbb{R})$, it is equipped with a relativistic metric, admits the isometric action of $SO_\uparrow^+(1,3,R)$ or of its universal covering group $Sl(2,C)$, and has constant negative curvature [13]. The Minkowski metric restricted to H^3 yields the notions of geodesics and their length, which carry over to M. The homotopy group $\pi_1(M)$ acts by path concatenation on M. Isomorphic to $\pi_1(M)$ is the holonomy group, [11, p. 141]. It acts as a discrete subgroup of $Sl(2,\mathbb{C})$ without fixpoints on the universal covering manifold H^3 and generates a tesselation by copies of M. Pairs of points on different copies of M are equivalent if there is an element g of the holonomy group which has this pair as preimage and image. We shall use the same symbol $\pi_1(M)$ for the homotopy and for the holonomy group which is our main tool. Equivalent points of H^3 are identified on M. Any geodesic section on H^3 between equivalent points when mapped to M closes by self-intersection.

The classification of closed geodesics on M can now be seen under two aspects: (i) Find the variety of closed geodesics for a given fixed element g of the holonomy group. (ii) Compare closed geodesics for different elements of the holonomy group, leading to a length spectrum [4]. In what follows we consider mainly the aspect (i) for a general compact hyperbolic manifold M. We illustrate it on the dodecahedral Weber–Seifert manifold. In Section 2 we briefly describe the groups, in Section 3 the geometry of the hyperbolic space H^3 and hyperbolic ball B^3, and the group actions. In Section 4 we develop the classification of closed geodesics. Our main results are:

(1) The starting points on M for closed geodesic sections of equal length are classified by orbit surfaces on H^3 under the continuous commutative two-parameter normalizer N_g associated with g.
(2) For given holonomy g, the various length values of the closed geodesic sections are explicitly expressed by the orbit and group parameters.
(3) The shortest closed geodesic length for fixed g is unique, determined by the character $\chi(g)$, and so independent of conjugations. Closed geodesics for different elements $g \neq g'$ of $\pi_1(M)$ can be compared by means of their characters.
(4) The defect angle at the self-intersection of any closed geodesic is explicitly expressed by orbit and group parameters. The shortest closed geodesics have vanishing defect angle and so are smooth.

2. The Lorentz group $SO_\uparrow^+(1,3,\mathbb{R})$ and its covering

The universal covering for compact hyperbolic 3-manifolds is the hyperbolic space H^3. It is a homogeneous space under the Lorentz group $SO_\uparrow^+(1,3,\mathbb{R})$. In this section we collect results and notations for the action of this and related continuous groups.

2.1. Class and in-class structure of $Sl(2,\mathbb{C})$, adjoint representation.

The universal covering of the proper time-preserving Lorentz group $SO_\uparrow^+(1,3,\mathbb{R})$ is the unimodular group $Sl(2,\mathbb{C})$. Both groups have 6 real parameters.

We use exponential parameters to display the class structure of Sl(2, $\mathbb{C}$). Any class representative can be chosen in diagonal form,

$$(2.1) \qquad g_0 = \begin{bmatrix} \exp(c+i\gamma) & 0 \\ 0 & \exp(-(c+i\gamma)) \end{bmatrix}, \quad -\infty < c < \infty, \ 0 \leq \gamma < 2\pi.$$

with two real class parameters c, γ. The elements (2.1) belong to a subgroup H isomorphic to $SO(1,1,\mathbb{R}) \times SO(2,\mathbb{R})$ but in diagonal, not in standard form. Elements of type (2.1) and its conjugates with $|c| > 0$ will be called hyperbolic. The corresponding Lorentz transformations, see the next subsection, have no fixpoints on H^3 and so are candidates for elements of the discrete holonomy group of M. Given a general element $g \in Sl(2,\mathbb{C})$, we can determine its class parameters by use of the character,

$$(2.2) \qquad \tfrac{1}{2}\chi(g) = \cosh(c+i\gamma) = \cosh(c)\cos(\gamma) + i\sinh(c)\sin(\gamma).$$

In addition to (2.1) we define the elements

$$(2.3) \qquad \begin{aligned} g_1 = g_1(a,\alpha) &= \begin{bmatrix} \cosh(a+i\alpha) & \sinh(a+i\alpha) \\ \sinh(a+i\alpha) & \cosh(a+i\alpha) \end{bmatrix}, \quad (a,\alpha) \text{ real}, \\ g_2 = g_0(b,\beta) &= \begin{bmatrix} \exp(b+i\beta) & 0 \\ 0 & \exp(-(b+i\beta)) \end{bmatrix}, \quad (b,\beta) \text{ real}. \end{aligned}$$

By use of complex Euler angles it can easily be shown that the products $g_2 g_1$ parametrize the cosets $Sl(2,\mathbb{C})/H$. By $\sigma_1, \sigma_2, \sigma_3$ we denote the standard Hermitian Pauli matrices and by e the 2×2 unit matrix. We claim: The general element g of the class with representative (2.1) may be written as

$$(2.4) \qquad \begin{aligned} g &= (g_2 g_1) g_0 (g_1 g_2)^{-1} \\ &= \cosh(c+i\gamma) + \sinh(c+i\gamma) \sum_{i=1}^{3} \eta_i \sigma_i, \quad \sum_{i=1}^{3} (\eta_i)^2 = 1, \\ \begin{bmatrix} \eta_1 \\ \eta_2 \\ \eta_3 \end{bmatrix} &= \mathrm{Ad}(g_2 g_1) \begin{bmatrix} 0 \\ 0 \\ 1 \end{bmatrix}, \\ \mathrm{Ad}(g_2 g_1) &= \begin{bmatrix} \cosh(2(b+i\beta)) & -i\sinh(2(b+i\beta)) & 0 \\ i\sinh(2(b+i\beta)) & \cosh(2(b+i\beta)) & 0 \\ 0 & 0 & 1 \end{bmatrix} \\ &\quad \times \begin{bmatrix} 1 & 0 & 0 \\ 0 & \cosh(2(a+i\alpha)) & -i\sinh(2(a+i\alpha)) \\ 0 & i\sinh(2(a+i\alpha)) & \cosh(2(a+i\alpha)) \end{bmatrix}. \end{aligned}$$

We call a, α, b, β the 4 in-class parameters of g. For $a = b = c = 0$, (2.4) becomes a variant of the familiar parametrization of SU(2). By Ad we denote the adjoint representation of Sl(2, $\mathbb{C}$) generated by $\mathrm{Ad}(g_2)$, $\mathrm{Ad}(g_1)$ in (2.4).

2.2. The proper time-preserving Lorentz group. We obtain in the usual fashion the two-to-one homomorphism from Sl(2, $\mathbb{C}$) to the Lorentz group with transformations $L(g) \in SO_\uparrow^+(1,3,\mathbb{R})$. We introduce in Minkowski space M(1,3) the coordinates (x_0, x_1, x_2, x_3). The scalar product is taken as

$$(2.5) \qquad \langle x, y \rangle = x_0 y_0 - x_1 y_1 - x_2 y_2 - x_3 y_3.$$

We introduce the 2×2 Hermitian matrix

$$\tilde{x} = x_0 e + \sum_{i=1}^{3} x_i \sigma_i. \tag{2.6}$$

with $det(\tilde{x}) = \langle x, x \rangle$. The linear Lorentz action $L(g) = L(-g)$ of $\mathrm{Sl}(2,\mathbb{C})$ on $\mathrm{M}(1,3)$ is

$$g \in \mathrm{Sl}(2,C) \colon \tilde{x} \to \tilde{x}' = g \tilde{x} g^\dagger,$$
$$x'_\mu = \sum_{0}^{3} L_{\mu\nu}(g) x_\nu. \tag{2.7}$$

In particular one finds from (2.1), (2.3), (2.7)

$$L(g_0(c,\gamma)) := \begin{bmatrix} \cosh(2c) & 0 & 0 & \sinh(2c) \\ 0 & \cos(2\gamma) & \sin(2\gamma) & 0 \\ 0 & -\sin(2\gamma) & \cos(2\gamma) & 0 \\ \sinh(2c) & 0 & 0 & \cosh(2c) \end{bmatrix}$$

$$L(g_1(a,\alpha)) := \begin{bmatrix} \cosh(2a) & \sinh(2a) & 0 & 0 \\ \sinh(2a) & \cosh(2a) & 0 & 0 \\ 0 & 0 & \cos(2\alpha) & \sin(2\alpha) \\ 0 & 0 & -\sin(2\alpha) & \cos(2\alpha) \end{bmatrix}, \tag{2.8}$$

$$L(g_2) := L(g_0(b,\beta)).$$

3. The hyperbolic space H^3 and ball B^3

The hyperbolic space H^3 arises as the universal covering of compact hyperbolic manifolds. For details on the hyperbolic space and ball we refer to Ratcliffe [10] pp. 56–104 and pp. 127–135 respectively.

The hyperbolic space is the coset space $\mathrm{H}^3 := \mathrm{SO}^+_\uparrow(1,3,\mathbb{R})/\mathrm{SO}(3,\mathbb{R})$. In $\mathrm{M}(1,3)$ its points form the hyperboloid

$$\mathrm{H}^3 = \langle x \mid \langle x, x \rangle = 1, \quad x_0 \geq 1 \rangle \tag{3.1}$$

We follow [10] up to a sign and rewrite the scalar product (2.5) in $\mathrm{M}(1,3)$ as

$$\langle x, y \rangle = x_0 y_0 - (x, y),$$
$$(x, y) := x_1 y_1 + x_2 y_2 + x_3 y_3. \tag{3.2}$$

Given two points $x, y \in \mathrm{H}^3$, we take their scalar product as the restriction of the scalar product $\langle x, y \rangle$ from $\mathrm{M}(1,3)$ to H^3. The restricted scalar product on H^3 is positive definite. H^3 with this metric is a space of constant negative curvature.

The conformal ball model B^3 for H^3 is obtained from the points of this hyperboloid by the nonlinear map from H^3 to a Euclidean space E^3,

$$x_i \to \xi_i = \frac{x_i}{1+x_0}, \quad i = 1,2,3, \quad (\xi,\xi) < 1,$$
$$x_i = \frac{2\xi_i}{1-(\xi,\xi)}, \quad i = 1,2,3. \tag{3.3}$$

Although all points of H^3 from (3.3) map bijectively into points of B^3, we shall need points in E^3 but outside B^3 to characterize its symmetries. We shall use the scalar product $(\,,\,)$ with respect to the space-like components of vectors from $\mathrm{M}(1,3)$, for vectors in B^3, and in E^3 embedding B^3. Consider a space-like vector $k \in \mathrm{M}(1,3)$,

$k_0 \neq 0$, and a point $x : \langle x, x \rangle = 1$ of H^3. Using for x the coordinates (3.3) on B^3 one finds

(3.4)
$$\langle k, x \rangle = k_0 (1 + x_0) \tfrac{1}{2} \big(1 + (\xi, \xi) - 2(q, \xi) \big),$$
$$k \to q = k_0^{-1}(k_1, k_2, k_3).$$

LEMMA 3.1. *The intersection of the hyperplane $\langle k, x \rangle = 0$, $k_0 \neq 0$ in $M(1, 3)$ with the hyperboloid H^3 in the conformal ball model becomes the intersection of a sphere $S^2(q, R)$ of center q, $(q, q) > 1$ and radius R with $B^3 \subset E^3$,*

(3.5)
$$(\xi - q, \xi - q) - R^2 = 0,$$
$$q = k_0^{-1}(k_1, k_2, k_3), \quad R = \sqrt{(q, q) - 1}.$$

PROOF. It suffices to rewrite part of (3.5) as

(3.6) $\quad \big(1 + (\xi, \xi) - 2(q, \xi) \big) = (\xi - q, \xi - q) - (q, q) + 1 = (\xi - q, \xi - q) - R^2.$

The condition $R^2 = (q, q) - 1$ insures that the sphere $S^2(q, R)$ has orthogonal intersections with the surface ∂B^3. □

The sphere $S^2(q, R)$ separates the points of B^3 into two disjoint parts. In case of a vector $k : (0, k_1, k_2, k_3)$ when q in (3.5) is not well-defined we replace the sphere in E^3 by a plane through the origin and perpendicular to (k_1, k_2, k_3).

4. Geodesics on H^3 and B^3

As pointed out in the introduction, for any closed geodesic on M there is a unique element g of the holonomy group acting on H^3. We wish to characterize the variety of closed geodesics associated with a fixed g. To this purpose we determine the continuous normalizer of $g \in Sl(2, \mathbb{C})$ and characterize its orbits on H^3. The orbits classify closed geodesics by length and direction and yield explicit expressions for them. A geodesic which closes on M must intersect itself. From H^3 we compute the defect angle at the self-intersection. For given g, there is a unique set of closed geodesic lines which have minimum length and are smooth, i.e., have vanishing defect angle.

4.1. Two-parameter normalizer subgroups of $Sl(2, \mathbb{C})$ and their orbits.

DEFINITION 4.1. Consider a hyperbolic element $g \in Sl(2, \mathbb{C})$ in diagonal form, $g = g_0(c, \gamma)$, $|c| > 0$, (2.1) with fixed exponential parameters c, γ. Define the group N_{g_0} with elements

(4.1) $\quad h(\lambda, \phi) := g_0(\lambda c/2, \lambda \gamma/2 + \phi), \quad -\infty < \lambda < \infty, \ 0 \leq \phi < 2\pi.$

This is the two-parameter commutative normalizer N_{g_0} (which coincides with the centralizer) whose elements commute with g_0, (2.1).

N_{g_0} is isomorphic to $SO(1, 1, \mathbb{R}) \times SO(2, \mathbb{R})$. With $L\big(g_0(\lambda c/2, \lambda \gamma/2 + \phi)\big)$ we represent N_{g_0} by Lorentz transformations. The reasons for our choice of parameters will appear from the actions.

Turn to the action of N_{g_0} on $M(1,3)$, H^3, B^3. On H^3 we can choose the hyperplane $\langle e_3, x \rangle = 0$ for the orbit representatives on H^3 under N_{g_0}. We incorporate the second parameter ϕ of N_{g_0} into the representatives and write them as

(4.2) $\quad x(\rho, \phi) = (\cosh(2\rho), \sinh(2\rho)\cos(2\phi), \sinh(2\rho)\sin(2\phi), 0)$,
$$0 \leq \rho < \infty, \ 0 \leq 2\phi < 2\pi.$$

The points on H^3, B^3 under the action of N_{g_0} we call orbit surfaces. Each one is determined by the value of ρ. We call orbit lines the points from the action of N_{g_0} on B^3 for varying λ and fixed ϕ. The full orbit surface is obtained by rotating this orbit line with the angle ϕ.

LEMMA 4.2. *Preimages on a fixed orbit surfaces under g_0 are mapped into images on the same orbit surface. The relative geodesic distance between preimage and image point is independent of the starting point on the orbit surface.*

PROOF. The first part follows from the fact that N_{g_0} commutes with g_0. Any point on an orbit surface may be written as $L\bigl(g_0(\lambda c/2, \lambda\gamma/2)\bigr)x(\rho, \phi)$. With $g_0 = g_0(c, \gamma)$ we obtain from the scalar product for the geodesic distance between preimage and image points under $L\bigl(g_0(c, \gamma)\bigr)$

(4.3) $\quad \langle L\bigl(g_0(-\lambda c/2, -\lambda\gamma/2)\bigr)x(\rho, \phi), L\bigl(g_0(c, \gamma)\bigr)L\bigl(g_0(-\lambda c/2, -\lambda\gamma/2)\bigr)x(\rho, \phi)\rangle$
$$= \langle x(\rho, \phi), L\bigl(g_0(c, \gamma)\bigr)x(\rho, \phi)\rangle = \langle x(\rho, 0), L\bigl(g_0(c, \gamma)\bigr)x(\rho, 0)\rangle. \quad \square$$

The Lorentz invariance of the scalar product N_{g_0} and the commutativity of N_{g_0} are crucial for this result.

We choose as specific preimages on H^3 for the orbit lines the points $L\bigl(g_0(-c/2, -\gamma/2)\bigr)x(\rho, \phi)$. The images then are of the form $L\bigl(g_0(c/2, \gamma/2)\bigr)x(\rho, \phi)$. Preimages and images correspond to parameter values $\lambda = \mp 1$ in (4.1) respectively. For fixed λ, all orbit lines pass the intersection of the hyperplane $\langle k, x\rangle = 0$, $k = L\bigl(g_0(\lambda c/2, \lambda\gamma/2)\bigr)e_3$ with H^3.

On H^3 the orbit lines take the form

(4.4) $\quad L\bigl(g_0(\lambda c/2, \lambda\gamma/2)\bigr)x(\rho, \phi) = \begin{bmatrix} \cosh(2\rho)\cosh(\lambda c) \\ \sinh(2\rho)\cos(2\phi - \lambda\gamma) \\ \sinh(2\rho)\sin(2\phi - \lambda\gamma) \\ \cosh(2\rho)\sinh(\lambda c) \end{bmatrix}.$

We compute the orbit lines in B^3 and find by applying (3.3) to (4.4)

(4.5) $\quad \xi(\lambda) = (1 + \cosh(2\rho)\cosh(\lambda c))^{-1} \begin{bmatrix} \sinh(2\rho)\cos(2\phi - \lambda\gamma) \\ \sinh(2\rho)\sin(2\phi - \lambda\gamma) \\ \cosh(2\rho)\sinh(\lambda c) \end{bmatrix}.$

For fixed λ, the orbit lines $\xi(\lambda)$ intersect the Möbius sphere $S^2\bigl(q(\lambda), R(\lambda)\bigr)$ with

(4.6) $\quad q(\lambda) = \bigl(0, 0, \cotanh(\lambda c)\bigr), \quad R(\lambda) = \bigl(\sinh(\lambda c)\bigr)^{-1}.$

Since the action of $\mathrm{Sl}(2, \mathbb{C})$ is conformal we have

LEMMA 4.3. *The orbit lines on B^3 for fixed parameter λ intersect the Möbius sphere (4.6). The angle between tangents to orbit lines and the normal of the Möbius sphere depends on the starting value of ρ but is independent of λ, ϕ.*

All orbit lines for $\lambda = 0$ cross the plane $\xi_3 = 0$. This choice allows a simple visualization of the orbit lines and geodesics in Figure 1. The full orbit surfaces are obtained by rotating the orbit lines to any angle ϕ around the 3-axis.

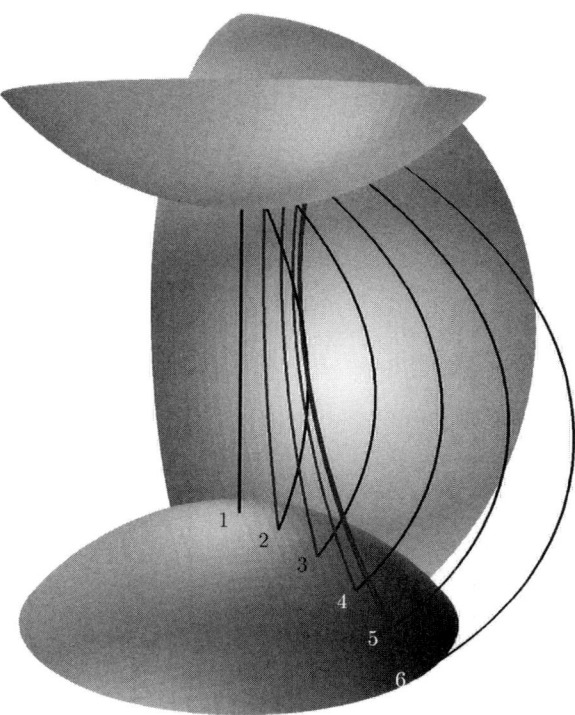

FIGURE 1. A sector of area π from the boundary of B^3 and its intersection with two Möbius spheres $S^2(q, R)$. The generator of the holonomy $g = C_1 \in \pi_1(M)$ for the Weber-Seifert manifold maps the lower into the upper Möbius sphere. Shown are six orbit lines $1, \ldots, 6$ for this map (see 4.1), and six geodesics sections between preimages and images under C_1. Only the straight orbit line 1 yields shortest closed geodesics 1 under C_1. The orbit surface arises from each orbit line by a rotation around the vertical 3-axis.

4.2. Geodesic lines.

LEMMA 4.4. *The geodesic connecting two points ξ, η, $\xi \neq \eta$ on B^3 is a section on the circle $S^1(q, R)$ in the plane containing the points $\langle 0, \xi, \eta \rangle$. The circle has two perpendicular intersections with the surface ∂B^3.*

We compute the geodesic circle $S^1(q, R)$ explicitly. The conditions that the two points be on the same circle with center q and radius $R = \sqrt{(q,q) - 1}$ in a plane containing $\xi = 0$ are

$$(4.7) \qquad q = \mu_1 \xi + \mu_2 \eta, \quad 2(q, \xi) = (\xi, \xi) + 1, \quad 2(q, \eta) = (\eta, \eta) + 1.$$

Solving these equations for μ_1, μ_2 yields

$$(4.8) \qquad \begin{bmatrix} \mu_1 \\ \mu_2 \end{bmatrix} = \frac{1}{2((\xi,\xi)(\eta,\eta) - (\xi,\eta)^2)} \begin{bmatrix} (\eta,\eta) & -(\xi,\eta) \\ -(\xi,\eta) & (\xi,\xi) \end{bmatrix} \begin{bmatrix} (\xi,\xi) + 1 \\ (\eta,\eta) + 1 \end{bmatrix}.$$

4.3. Geodesics between preimage and image points on orbit surfaces.
First we compute the geodesic distance of the preimage and image points under class representatives (2.1) from the scalar products.

LEMMA 4.5. *The length of all geodesic sections for fixed hyperbolic $g_0 = g_0(c, \gamma)$ and orbit parameter ρ is given by*

$$
\begin{aligned}
(4.9) \quad &\langle L(g_0(-c/2, -\gamma/2))x(\rho, \phi), L(g_0(c/2, \gamma/2))x(\rho, \phi)\rangle \\
&= \cosh(2c) + (\cosh(2c) - \cos(2\gamma))(\sinh(2\rho))^2 \geq \cosh(2c).
\end{aligned}
$$

The minimal geodesic length under $L(g_0(c, \gamma))$ is reached for points on the straight geodesic orbit line with representative point $\rho = 0$, $x = (1, 0, 0, 0)$.

PROOF. The scalar product between the points $\lambda = \pm 1$ on the orbit line by evaluation of (4.3) yields (4.9). This expression depends on the parameters c, γ of g_0 and on the parameter ρ which characterizes the orbit but is independent of the parameters λ, ϕ of the starting point on the orbit. The geodesic distance takes its minimum value $\cosh(2c)$ for $\rho = 0$. Its length is determined by the character $\chi(g)$ (2.2). □

Next we characterize the geodesics between the preimage and image points. We particularize the general construction of Lemma 4.4 to the geodesic between $\xi(\lambda)$, $\lambda = \mp 1$. Both vectors have the same length. From (4.8) it can be shown that this geodesic is a section on the circle $S^1(q(-1, +1), R(-1, +1))$ with center and radius

$$
\begin{aligned}
(4.10) \quad & q(-1, +1) = \coth(2\rho)\frac{\cosh(c)}{\cos(\gamma)}(\cos(2\phi), \sin(2\phi), 0), \\
& R(-1, +1) = \sqrt{(q(-1, +1), q(-1, +1)) - 1}.
\end{aligned}
$$

In Figure 1 we show the orbit lines and geodesics for the Lorentz transformation $C_1 = L(g_0(c, \gamma))$ which is a generator of the holonomy group $\pi_1(M)$ of the dodecahedral Weber–Seifert manifold. For the orbit parameters we use six values $\operatorname{arc} \cosh(2\rho) = (0.0, 0.2, 0.4, 0.6, 0.8, 1.0)$, $\phi = 0$. In Figure 2 we show the orbit and geodesic lines between two vertices of the Weber–Seifert dodecahedron.

4.4. Defect angle of closed geodesics.
A closed geodesic starting at a point $P \in M$ will intersect itself at P. Denote by Δ the defect angle between the starting geodesic and its continuation after return and intersection. For a smooth closed geodesic we must have $\Delta = 0$. We compute Δ on the universal covering B^3 by considering along with P the preimage and image $g_0^{-1}P$, $g_0 P$ under a fixed diagonal holonomy g_0 with parameters c, γ. As orbit representative P we choose on the plane $\xi_3 = 0$ from (4.5) the point $\xi(\lambda) \in B^3$, $\lambda = 0$ with $\phi = 0$. The images of this plane under g_0^{-1} and g_0 are two Möbius spheres for the parameter values $\lambda = \mp 2$. With the abbreviations

$$
\begin{aligned}
(4.11) \quad & u := \cosh(2c), \quad v := \sinh(2c), \\
& c := \cos(2\gamma), \quad s := \sin(2\gamma), \\
& \tau := \cosh(2\rho), \quad \sigma := \sinh(2\rho),
\end{aligned}
$$

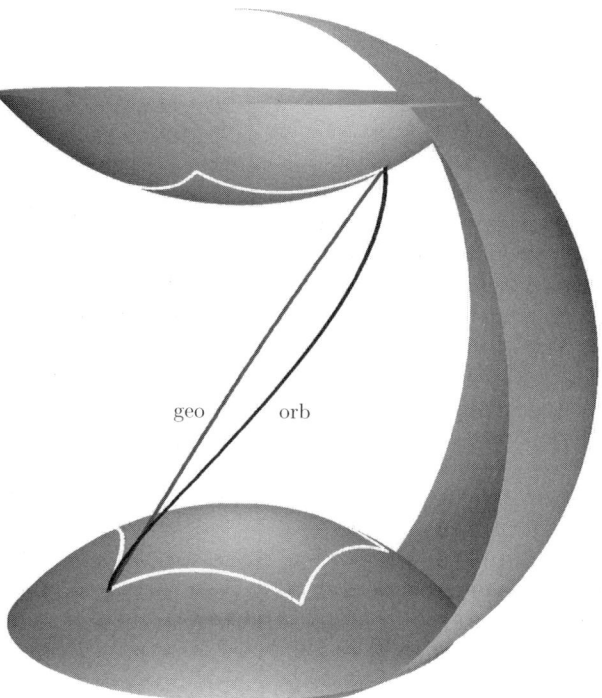

FIGURE 2. View of part of B^3 and the same two Möbius spheres as in Figure 1. An orbit line (orb) of $g = C_1$ connects vertices of opposite pentagonal faces of the Weber-Seifert dodecahedron, the geodesic section between them is marked (geo).

we compute from (4.5) the three points $\xi(-2), \xi(0), \xi(2)$ in terms of the parameters of (4.11),

(4.12)
$$\xi(-2) = (1+\tau u)^{-1} \begin{bmatrix} \sigma c \\ \sigma s \\ -\tau v \end{bmatrix}, \quad \xi(0) = (1+\tau)^{-1} \begin{bmatrix} \sigma \\ 0 \\ 0 \end{bmatrix},$$

$$\xi(2) = (1+\tau u)^{-1} \begin{bmatrix} \sigma c \\ -\sigma s \\ \tau v \end{bmatrix}.$$

All three points are on a single orbit of the continuous normalizer. We observe from (4.12) that, by the planar rotation

(4.13)
$$R_{2,3} = \begin{bmatrix} c' & s' \\ -s' & c' \end{bmatrix},$$
$$c' := \tau v/\omega, \quad s' := \sigma s/\omega, \quad \omega := \sqrt{\sigma^2 s^2 + \tau^2 v^2},$$

applied to the components 2, 3 of all three vectors in eq. 4.12, their new coordinate expressions become

(4.14)
$$\xi(-2) = (1+\tau u)^{-1} \begin{bmatrix} \sigma c \\ 0 \\ -\omega \end{bmatrix}, \quad \xi(0) = (1+\tau)^{-1} \begin{bmatrix} \sigma \\ 0 \\ 0 \end{bmatrix},$$

$$\xi(2) = (1+\tau u)^{-1} \begin{bmatrix} \sigma c \\ 0 \\ \omega \end{bmatrix},$$

and so all three vectors are in the single new $1,3$ plane of B^3.

Now we construct from (4.7) the two geodesic sections between the pairs of points corresponding to $\lambda = (0,2)$ and $\lambda = (-2,0)$ respectively. Clearly both geodesic sections run on the new $1,3$ plane and intersect at $\xi(0)$. On B^3 let the starting geodesic run on the first, the returning geodesic on the second section. Evaluating (4.7), (4.8) for pairs of points we find for the center vectors $q(0,2)$, $q(-2,0)$ of the two geodesic circles

(4.15)
$$q(0,2) = \mu_2 \xi(0) + \mu_1 \xi(2), \quad q(-2,0) = \mu_1 \xi(-2) + \mu_2 \xi(0),$$
$$\mu_1 = (1+\tau u)\frac{\tau(u-c)}{\omega^2}, \quad \mu_2 = (1+\tau)\frac{\tau[-\sigma^2(cu-1)+\tau^2 v^2]}{\sigma^2 \omega^2}.$$

The radius vectors $\xi(0) - q(0,2)$, $\xi(0) - q(-2,0)$ are perpendicular respectively to the starting and returning geodesics at $\xi(0)$. Therefore their angle of intersection determines the defect angle Δ. From (4.14), (4.15) we find in the new coordinates

(4.16) $\xi(0) - q(-2,0) = \begin{bmatrix} -\sigma^{-1} \\ 0 \\ \tau(u-c)/\omega \end{bmatrix}, \quad \xi(0) - q(0,2) = \begin{bmatrix} -\sigma^{-1} \\ 0 \\ -\tau(u-c)/\omega \end{bmatrix}.$

The two vectors differ only by a reflection in the new 1-coordinate axis and so we get for half the defect angle the expression

(4.17) $\tan\left(\frac{1}{2}\Delta\right) = \frac{(\xi(0) - q(0,2))_3}{(\xi(0) - q(0,2))_1} = \sigma \frac{\tau(u-c)}{\omega}$
$$= \sinh(2\rho) \frac{\cosh(2\rho)(\cosh(2c) - \cos(2\gamma))}{\sqrt{\sinh(2\rho)^2 \sin(2\gamma)^2 + \cosh(2\rho)^2 \sinh(2c)^2}}.$$

In the final expression we replaced the abbreviations from (4.11). We analyze the terms in the last expression. The exponential parameters for a fixed non-trivial hyperbolic g_0 must obey $(\cosh(2c) - \cos(2\gamma)) > 0$, $|\sinh(2c)| > 0$. Therefore the factor of $\sinh(2\rho)$ in (4.17) is always different from zero. The value $\Delta = 0$ of the defect angle for a smooth closed geodesic therefore enforces the unique orbit line with representative $\sinh(2\rho) = 0 \to \rho = 0$.

LEMMA 4.6. *The defect angle Δ for geodesic sections associated with g_0 is given as the function of group and orbit parameters by (4.17). The only smooth closed geodesics associated with g_0 are sections of hyperbolic length $2c$ on the geodesic line $\xi = (0, 0, \xi_3)$, $-1 \leq \xi_3 \leq 1$.*

4.5. Orbit surfaces and geodesics for general holonomies. So far we dealt only with the action of a diagonal representative g_0 of a holonomy on general points of H^3. A general holonomy g is given from (2.4) by g_0 as in (2.1) conjugated with $g_2 g_1$ as $g = (g_2 g_1) g_0 (g_2 g_1)^{-1}$. Let c, γ again denote the exponential parameters of the diagonal form g_0.

By conjugation with $g_2 g_1$ we introduce the two-parameter commutative general normalizer $N_g := (g_2 g_1) N_{g_0} (g_2 g_1)^{-1}$ which is conjugate to $SO(1,1,\mathbb{R}) \times SO(2,\mathbb{R})$

and commutes with g. We let this normalizer act on H^3 or B^3 and obtain orbit surfaces. Lemma 4.2 is easily generalized to these orbit surfaces. The geometric results given in Lemmas 4.3–4.6 for $L(g_0)$ can be transcribed to $L(g)$ if before we pass on H^3, B^3 from initial coordinates x to new ones defined as $y = L\big((g_2 g_1)^{-1}\big)x$. For general $L(g)$ it follows from Lemma 4.5 that the unique shortest closed geodesic sunder $L(g)$ occurs between points on the image of a straight orbit line under the map $L\big((g_2 g_1)^{-1}\big)$ acting on the coordinates y. This action is isometric and conformal. Therefore the expressions (4.9) for the geodesic length and (4.17) for the defect angle remain true in terms of the parameters for the diagonal form g_0 of g. For given g there is a unique set of shortest and smooth geodesics.

Since $L(g_2 g_1)$ maps geodesics into geodesics of the same length, the unique minimum length for the action of a general $L(g)$ is determined from Lemma 4.5 by the character $\chi(g) = \chi(g_0(c, \gamma))$ as $\cosh(2c)$. Similarly, the defect angle Δ (4.17) keeps the same value independent of conjugation.

5. The holonomy group of the Weber–Seifert manifold

Symmetries of compact manifolds play an important part in their classification, see [9, pp. 266–279]. The Weber–Seifert hyperbolic dodecahedral manifold [12] illuminates the role played by symmetries. It is related to the hyperbolic Coxeter group with the Dynkin diagram, [10, p. 284], o—^{5}—o—^{3}—o—^{5}—o. A representation by reflections of the full Coxeter group can be constructed in the hyperbolic space H^3. In B^3 the reflections become Möbius inversions. The subgroup o—^{5}—o—^{3}—o is the standard icosahedral Coxeter group. It generates the Weber–Seifert dodecahedron centered at $\xi = (0, 0, 0)$. The holonomy group $\pi_1(M)$ from [1, 12] has six generators $C_1, \ldots, C_6$ with six relations [1, 12]. Each one of them moves one face of the dodecahedron to the opposite face while rotating it around the shortest diameter between the faces by an angle $3\pi/5$. If we choose one diameter along the 3-axis of B^3, the corresponding generator C_1 has the form $L(g_0)$ of (2.1), (2.8) with the parameters

(5.1) $\qquad C_1 = L\big(g_0(c, \gamma)\big), \quad c = \operatorname{arc cosh}(\tfrac{1}{2}\sqrt{4\tau + 3}), \quad \gamma = 3\pi/10.$

These values are used in Figures 1 and 2. The full Coxeter group can be written [7] as a semidirect product

(5.2) $\qquad \text{o—}^5\text{—o—}^3\text{—o—}^5\text{—o} \sim (\pi_1(M)) \times_s (\text{o—}^5\text{—o—}^3\text{—o}).$

In particular the holonomy group is normal in the hyperbolic Coxeter group. The icosahedral Coxeter group acts by conjugation and produces discrete symmetries of the Weber–Seifert manifold M. For the Weber–Seifert manifold, any word of $\pi_1(M)$ from (5.2) can be rewritten in terms of generators for the hyperbolic Coxeter group. Due to the simpler generators and relations for the Coxeter group, the word problem for the holonomy group can then be considered within the Coxeter group, see [6, p. 171]. Moreover it is shown in [7] that the Weyl reflections which generate the hyperbolic Coxeter group can act on the level of $\mathrm{Sl}(2, \mathbb{C})$.

5.1. Closed geodesic sections on M. We list the main aspects in the search for closed geodesics.

(i) Closed geodesic sections are classified by hyperbolic elements $g \in \pi_1(M)$, by their classes, by the continuous normalizer N_g and its orbit surfaces. The orbit and class parameters for the diagonal form g_0 of g determine the length of closed geodesics and their defect angle according to (4.9) and (4.17).

(ii) Let $(g_2 g_1)$ be the matrix which brings g to diagonal form g_0 and determines its in-class parameters. Construct the image of the straight geodesic line $\xi = (0, 0, \xi_3)$ under $(g_2 g_1)^{-1}$. The unique shortest and smooth closed geodesics under g are sections of this line, of length determined by the character $\chi(g)$ according to (2.2).

(iii) For the example of the compact dodecahedral Weber–Seifert manifold, the holonomy group is normal in a hyperbolic Coxeter group. Conjugations by the icosahedral Coxeter group map elements of the holonomy group to one another. They relate closed geodesics beyond the classes of the holonomy group.

(iv) To study the spectrum of shortest smooth closed geodesics on the Weber–Seifert manifold, one must analyze the words of the holonomy group $\pi_1(M)$. Upon rewriting these words in terms of the hyperbolic Coxeter group, techniques used for words in Coxeter groups [6, p. 171] can be applied.

6. Conclusion

Closed geodesic sections on a compact hyperbolic manifold M connect points on B^3 equivalent under a holonomy $g \in \pi_1(M)$. The continuous commutative normalizer N_g isomorphic to $\mathrm{SO}(1,1,R) \times \mathrm{SO}(2,R)$ and its orbit surfaces on M classify by length, direction and defect angle the closed geodesic sections arising from g. Any holonomy g determines a unique geodesic line on B^3. All the shortest and smooth closed geodesics under g are sections of length determined by the character $\chi(g)$ on this geodesic line.

References

1. L. A. Best, *On torsion-free discrete subgroups of* $\mathrm{PSL}(2,\mathbb{C})$ *with compact orbit space*, Canad. J. Math. **23** (1971), 451–460.
2. H. V. Fagundes, *Quasi-galaxy associations with discordant redshifts as a topological effect.* I. *Two-dimensional study.* Astrophys. J. **291** (1985), 450–459.
3. _____, *Quasi-galaxy associations with discordant redshifts as a topological effect.* II. *A closed hyperbolic model*, Astrophys. J. **338** (1989), 618–629.
4. M. C. Gutzwiller *Chaos in classical and quantum mechanics*, Interdiscip. Appl. Math., vol 1, Springer-Verlag, New York, 1990.
5. D. Hilbert and S. Cohn-Vossen, *Geometry and the imagination*, Chelsea Publishing, New York, 1952.
6. J. E. Humphreys, *Reflection groups and Coxeter groups*, Cambridge Stud. Adv. Math., vol. 29, Cambridge Univ. Press, Cambridge, 1990
7. P. Kramer, *Group actions, geodesic loops, and symmetries of compact hyperbolic manifolds* (submitted).
8. M. Lachièze-Rey and J.-P. Luminet, *Cosmic topology*, Phys. Rep. **254** (1995), 135–214.
9. J. Levin, *Topology and the cosmic microwave background*, Phys. Rep. **365** (2002), 251–333.
10. J. G. Ratcliffe, *Foundations of hyperbolic manifolds*, Grad. Texts in Math., vol. 149, Springer-Verlag, New York, 1994.
11. W. P. Thurston, *Three-dimensional geometry and topology*, Princeton Math. Ser., vol. 35, Princeton Univ. Press, Princeton, 1997.
12. C. Weber and H. Seifert, *Die beiden Dodekaederräume*, Math. Z. **37** (1933), 237–253.
13. J. A. Wolf, *Spaces of constant curvature*, 5th edition, Publish or Perish, Houston, TX, 1984.

Institut für Theoretische Physik, Universität Tübingen, 72076 Tübingen, Germany
E-mail address: peter.kramer@uni-tuebingen.de

Is There an Ultimate Symmetry in Physics?

C. S. Lam

ABSTRACT. A speculative discussion on the roles of various symmetries in physics.

1. The ever increasing symmetry

This is dedicated to the memory of Bob Sharp, the kindest and the most decent person in the McGill physics department. Our discussions often turned to group theory, the subject of his love, with him teaching me. This talk is in a sense a continuation of those discussions, and I wish Bob were here to tell me the answer to the question I am now posing, namely, whether there is an ultimate symmetry in physics which can replace all the dynamical laws.

History shows that we became aware of more and more symmetry in Nature as time went by, and that many important discoveries were spear headed by the discovery of a new symmetry. Symmetry can sometimes be used to replace dynamics, at least partially. Thus Newton's first law in mechanics is a reflection of translational symmetry in space, and energy conservation is a revelation of the invariance of translation in time. The huge coriolis force that drives storms on earth into cyclones and anti-cyclones is caused by the rotational invariance of our physics, with conservation of angular momentum being another manifestation. Einstein's special relativity is based on Lorentz invariance, and his general relativity is based on invariance under diffeomorphism. These well-known spacetime symmetries work from astronomical scale down to the tiniest scale we know today, around 10^{-18} metres. Do they continue to hold down to much smallest distances? We do not know the answer but we will come back to this kind of question later.

Then there is the internal symmetry. Conservation of charge as well as other additive quantum numbers can be traced back to the invariance under a phase change, namely a U(1) symmetry. The discovery of isospin by Heisenberg for the nucleon opened up another symmetry, the $SU(2)_F$ symmetry, with the subscript F standing for "flavour." Unlike the other symmetry we talked about, this symmetry is only approximate, being broken by the electromagnetic force. The subsequent discovery of associated production led Gell-Mann and Nishijima to postulate strangeness, which in due course led to the discovery of the $SU(3)_F$ symmetry by Gell-Mann

---

2000 *Mathematics Subject Classification.* 81R99.
This is the final form of the paper.

and Ne'eman. Nowadays we have the Standard Model of strong and electroweak interactions, whose dynamics is governed by the $SU(3)_c \times SU(2)_L \times U(1)_Y$ symmetry.

2. Local symmetry and dynamics

According to Noether's theorem, each continuous symmetry leads to a conserved current $j^\mu(x)$. A local symmetry such as the gauge symmetry of electromagnetism, or the diffeomorphism symmetry in Einstein's gravitational equations, contains an infinite number of symmetries, one at each spacetime point. This means we have an infinite number of conserved currents. What are they? We surely do not have an infinite number of conserved quantities in electromagnetism.

To understand this puzzle in the context of a U(1) local symmetry, let $F^{\mu\nu}(x)$ be the electromagnetic field strength, and $A^\mu(x)$ be its vector potential, coupled to a current density $J^\mu(x)$ through the Lagrangian density $\mathcal{L}(x) = -F^{\mu\nu}F_{\mu\nu}/4 + J^\mu A_\mu$. Under an infinitesimal gauge transformation $\delta A_\mu(x) = \partial_\mu \lambda(x)$, the Lagrangian density changes by an amount $\delta\mathcal{L} = J^\mu \partial_\mu \lambda$. Using Noether's procedure, we know that the current $j^\mu_\lambda(x) = -F^{\mu\nu}(x)\partial_\nu\lambda(x)$ is driven by a source $\delta\mathcal{L}$. In other words,

$$\partial_\mu j^\mu_\lambda(x) = \partial_\mu\bigl(-F^{\mu\nu}(x)\partial_\nu\lambda(x)\bigr) = \delta\mathcal{L}(x) = J^\nu(x)\partial_\nu\lambda(x).$$

Since $F^{\mu\nu}$ is antisymmetric in its two indices, it can be reduced to

$$\bigl(\partial_\mu F^{\mu\nu}(x) + J^\nu(x)\bigr)\partial_\nu\lambda(x) = 0,$$

or simply $\partial_\mu F^{\mu\nu}(x) + J^\nu(x) = 0$ because $\lambda(x)$ is arbitrary. This is the Maxwell equations. In other words, local gauge symmetry gives the full dynamics of Maxwell equations.

What are then the infinite number of conserved charges? j^μ is not conserved unless $J^\mu = 0$, so let us consider that case. The conserved charge is then

$$Q_\lambda(t) = \int d^3x\, j^0_\lambda(\vec{x},t)\, d^3x = -\int d^3x\, F^{0k}\partial_k\lambda.$$

Performing an integration by part and assuming as usual that the surface term vanishes, we come to the realization that $Q_\lambda(t) = 0$ for all $\lambda(x)$ as long as the Gauss law $\partial_k F^{0k} = 0$ is obeyed. So the infinite number of conserved currents simply reduce to the Maxwell equation, and nothing more.

Similar outcome happens in other local symmetries. This suggests that if we are to have an ultimate symmetry which can replace all dynamics, local symmetry is likely to be involved.

Local symmetries as we know them in the Standard Model and in Einstein's gravity are symmetry for the spin-1 gauge fields and the spin-2 gravitational fields. At this moment we know of no proven local symmetry involving fermions which can give us dynamical equations of motion for fermions. These would be necessary if an ultimate symmetry is to be found. Is local supersymmetry the answer to this prayer?

Even granting that there are local symmetries, or even supersymmetries, one might still ask what the particular symmetry is and why that particular symmetry. For that matter why do we live in a four-dimensional spacetime? Why is $SU(3)_c X S \times SU(2)_L \times U(1)_Y$ the Standard Model group and not something else? If there is to be an ultimate symmetry, presumably it should be unique and the symmetry itself should be determined somehow. At this moment the only theory we know of where the symmetry is self-determined is the string theory. It tells us that the spacetime is really 10-dimensional and the gauge group, if present, is either SU(32) or $E_8 \times E_8$. Is the string theory pointing us to the direction of an ultimate symmetry then? We shall come back to discuss this question in a later section.

If the universe possesses an ultimate symmetry, we are surely not seeing all of them in this world. That, we know, is because we are living in a world of low temperature in a late universe, where many of the big symmetries are spontaneously broken. So another criterion for an ultimate symmetry is that it possesses a mechanism to spontaneously breaking itself at low temperatures, and hopefully that mechanism is unique as well.

If there is an ultimate symmetry with these characteristics, we are surely very far from understanding it. String theory or M theory may or may not be the answer, but unless their correctness, and the uniqueness of their symmetry-breaking mechanism can be demonstrated, it will not fully satisfy our criterion here as being the ultimate symmetry.

3. An argument for a supersymmetric standard model

Physics requires three fundamental constants, to tie down the three independent units we are measuring everything with. It is commonly accepted that $\hbar$ and c are two of these constants. They turn all units into powers of energy. The third constant which determines the natural energy scale is often taken to be the Newtonian gravitational constant G, which gives a mass scale $M_P \sim 10^{19}$ GeV.

Accepting M_P to be the natural mass unit, a good question to ask is why all the particles we know have such tiny masses compared to M_P. These particles all interact gravitationally. What keeps them from acquiring a fraction of M_P in their masses? In comparison to M_P all the observed masses can be taken to be zero, and I shall do so in the following. If you are fussy I will explicitly assume that all the finite masses we know derive themselves from spontaneously broken symmetry so they are zero originally, at high temperature.

There is a simple mechanism Nature seems to make use of to keep masses zero, a mechanism which I shall refer to as the 'spin trick'. A particle with spin s must have $2s + 1$ spin degrees of freedom. The only exception is when the particle travels with the speed of light, which means that it is massless. So if we can somehow concoct a theory in which less than $2s + 1$ degrees of freedom enters, then the particle must be massless. For spin-1 particles, local gauge invariance does just that. It prevents the particle from having a physical longitudinal polarization, thus necessarily it must be massless. Conversely, this may be taken as an argument for the emergence of gauge theories in the Standard Model.

For spin-$\frac{1}{2}$ fermions, the spin trick demands every fermion to have only one helicity or chirality for it to be massless. This is precisely what the $SU(2)_L \times U(1)_Y$ symmetry of the Standard Model does. It gives different quantum numbers to different chiralities of the "same" fermion, making them effectively two different particles, each having only one chirality. Conversely, we can use the massless property of fermions to argue for the presence of a symmetry group, such as $SU(2)_L \times U(1)_Y$, which gives different quantum numbers to different chiralities.

What about spin-0 particles? So far, no spinless fundamental particle has been found, though the Higgs may be looming just around the corner. Suppose the Higgs is found and it is elementary. It has no spin, so the spin trick would not work. What keeps its mass small? Well, supersymmetry might. Supersymmetry links it to a spin-$\frac{1}{2}$ partner, which may be massless because of the spin trick. This will force the spinless particle to be massless as well.

We have been dealing with continuous symmetries, but in applying the spin trick to spin-$\frac{1}{2}$ particles, it is important for parity to be violated, for otherwise there is no way for a particle to have just one helicity and not two. So parity violation is an integral part of the game plan as well.

This line of thinking is encouraging for the enterprise of an ultimate symmetry. The very simple requirement of masslessness ties together various pieces of physics we know to be true, but taken by themselves it is hard to see why these pieces must all appear on stage in the same act. These are local gauge symmetry, the parity-breaking electroweak symmetry, and possibly supersymmetry as well. Maybe there is after all a simple grand design where everything occurs for a reason, and not just by chance? Or is that too optimistic?

4. An argument for a string theory

The universe started with a big bang, preceeded by a short inflationary era. What happens *before* all that and where is the universe going to be put when it is created?

The first is an age old question. In the ancient form, it asks what happens before God created the universe. That question led St. Augustine of Hippo to an important theological study, published in the book "Confessions." He essentially concluded that God is eternal, but He created time when He created the universe we live in, so it does not make sense to ask what happens before time is created. He could have said that God created space to put the universe in as well, but I am not aware of this statement by him.

Suppose we accept the premise that space and time are created at the beginning. Can we construct a theory capable of describing such a creation? Presumably we need a quantum field theory to do it, because that is the only theory we know of which deals with the creation and annihilation of objects.

Ordinary field amplitudes are functions of the spacetime coordinates x^μ. To be able to create space and time, these coordinates must be promoted into dynamical fields X^μ, so we need a replacement of the independent variables which fields depend on. Denote these new variables as σ_a and τ, and call them the *artifical* spacetime coordinates. We know nothing about these parameters, not even how many σ_a's there are. However, in this case knowing nothing is knowing a lot, because we are saying that these parameters are artificial and cannot be observed. Such a theory must involve a large symmetry as we shall see.

We encounter a similar situation in general relativity, where physics in independent of the choice of spacetime coordinates x^μ. But there the curvature of spacetime is observable as the gravitational field. Nobody has seen artificial gravity, and probably never will, so even the curvature of the artificial spacetime cannot be an observable. How can that be achieved?

Let n be the number of artificial spatial coordinates σ_a. If $n = 0$, there is no internal curvature, but it is not a quantum field theory so we do not have creation and annihilation either. If $n = 1$, the curvature at every point in the two-dimensional artificial spacetime has a dimension, os it can be affected by a local scale transformation. To make it unobservable, we need the theory to be invariant under local scale transformation. A local scale transformation in two dimensions is a conformal transformation, so the theory must be diffeomorphic and conformally invariant. This is the string theory. If $n > 2$, the curvature at every point is a tensor. We do not know of any local invariance which can transform all the components away. Such a transformation may or may not exist, but if it does it is certainly not simple because nobody has succeeded in constructing one. Barring that as a possibility, we are led to the string theory as the unique solution of St. Augustine's premise. That is amazing, especially because string theory demands the (real) spacetime has to be ten dimensional, and the gauge groups, if present, to be SO(32) or $E_8 \times E_8$. It is also the only candidate for a consistent quantum gravity theory. We have no idea whether strng theory is real physics, but certainly it is beautiful and it goes a long way towards satisfying the criteria I set out for an ultimate symmetry.

If the argument presented above has any merit, it also suggests that we need a more intimate connection between cosmology and the formulation of string theory. I would like to see a string theory which can describe everything, including the spacetime, and not just the present version where spacetime is already there for the strings to live in. Whether this view is misguided or not remains to be seen.

We think now that the five viable superstring theories are manifestations of a larger 11-dimensional "M theory," but we know very little about what that theory is. We do not even know what symmetry M theory has, though I expect it to be much larger than the conformal symmetry of the strings. To my mind that is a very important question to answer.

5. Is gravity the key?

Can gravity provide the missing link to propel us into an ultimate symmetry? There is no way to know at this moment, but certainly strong quantum gravity is a fairly unexplored virgin land which might give us new insights and bring us new wealth.

For example, if gravity is taken into account, the maximum number of degrees of freedom N that can be squeezed into a spherical volume of radius R is proportional to its surface area R^2, but not its volume R^3. This strange fact comes about in the following way.

Let us put N particles of mass m into the spherical volume. When the total mass $M = Nm$ reaches $Rc^2/2G$, with G being the Newtonian gravitational constant, the volume turns into a black hole. This is the maximum mass the original volume can accommodate, because adding more mass will cause the black hole to grow beyond the confines of the original volume. This maximum M leads to a maximum N, because a particle in a relativistic theory has a natural spread of the order of its Compton wavelength $\hbar/mc$. This number must be less than R to have the particle stay inside the volume, hence $N = M/m$ is bounded by a number of the order of $(R/\ell_P)^2$, where $\ell_P = \sqrt{\hbar G/c^3} \sim 10^{-33}$ cm is the Planck length.

Can that strange fact be hinting at us that our physical theory in the volume can be described by an equivalent theory on the surface of the sphere, with about one degree of freedom per unit Planck area ℓ_P^2? This possibility first pointed out by 't Hooft is known as the holographic principle. Some examples of that type is known in string theory but its full implication is perhaps not yet understood. Is that an indication for the presence of a larger symmetry to make it possible?

There are other puzzles. Why is the cosmological constant in the universe so small? When a black hole evaporates by emitting Hawking radiation, does it recover all the information that seems to have been lost when matter collapsed into the black hole?

We do not know whether these strange facts help us or not in the quest for an ultimate symmetry, and of course we do not even know that an ultimate symmetry exists or not.

DEPARTMENT OF PHYSICS, MCGILL UNIVERSITY, 3600 UNIVERSITY ST., MONTRÉAL, QC H3A 2T8, CANADA

E-mail address: `lam@physics.mcgill.ca`

Formal Characters and Resolution of Infinite-Dimensional Simple A_r-Modules of Finite Degree

Daniel Britten and *Frank Lemire*

In memory of Bob Sharp

ABSTRACT. In this paper we construct resolutions of infinite-dimensional simple A_r-modules having bounded weight multiplicities in terms of Generalized Verma Modules induced parabolically from simple torsion free modules. These resolutions then provide formulas for the formal characters of these simple modules in terms of alternating sums of partition functions arising from the formal characters of Generalized Verma Modules.

1. Introduction

In 1975 Bernstein, Gelfand and Gelfand [2] established resolutions for finite-dimensional simple Lie modules $L(\nu)$ in terms of Verma modules. These resolutions yield formal character formulas for simple finite-dimensional Lie modules $L(\nu)$ in terms of formal characters of Verma modules and hence produce formulas for weight multiplicities of $L(\nu)$ in terms of partition functions describing the weight multiplicities in the Verma modules. There have been a number of generalizations of these results. Lepowsky [6] provided a sequence of resolutions of simple finite-dimensional Lie modules in terms of Generalized Verma modules. Futorny and Mazorchuk [5] have provided resolutions for simple alpha-stratified Lie modules.

In this paper we consider analogous resolutions and hence formal character formulas for certain simple infinite-dimensional Lie modules having bounded multiplicities. The least upper bound of the multiplicities is called the *degree* of the module. It follows from [1] that the only simple Lie algebras which admit infinite-dimensional simple modules of finite degree are the algebras of types A and C. In this paper we provide resolutions and formal character formulas for *nonintegral* A_r-modules of bounded multiplicities.

In Section 2 we briefly outline Lepowsky's generalization of the BGG resolutions for finite-dimensional simple Lie modules. The third section presents Mathieu's results classifying the simple A_r modules which have bounded weight multiplicities and provide an alternate realization of these modules in terms of Generalized Verma

2000 *Mathematics Subject Classification.* 17B10.
This is the final form of the paper.

modules parabolically induced from simple finite-dimensional modules. In Section 4 we extend Lepowsky's resolutions to infinite-dimensional simple A_r-modules in terms of Generalized Verma modules parabolically induced from simple torsion free modules.

2. Resolutions of finite-dimensional simple Lie modules

In this section we briefly review Lepowsky's paper [6] on resolutions of finite-dimensional simple Lie modules. Let $\mathcal{G}$ be a simple Lie algebra over the complex numbers; $\mathfrak{h}$ a Cartan subalgebra of $\mathcal{G}$; $\Delta \subset \mathfrak{h}^*$ the root system of $\mathcal{G}$ with respect to $\mathfrak{h}$; $\Delta_+ \subset \Delta$ a set of positive roots; $\mathfrak{n}$ the subalgebra $\sum_{\alpha \in \Delta_+} \mathcal{G}^\alpha$ of $\mathcal{G}$ where $\mathcal{G}^\alpha = \mathbb{C} X_\alpha$ denotes the α root space; $\mathfrak{B}$ the Borel subalgebra $\mathfrak{h} + \mathfrak{n}$; $\rho = \frac{1}{2} \sum_{\alpha \in \Delta_+} \alpha$; $U(\cdot)$ the universal enveloping algebra functor. A $\mathcal{G}$-module V is said to be a *weight module* provided $V = \sum \oplus V_\lambda$ where $V_\lambda = \{v \in V \mid hv = \lambda(h)v \ (\forall h \in \mathfrak{h})\}$.

For each $\lambda \in \mathfrak{h}^*$ let $\mathbb{C}_\lambda$ be the 1-dimensional $\mathfrak{B}$-module on which $\mathfrak{h}$ acts according to λ and $\mathfrak{n}$ acts trivially. The *Verma module* $M(\lambda)$ is defined to be the induced $\mathcal{G}$-module

$$M(\lambda) = U(\mathcal{G}) \otimes_{U(\mathfrak{B})} \mathbb{C}_\lambda.$$

Clearly each Verma module is a weight module and is cyclic generated by the vector $e_\lambda = 1 \otimes 1$.

For all $\alpha \in \Delta$, define $h_\alpha \in \mathfrak{h}$ by $h_\alpha \in [\mathcal{G}^\alpha, \mathcal{G}^{-\alpha}]$ and $\alpha(h_\alpha) = 2$. Let $r_\alpha \in \text{Aut } \mathfrak{h}^*$ be the Weyl reflection with respect to α where $r_\alpha(\lambda) = \lambda - \lambda(h_\alpha)\alpha$ for all $\lambda \in \mathfrak{h}^*$. The Weyl group W of $\mathcal{G}$ with respect to $\mathfrak{h}$ is the subgroup of $\text{Aut } \mathfrak{h}^*$ generated by $\{r_\alpha \mid \alpha \in \Delta_+\}$. Let $r = \dim \mathfrak{h}$ and $\Delta_{++} = \{\alpha_1, \ldots, \alpha_r\}$ be the simple roots of $\mathcal{G}$ with respect to $\mathfrak{h}$ and Δ_+. Then W is generated by the simple reflections $\{r_i = r_{\alpha_i} \mid i = 1, \ldots, r\}$. The length $\ell(w)$ of $w \in W$ is the number of elements in the set $\Delta_+ \cap w(-\Delta_+)$. The *dot action* of W on $\mathfrak{h}^*$ is defined by $w \cdot \mu = w(\mu + \rho) - \rho$. If $w, w' \in W$ we write $w \xrightarrow{\alpha} w'$ provided there exists a root $\alpha \in \Delta_+$ such that $w = r_\alpha \cdot w'$ and $\ell(w) = \ell(w') + 1$. We also write $w \leq w'$ provided there exist $w_1, \ldots, w_n \in W$ such that $w = w_1 \to w_2 \to \cdots \to w_n = w'$. According to [2] there exists a $\mathcal{G}$-module injection from $M(\lambda)$ into $M(\mu)$ if and only if there exists $w \in W$ with $w \leq \text{id}$ and $w \cdot \mu = \lambda$. This injection is necessarily unique up to a scalar multiple.

The concept of a Verma module has been generalized in the following manner. If P is a parabolic subalgebra of $\mathcal{G}$ which contains $\mathfrak{B}$ and V is any simple P-module then the $\mathcal{G}$-module formed by $U(\mathcal{G}) \otimes_{U(P)} V$ is called a *Generalized Verma module*, denoted GVM.

Let S be an arbitrary subset of $\{1, \ldots, r\}$; $\mathfrak{h}_S$ denotes the span of the h_{α_i} with $i \in S$; $\mathcal{G}_S$ the subalgebra of $\mathcal{G}$ generated by $\mathfrak{h}_S$ and $\mathcal{G}^{\pm \alpha_i}$ with $i \in S$; $\Delta^S = \Delta \cap \sum_{i \in S} \mathbb{Z}\alpha_i$ and $\Delta_+^S = \Delta^S \cap \Delta_+$. We also define the following subalgebras of $\mathcal{G}$: $\mathfrak{n}_S = \sum_{\alpha \in \Delta_+^S} \mathcal{G}^\alpha$; $\mathfrak{n}_S^- = \sum_{\alpha \in \Delta_+^S} \mathcal{G}^{-\alpha}$; $\mathfrak{u} = \sum_{\alpha \in \Delta_+ \setminus \Delta_+^S} \mathcal{G}^\alpha$; $\mathfrak{u}^- = \sum_{\alpha \in \Delta_+ \setminus \Delta_+^S} \mathcal{G}^{-\alpha}$; $\mathfrak{r} = \mathcal{G}_S + \mathfrak{h}$; and $\mathfrak{p}_S = \mathfrak{r} \oplus \mathfrak{u}$. Clearly $\mathcal{G}_S$ is a semisimple Lie algebra with Cartan subalgebra $\mathfrak{h}_S$ with root system $\{\alpha \downarrow \mathfrak{h}_S \mid \alpha \in \Delta^S\}$ and positive roots $\{\alpha \downarrow \mathfrak{h}_S \mid \alpha \in \Delta_+^S\}$. $\mathfrak{r}$ is a reductive subalgebra in $\mathcal{G}$ with $\mathcal{G}_S = [\mathfrak{r}, \mathfrak{r}]$ and the center of $\mathfrak{r}$ is a subalgebra of $\mathfrak{h}$. $\mathfrak{p}_S$ is the most general parabolic subalgebra of $\mathcal{G}$ containing $\mathcal{G}_S$ and the Borel subalgebra $\mathfrak{B}$; the reductive part of $\mathfrak{p}_S$ is $\mathfrak{r}$ and the nilpotent part is $\mathfrak{u}$.

Let $\Lambda_S = \{\lambda \in \mathfrak{h}^* \mid \lambda(h_{\alpha_i}) \in \mathbb{Z}_{\geq 0} \ (\forall i \in S)\}$. Then Λ_S indexes the set of finite-dimensional simple $\mathfrak{r}$-modules which remain simple as $\mathcal{G}_S$-modules. To each $\lambda \in \Lambda_S$ we associate the simple $\mathfrak{r}$-module $L(\lambda)$ having highest weight λ and define the GVM by setting
$$V^{L(\lambda)} = U(\mathcal{G}) \otimes_{U(\mathfrak{p}_S)} L(\lambda)$$
after the action of $\mathfrak{r}$ on $L(\lambda)$ is extended to $\mathfrak{p}_S$ by letting $\mathfrak{u}$ act trivially. In [**6**, Proposition 3.2] it is shown that
$$V^{L(\lambda)} \simeq M(\lambda)/K_S(\lambda)$$
where $K_S(\lambda) = \sum_{i \in S} M(r_i \cdot \lambda)$. Further if $\lambda \to \mu$ and $f \colon M(\lambda) \to M(\mu)$ is a nonzero $\mathfrak{p}_S$-module homomorphism we obtain a *standard* $\mathcal{G}$-module homomorphism $\hat{f} \colon V^{L(\lambda)} \to V^{L(\mu)}$ by composing f with the natural quotient map. This map $\hat{f}$ is zero if and only if $M(\lambda) \subset M(r_i \cdot \mu)$ for some $i \in S$.

Let W_S denote the subgroup of the Weyl group W generated by $\{r_i \mid i \in S\}$. Clearly W_S is the Weyl group of $\mathfrak{r}$. Let $W^S = \{w \in W \mid \Delta_+^S \subset w(\Delta_+)\}$. It is easily seen that every element $w \in W$ can be uniquely expressed as a product $w_1 w^1$ where $w_1 \in W_S$ and $w^1 \in W^S$ and further for any $w^1 \in W^S$ and $\lambda \in \Lambda_S$ we have $w^1 \cdot \lambda \in \Lambda_S$.

Fix a dominant integral weight $\mu \in \mathfrak{h}^*$ and for each $j = 0, \ldots, \dim \mathfrak{u}$ define
$$C_j^S = \bigoplus_{\substack{w \in W^S \\ \ell(w) = j}} V^{L(w \cdot \mu)}.$$

Construct the $\mathcal{G}$-module homomorphisms $d_j^S \colon C_j^S \to C_{j-1}^S$ for all $j = 1, \ldots, \dim \mathfrak{u}$ as follows. By [**6**, Proposition 3.7] if $w, w' \in W^S$ with $\ell(w) = \ell(w') + 1$ then there exists a nonzero $\mathcal{G}$-module homomorphism $V^{L(w \cdot \mu)} \to V^{L(w' \cdot \mu)}$ if and only if $w \to w'$ and in this case the standard map is nonzero. On the other hand, if $w \in W$, $w \notin W^S$, $w' \in W^S$ with $w \to w'$ then $w \xrightarrow{\alpha_i} w'$ for some $i \in S$ and hence if $w_1, w_2 \in W$, $w \in W \setminus W^S$ and $w_1 \leq w \to w_2$ then the standard map $V^{L(w_1 \cdot \mu)} \to V^{L(w_2 \cdot \mu)}$ is zero.

For all $w \in W$, fix an embedding $M(w \cdot \mu) \subset M(\mu)$. Then for all $w, w' \in W$ with $w \leq w'$, we have a fixed embedding $f_{w,w'} \colon M(w \cdot \mu) \to M(w' \cdot \mu)$. Whenever, $w_1, w_2, w_3 \in W$ satisfy the condition $w_1 \leq w_2 \leq w_3$, we have a commutative triangle of embeddings. If $w_1, w_2, w_3 \in W^S$, then the three associated standard maps among the GMVs $V^{L(w_i \cdot \mu)}$ also form a commutative triangle. In particular by [**6**, Lemma 4.1] we have that for each arrow $w_1 \to w_2$ we can assign a number $s(w_1, w_2) \in \{\pm 1\}$ such that for every quadruple $w_1, w_2, w_3, w_4 \in W$ with $w_1 \to w_2 \to w_4$ and $w_1 \to w_3 \to w_4$ the product of the numbers assigned to the four arrows is equal to -1. Finally for any $w, w' \in W^S$ such that $\ell(w) = \ell(w') + 1$ we define the $\mathcal{G}$-module homomorphism
$$h_{w,w'} \colon V^{L(w \cdot \mu)} \to V^{L(w' \cdot \mu)}$$
to be the standard map associated with the map $s(w, w') f_{w,w'}$ if $w \to w'$ and equal to zero otherwise. The array of maps $h_{w,w'}$ then provides $\mathcal{G}$-module homomorphisms $d_j^S \colon C_j^S \to C_{j-1}^S$ for $j = 1, \ldots, \dim \mathfrak{u}$. When these maps are combined with a surjective map $\varepsilon \colon C_0^S \to L(\mu)$ Lepowsky shows that we obtain an exact sequence
$$0 \to C_{\dim \mathfrak{u}}^S \xrightarrow{d_{\dim \mathfrak{u}}^S} \cdots \xrightarrow{d_1^S} C_0^S \xrightarrow{\varepsilon} L(\mu) \to 0.$$

3. Infinite-dimensional simple A_r-modules with bounded multiplicities

In this section we review the work of Mathieu on constructing the infinite-dimensional simple A_r-modules having bounded weight multiplicities and provide an alternate construction of these modules. Let $\mathcal{G} = \mathrm{sl}(r+1,\mathbb{C})$ and fix a Cartan subalgebra $\mathfrak{h}$ consisting of all diagonal elements in $\mathcal{G}$. If $\epsilon_i \in \mathfrak{h}^*$ denotes the projection onto the (i,i) entry, then the root system of $\mathcal{G}$ is given by $\Delta = \pm\{\epsilon_i - \epsilon_j \mid 1 \leq i < j \leq r+1\}$, $\Delta_+ = \{\epsilon_i - \epsilon_j \mid 1 \leq i < j \leq r+1\}$ is a positive system of roots and a base of simple roots is given by $\Delta_{++} = \{\alpha_i = \epsilon_i - \epsilon_{i+1} \mid i = 1, \ldots, r\}$. Let $\langle \cdot, \cdot \rangle \colon \mathfrak{h}^* \times \mathfrak{h}^* \to \mathbb{C}$ denote the inner product on $\mathfrak{h}^*$ determined by setting $\langle \epsilon_i, \epsilon_j \rangle = \delta_{ij}$. The fundamental weights $\{\omega_1, \ldots, \omega_r\}$ provide a basis of $\mathfrak{h}^*$ which is dual to Δ_{++} in the sense that $\langle \omega_i, \alpha_j \rangle = \delta_{ij}$. The Weyl group W of $\mathcal{G}$ can be identified with the permutations of $\{1, \ldots, r+1\}$ acting on the subscripts of the ϵ_i.

Fix an integer ℓ with $0 \leq \ell \leq r$ and denote by $\hat{\mathcal{G}}$ the regular subalgebra of $\mathrm{sl}(r+1,\mathbb{C})$ generated by $\{E_{ij}, E_{ji} \mid 1 \leq i < j \leq \ell+1\}$ where E_{ij} denote the standard matrix units in $\mathrm{sl}(r+1,\mathbb{C})$. Clearly $\hat{\mathcal{G}}$ is isomorphic to $\mathrm{sl}(\ell+1,\mathbb{C})$ Let $\hat{\mathfrak{h}}_\ell = \hat{\mathcal{G}} \cap \mathfrak{h}$; $\mathfrak{h}_\ell^\perp = \{h \in \mathfrak{h} \mid [h, \hat{\mathcal{G}}] = 0\}$; $\mathfrak{u}_\ell^+ = \sum_{j > \ell+1; 1 \leq i < j \leq r+1} \mathbb{C}E_{ij}$; $\mathfrak{u}_\ell^- = \sum_{i > \ell+1; 1 \leq j < i \leq r+1} \mathbb{C}E_{ji}$; $\Delta_\ell^\pm = $ the roots of $\mathfrak{u}_\ell^\pm$ and $P_\ell = \hat{\mathcal{G}} \oplus \mathfrak{h}_\ell^\perp \oplus \mathfrak{u}_\ell^+$. Let $Q = \sum_{i=1}^\ell \mathbb{Z}\alpha_i$ be the integral root lattice determined by the simple roots of $\hat{\mathcal{G}}$.

According to Mathieu's classification, a simple torsion free $\hat{\mathcal{G}}$-module, denoted $F(\lambda, \zeta)$, of finite degree is uniquely labeled by a pair of weights, $\lambda, \zeta \in \hat{\mathfrak{h}}_\ell^*$. The first weight λ is of the form $\lambda = a\omega_1 + n_2\omega_2 + \cdots + n_\ell\omega_\ell$ with $a \in \mathbb{C} \setminus \mathbb{Z}_{\geq 0}$ and $n_j \in \mathbb{Z}_{\geq 0}$ for $j = 2, \ldots, \ell$ and $F(\lambda, \zeta)$ has central character equal to the central character of the Verma module $M(\lambda)$. This central character is denoted by χ_λ. The second weight ζ (see [7, Section 5]) determines the weight lattice $\lambda + \zeta + Q$. The modules $F(\lambda, \zeta)$ are constructed from the simple $\hat{\mathcal{G}}$-module $L(\lambda)$ with highest weight λ and highest weight vector v^+ by the so-called *Mathieu's twisting functor* briefly described as follows. Note that $\Sigma_\ell = \{X_k = E_{k+1,1} \mid k = 1, \ldots, \ell\}$ is a set of pairwise commuting root elements of $\hat{\mathcal{G}}$. Denote by $U_{\Sigma_\ell}(\hat{\mathcal{G}})$ the localization of $U(\hat{\mathcal{G}})$ with respect to the Ore multiplicative subset generated by Σ_ℓ. Then the algebra $U_{\Sigma_\ell}(\hat{\mathcal{G}})$ has an n-parameter family of automorphisms $\Theta_{(t_1,\ldots,t_\ell)}$ such that $\Theta_{(t_1,\ldots,t_\ell)}(u) = X_1^{t_1} \cdots X_n^{t_\ell} u X_n^{-t_\ell} \cdots X_1^{-t_1}$ provided that all t_i's are integers. Since the map $(t_1, \ldots, t_\ell) \mapsto \Theta_{(t_1,\ldots,t_\ell)}(u)$ is polynomial in $(t_1, \ldots, t_\ell)$ for every $u \in U_{\Sigma_\ell}(\hat{\mathcal{G}})$, we may extend this class of automorphisms so that $\Theta_{(t_1,\ldots,t_\ell)}$ is defined for all $(t_1, \ldots, t_\ell) \in \mathbb{C}^\ell$. The weight ζ, in the label of $F(\lambda, \zeta)$, has the form $\zeta = \sum_{k=1}^\ell t_k(\alpha_1 + \cdots + \alpha_k) \in \hat{\mathfrak{h}}_\ell^*$ and must satisfy certain conditions set down by Mathieu. Now,

$$F(\lambda, \zeta) = [U_{\Sigma_\ell}(\hat{\mathcal{G}}) \otimes_{U(\hat{\mathcal{G}})} L(\lambda)]_\zeta,$$

where the action of $U(\hat{\mathcal{G}})$ on $F(\lambda, \zeta)$ is simply the action on $U_{\Sigma_\ell}(\hat{\mathcal{G}}) \otimes_{U(\hat{\mathcal{G}})} L(\lambda)$ twisted by the automorphism $\Theta_\zeta = \Theta_{(t_1,\ldots,t_\ell)}$.

Our subjects of study are the GVMs starting with the $\hat{\mathcal{G}}$-module $F(\lambda, \zeta)$. Fix any weight $\mu \in \mathfrak{h}^*$ such that $\mu \downarrow \hat{\mathfrak{h}}_\ell = \lambda$ and set $\mu \downarrow \mathfrak{h}_\ell^\perp = \tau$. $F(\lambda, \zeta)$ is a P_ℓ-module once we have defined

$$(x + h + z)v = xv + \tau(h)v$$

where $x \in \hat{\mathcal{G}}$, $h \in \mathfrak{h}_\ell^\perp$, $z \in \mathfrak{u}_\ell^+$ and $v \in F(\lambda, \zeta)$. The resultant P_ℓ-module is denoted $F(\lambda, \zeta)_\tau$, and our generalized Verma $\mathcal{G}$-module is

$$M(\ell, \mu, \zeta) = U(\mathcal{G}) \otimes_{U(P_\ell)} F(\lambda, \zeta)_\tau.$$

Clearly, $M(\ell, \mu, \zeta)$ is a cyclic $U(\mathcal{G})$-module generated by the element $[1 \otimes v^+]_\zeta$ and $M(\ell, \mu, \zeta)$ is a free $U(\mathfrak{u}_\ell^-)$-module with a free basis given by any basis of $F(\lambda, \zeta)$. It follows that the weight lattice of this module consists of all weights of the form $\nu = \mu + \zeta + \sum_{i=1}^\ell k_i \alpha_i - \sum_{i=\ell+1}^r k_i \alpha_i$ where $k_1, \ldots, k_\ell \in \mathbb{Z}$ and $k_{\ell+1}, \ldots, k_r \in \mathbb{Z}_{\geq 0}$. In particular, the multiplicity of such a weight ν in $M(\ell, \mu, \zeta)$ is equal to the degree of the module $F(\lambda, \zeta)$ times the number of ways that a sum of roots from Δ_ℓ^+ minus $\sum_{i=\ell+1}^r k_i \alpha_i$ is in $\sum_{i=1}^\ell \mathbb{Z} \alpha_i$. The latter number is evaluated by a modified partition function.

From the basic results on generalized Verma modules (see, e.g., [**8**]), every generalized Verma module admits a unique maximal submodule and hence a unique simple quotient module. Let $L(\ell, \mu, \zeta)$ denote the unique simple quotient of the module $M(\ell, \mu, \zeta)$.

The main goal of this paper is to establish a resolution for the modules $L(\ell, \mu, \zeta)$ in terms of generalized Verma modules when $\mu = a\omega_1 + n_2\omega_2 + \cdots + n_r\omega_r \in \mathfrak{h}^*$ where $a \in \mathbb{C} \setminus \mathbb{Z}_{\geq 0}$ and the n_i are nonnegative integers for $i = 2, \ldots, r$. These constitute all simple infinite-dimensional modules of bounded multiplicities having a nonintegral weight. As a first step in this direction we provide an alternate realization of the modules $L(\ell, \mu, \zeta)$ and $M(\ell, \mu, \zeta)$ connecting these modules to the generalized Verma modules considered by Lepowsky.

Let $\bar{\mathcal{G}}$ be the regular simple subalgebra of $\mathcal{G}$ having root system $\{\alpha_2, \ldots, \alpha_r\}$. Clearly $\bar{\mathcal{G}} \simeq A_{r-1}$. If $a \in \mathbb{C} \setminus \mathbb{Z}$, the multiplicities of the simple module $L(\ell, \mu, \zeta)$ are bounded by the dimension of the simple $\bar{\mathcal{G}}$-module having highest weight $\bar{\mu} = \sum_{i=2}^r n_i \omega_i$.

Let $\bar{\Delta} \subset \Delta$ be the root system of $\bar{\mathcal{G}}$; $\bar{\Delta}_+$ the positive root system; and $\bar{\Delta}_+^\ell$ the positive roots generated by $\{\alpha_2, \ldots, \alpha_\ell\}$; $\bar{\mathfrak{n}}^\pm = \bar{\mathcal{G}} \cap \mathfrak{n}^\pm$. Following Lepowsky, we can then introduce the usual subalgebras of $\bar{\mathcal{G}}$ where, in this case, the set $S = \{2, \ldots, \ell\} \subset \{2, \ldots, r\}$. In particular, we set $\bar{\mathfrak{u}} = \sum_{\alpha \in \bar{\Delta}_+ \setminus \bar{\Delta}_+^\ell} \bar{\mathcal{G}}^\alpha$. Denote $\bar{\mathfrak{h}} = \mathfrak{h} \cap \bar{\mathcal{G}}$; $\bar{\mathfrak{h}}^\perp = \{h \in \mathfrak{h} \mid \alpha_i(h) = 0 \text{ for } i = 2, \ldots, r\}$ and $\bar{P} = \bar{\mathcal{G}} \oplus \bar{\mathfrak{h}}^\perp \oplus \bar{\mathfrak{u}}$. Set $\bar{\mu} = \mu \downarrow \bar{\mathfrak{h}}$ and $\theta = \mu \downarrow \bar{\mathfrak{h}}^\perp$—i.e., $\mu = \bar{\mu} \oplus \theta$. Define $U_{\Sigma_\ell}(\mathcal{G})$ to be the localization of $U(\mathcal{G})$ with respect to the multiplicatively closed set generated by Σ_ℓ and fix $\zeta \in \mathfrak{h}_\ell^*$. The automorphism Θ_ζ on $U_{\Sigma_\ell}(\mathcal{G})$ can then be extended to an automorphism of $U_{\Sigma_\ell}(\bar{\mathcal{G}})$ using the same definition.

For any $\bar{\mathcal{G}}$ module V we can consider V as a $\bar{P}$-module where each element of $\bar{\mathfrak{u}}$ acts trivially and each element $h \in \bar{\mathfrak{h}}^\perp$ acts as scalar multiplication by the scalar $\theta(h)$. The resultant $\bar{P}$-module is denoted V_θ. By parabolic induction we construct a $U_{\Sigma_\ell}(\mathcal{G})$-module $U_{\Sigma_\ell}(\mathcal{G}) \otimes_{U(\bar{P})} V_\theta$, and finally we construct an associated $U(\mathcal{G})$-module $[U_{\Sigma_\ell}(\mathcal{G}) \otimes_{U(\bar{P})} V_\theta]_\zeta$ where the action is twisted by the automorphism Θ_ζ—i.e., for any $u \in U(\mathcal{G})$, $u' \in U_{\Sigma_\ell}(\mathcal{G})$, and any $v \in V$ we have $u[u' \otimes v]_\zeta = [\Theta_\zeta(u)u' \otimes v]_\zeta$.

We now restrict attention to the case where the weight $\mu \in \mathfrak{h}^*$ satisfies $\langle \mu, \alpha_1 \rangle = a$ is not an integer and $\langle \mu, \alpha_i \rangle \in \mathbb{Z}_{\geq 0}$ for $i = 2, \ldots, \ell$. In this case we have

THEOREM 3.1. *Let $\bar{v}^+$ be a highest weight vector of $M(\bar{\mu})_\theta$. Using the notation above we have*

$$M(\ell, \mu, \zeta) \simeq [U_{\Sigma_\ell}(\mathcal{G}) \otimes_{U(\bar{P})} (M(\bar{\mu})/K_\ell(\bar{\mu}))_\theta]_\zeta$$

and

$$L(\ell, \mu, \zeta) \simeq [U_{\Sigma_\ell}(\mathcal{G}) \otimes_{U(\bar{P})} L(\bar{\mu})_\theta]_\zeta$$

where $K_\ell(\bar{\mu}) = \sum_{i=2}^\ell M(r_i \cdot \bar{\mu}) = \sum_{i=2}^\ell U(\bar{\mathfrak{n}}^-) X_{\alpha_i}^{\langle r_i \cdot \bar{\mu}+\rho, \alpha_i \rangle} \bar{v}^+$ and $L(\bar{\mu})$ is the simple $U(\bar{\mathcal{G}})$-module having highest weight $\bar{\mu}$.

PROOF. Since $\left(U_{\Sigma_\ell}(\hat{\mathcal{G}}) \otimes_{U(\hat{\mathcal{G}})} M(\lambda)\right)_\tau \simeq U_{\Sigma_\ell}(P_\ell) \otimes_{U(P_\ell)} M(\lambda)_\tau$ as $U(P_\ell)$-modules, and $U(\mathcal{G}) \otimes_{U(P_\ell)} U_{\Sigma_\ell}(P_\ell) \simeq U_{\Sigma_\ell}(\mathcal{G})$ as $U(\mathcal{G})$-modules, we have the following $U(\mathcal{G})$-modules isomorphisms.

$$\begin{aligned} U(\mathcal{G}) \otimes_{U(P_\ell)} \left(U_{\Sigma_\ell}(\hat{\mathcal{G}}) \otimes_{U(\hat{\mathcal{G}})} M(\lambda)\right)_\tau &\simeq U(\mathcal{G}) \otimes_{U(P_\ell)} U_{\Sigma_\ell}(P_\ell) \otimes_{U(\hat{\mathcal{G}})} M(\lambda)_\tau \\ &\simeq U_{\Sigma_\ell}(\mathcal{G}) \otimes_{U(P_\ell)} M(\lambda)_\tau \\ &\simeq U_{\Sigma_\ell}(\mathcal{G}) \otimes_{U(\mathcal{G})} U(\mathcal{G}) \otimes_{U(P_\ell)} M(\lambda)_\tau \\ &\simeq U_{\Sigma_\ell}(\mathcal{G}) \otimes_{U(\mathcal{G})} M(\mu) \\ &\simeq U_{\Sigma_\ell}(\mathcal{G}) \otimes_{U(\mathcal{G})} U(\mathcal{G}) \otimes_{U(\bar{P})} M(\bar{\mu})_\theta \\ &\simeq U_{\Sigma_\ell}(\mathcal{G}) \otimes_{U(\bar{P})} M(\bar{\mu})_\theta. \end{aligned}$$

Let $\hat{v}^+$ be a highest weight vector of the $U(P_\ell)$-module $M(\lambda)_\tau$ and $\bar{v}^+$ be a highest weight vector of the $U(\bar{P})$-module $M(\bar{\mu})_\theta$. If $\mathbf{B}^-$ is a fixed PBW basis for $U(\bar{\mathfrak{u}}^-)$ and $\mathbf{B}_{\Sigma_\ell}$ is a fixed PBW basis for $U(\sum_{i=1}^\ell \mathbb{C} X_i)$. Note that $\sum_{i=1}^\ell \mathbb{C} X_i$ is an abelian subalgebra of $\mathcal{G}$. A basis for $U(\mathcal{G}) \otimes_{U(P_\ell)} \left(U_{\Sigma_\ell}(\hat{\mathcal{G}}) \otimes_{U(\hat{\mathcal{G}})} M(\lambda)\right)_\tau$ is given by

$$\{u_1 \otimes u_2^{-1} \otimes \hat{v}^+ \mid u_1 \in \mathbf{B}^- \text{ and } u_2 \in \mathbf{B}_{\Sigma_\ell} \text{ with } u_1 u_2^{-1} \text{ being reduced}\}$$

and a basis for $U_{\Sigma_\ell}(\mathcal{G}) \otimes_{U(\bar{P})} M(\bar{\mu})_\theta$ is given by

$$\{u_1 u_2^{-1} \otimes \bar{v}^+ \mid u_1 \in \mathbf{B}^- \text{ and } u_2 \in \mathbf{B}_{\Sigma_\ell} \text{ with } u_1 u_2^{-1} \text{ being reduced}\}$$

where the phrase "$u_1 u_2^{-1}$ being reduced" simply means that for each $X_i \in \Sigma_\ell$ we do not have both X_i and X_i^{-1} occurring in the monomial $u_1 u_2^{-1}$. An isomorphism between these two $U(\mathcal{G})$–modules is determined by mapping $u_1 \otimes u_2^{-1} \otimes \hat{v}^+$ to $u_1 u_2^{-1} \otimes \bar{v}^+$. This isomorphism determines a natural epimorphism

$$\psi \colon U(\mathcal{G}) \otimes_{U(P_\ell)} \left(U_{\Sigma_\ell}(\hat{\mathcal{G}}) \otimes_{U(\hat{\mathcal{G}})} M(\lambda)\right)_\tau \to U_{\Sigma_\ell}(\mathcal{G}) \otimes_{U(\bar{P})} \left(M(\bar{\mu})/K_\ell(\bar{\mu})\right)_\theta$$

and the kernel of ψ is $\mathcal{K} = U(\mathcal{G}) \otimes_{U(P_\ell)} \left(U_{\Sigma_\ell}(\hat{\mathcal{G}}) \otimes_{U(\hat{\mathcal{G}})} K_\ell(\lambda)\right)_\tau$. Hence

$$\begin{aligned} U_{\Sigma_\ell}(\mathcal{G}) \otimes_{U(\bar{P})} \left(M(\bar{\mu})/K_\ell(\bar{\mu})\right)_\theta &\simeq [U(\mathcal{G}) \otimes_{U(P_\ell)} \left(U_{\Sigma_\ell}(\hat{\mathcal{G}}) \otimes_{U(\hat{\mathcal{G}})} M(\lambda)\right)_\tau]/\mathcal{K} \\ &\simeq U(\mathcal{G}) \otimes_{U(P_\ell)} \left(U_{\Sigma_\ell}(\hat{\mathcal{G}}) \otimes_{U(\hat{\mathcal{G}})} L(\lambda)\right)_\tau. \end{aligned}$$

Therefore, by [3, Theorem 2], we have

$$U(\mathcal{G}) \otimes_{U(P_\ell)} [\left(U_{\Sigma_\ell}(\hat{\mathcal{G}}) \otimes_{U(\hat{\mathcal{G}})} L(\lambda)\right)_\tau]_\zeta \simeq [U(\mathcal{G}) \otimes_{U(P_\ell)} \left(U_{\Sigma_\ell}(\hat{\mathcal{G}}) \otimes_{U(\hat{\mathcal{G}})} L(\lambda)\right)_\tau]_\zeta$$

and hence

$$\begin{aligned} M(\ell, \mu, \zeta) &\simeq U(\mathcal{G}) \otimes_{U(P_\ell)} [\left(U_{\Sigma_\ell}(\hat{\mathcal{G}}) \otimes_{U(\hat{\mathcal{G}})} L(\lambda)\right)_\tau]_\zeta \\ &\simeq [U(\mathcal{G}) \otimes_{U(P_\ell)} \left(U_{\Sigma_\ell}(\hat{\mathcal{G}}) \otimes_{U(\hat{\mathcal{G}})} L(\lambda)\right)_\tau]_\zeta \\ &\simeq [U_{\Sigma_\ell}(\mathcal{G}) \otimes_{U(\bar{P})} \left(M(\bar{\mu})/K_\ell(\bar{\mu})\right)_\theta]_\zeta. \end{aligned}$$

The second isomorphism of the theorem follows immediately from the first isomorphism and the submodule lattice epimorphism established in [3]. $\square$

4. Resolutions of $L(\ell, \mu, \zeta)$

We assume throughout this section that $\mu = a\omega_1 + n_2\omega_2 + \cdots + n_r\omega_r$ where $a \in \mathbb{C} \setminus \mathbb{Z}$ and $n_i \in \mathbb{Z}_{\geq 0}$ for $i = 2, \ldots, r$. We now prove the main theorem of this paper. The idea is to "lift" the Lepowsky resolutions of finite-dimensional simple Lie modules to resolutions of simple A_r-modules of bounded multiplicities using the alternate realizations of the modules $L(\ell, \mu, \zeta)$ and $M(\ell, \mu, \zeta)$ provided in the previous section.

Let $\overline{W}$ be the subgroup of the Weyl group W of $\mathcal{G}$ generated by $\{r_{\alpha_i} \mid i = 2, \ldots, r\}$. Clearly $\overline{W}$ is equivalent to the Weyl group of $\overline{\mathcal{G}}$. Set $\overline{W}_\ell$ to be the subgroup of $\overline{W}$ generated by $\{r_{\alpha_i} \mid i = 2, \ldots, \ell\}$ and define

$$\overline{W}^\ell = \{w \in \overline{W} \mid \overline{\Delta}_+^\ell \subset w(\overline{\Delta}_+)\}.$$

This corresponds to the decomposition of $\overline{W}$ given by Lepowsky where the set $S = \{2, \ldots, \ell\}$.

LEMMA 4.1. *For any $w \in \overline{W}^\ell$ we have*

$$M(\ell, w \cdot \mu, \zeta) \simeq [U_{\Sigma_\ell}(\mathcal{G}) \otimes_{U(\bar{P})} (M(w \cdot \bar{\mu})/K_\ell(w \cdot \bar{\mu}))_\theta]_\zeta.$$

PROOF. It is easily checked that for any $w \in \overline{W}^\ell$ we have $w \cdot \mu = w \cdot \bar{\mu} \oplus \theta$ where $<w \cdot \mu, \alpha_1> \in \mathbb{C} \setminus \mathbb{Z}$ and $<w \cdot \mu, \alpha_i> \in \mathbb{Z}_{\geq 0}$ for $i = 2, \ldots, \ell$. Hence the proof of the first isomorphism of Theorem 3.1 also implies this isomorphism. □

For each $j = 0, \ldots, \dim \bar{\mathfrak{u}}$ we define the $\mathcal{G}$-modules

$$D_j^\ell = \bigoplus_{\substack{w \in \overline{W}^\ell \\ \ell(w) = j}} M(\ell, w \cdot \mu, \zeta).$$

We first note that these modules are closely related to the generalized Verma modules used in Lepowsky's resolution of simple finite-dimensional $\overline{\mathcal{G}}$-modules. In fact, by Theorem 3.1 and Lemma 4.1 we have

$$D_j^\ell = \bigoplus_{\substack{w \in \overline{W}^\ell \\ \ell(w) = j}} \left[U_{\Sigma_\ell}(\mathcal{G}) \otimes_{U(\bar{P})} (M(w \cdot \bar{\mu})/K_\ell(w \cdot \bar{\mu}))_\theta\right]_\zeta$$

$$= \left[U_{\Sigma_\ell}(\mathcal{G}) \otimes_{U(\bar{P})} \bigoplus_{\substack{w \in \overline{W}^\ell \\ \ell(w) = j}} (M(w \cdot \bar{\mu})/K_\ell(w \cdot \bar{\mu}))_\theta\right]_\zeta$$

$$= [U_{\Sigma_\ell}(\mathcal{G}) \otimes_{U(\bar{P})} (C_j^\ell)_\theta]_\zeta$$

where C_j^ℓ is the $\overline{\mathcal{G}}$-module C_j^S defined by Lepowsky with $S = \{2, \ldots, \ell\}$.

Recall that in Section 2 for each $w, w' \in \overline{W}^\ell$ with $\ell(w) = \ell(w') + 1$ and $w \to w'$ there exists a nonzero $\overline{\mathcal{G}}$-module homomorphism

$$h_{w,w'} : \bigl(M(w \cdot \mu)/K(w \cdot \mu)\bigr) \to \bigl(M(w' \cdot \mu)/K_\ell(w' \cdot \mu)\bigr).$$

Since the module $[U_{\Sigma_\ell}(\mathcal{G}) \otimes_{U(\bar{P})} (M(w \cdot \mu)/K_\ell(w \cdot \mu))_\theta]_\zeta$ is a cyclic $\mathcal{G}$-module which is generated by the element $[1 \otimes \bar{v}]_\zeta$ where $\bar{v}$ is a generator of the module $M(w \cdot \mu)/K_\ell(w \cdot \mu)$ we can extend $h_{w,w'}$ to a $U_{\Sigma_\ell}(\mathcal{G})$-module homomorphism

$k_{w,w'}$ from the module $\left[U_{\Sigma_\ell}(\mathcal{G}) \otimes_{U(\bar{P})} \left(M(w \cdot \bar\mu)/K_\ell(w \cdot \bar\mu)\right)_\theta\right]_\zeta$ to the module $\left[U_{\Sigma_\ell}(\mathcal{G}) \otimes_{U(\bar{P})} \left(M(w' \cdot \bar\mu)/K_\ell(w' \cdot \bar\mu)\right)_\theta\right]_\zeta$ by setting

$$k_{w,w'}([1 \otimes \bar v]_\zeta) = [1 \otimes h_{w,w'}(\bar v)]_\zeta.$$

This array of homomorphisms can then be used to determine $\mathcal{G}$-module homomorphisms for $j = 1 \ldots, \dim \bar u$

$$f_j^\ell : D_j^\ell \to D_{j-1}^\ell.$$

Since $L(\ell,\mu,\zeta)$ is the unique simple quotient module of $M(\ell,\mu,\zeta)$ there exists a surjective $\mathcal{G}$-module homomorphism $\varepsilon : D_0^\ell \to L(\ell,\mu,\zeta)$. We are now in position to state the main theorem.

THEOREM 4.1. *With the notation above we have that*

$$0 \to D_{\dim \bar u}^\ell \xrightarrow{f_{\dim \bar u}^\ell} \cdots \xrightarrow{f_1^\ell} D_0^\ell \xrightarrow{\varepsilon} L(\ell,\mu,\zeta) \to 0$$

is an exact resolution of the module $L(\ell,\mu,\zeta)$.

PROOF. We first observe that the image of the map f_j^ℓ is determined by the image of the map d_j^ℓ, in fact, we claim

$$\mathrm{Im}(f_j^\ell) = \left[U_\Sigma(\mathcal{G}) \otimes_{U(\bar P)} \left(\mathrm{Im}(d_j^\ell)\right)_\theta\right]_\zeta.$$

It suffices to observe that if $\bar v$ is a generator of the $U(\bar{\mathcal{G}})$-module $M(w \cdot \mu)/K(w \cdot \mu)$ and for any elements $u \in U_{\Sigma_\ell}(\mathcal{G})$ $u' \in U(\bar{\mathcal{G}})$ we have

$$\begin{aligned}k_{w,w'}([u \otimes u'\bar v]_\zeta) &= k_{w,w'}(\Theta_\zeta^{-1}(uu')[1 \otimes \bar v]_\zeta) \\ &= \Theta_\zeta^{-1}(uu')k_{w,w'}([1 \otimes \bar v]_\zeta) \\ &= \Theta_\zeta^{-1}(uu')([1 \otimes h_{w,w'}(\bar v)]_\zeta) \\ &= [u \otimes h_{w,w'}(u'\bar v)]_\zeta\end{aligned}$$

It follows that $\mathrm{Im}\, k_{w,w'}$ is equal to $[U_{\Sigma_\ell}(\mathcal{G}) \otimes_{U(\bar P)} (Im(h_{w,w'}))_\theta]_\zeta$. Since the maps f_j^ℓ are defined by the array of maps $k_{w,w'}$ we conclude that

$$\mathrm{Im}(f_j^\ell) = \left[U_\Sigma(\mathcal{G}) \otimes_{U(\bar P)} \left(\mathrm{Im}(d_j^\ell)\right)_\theta\right]_\zeta.$$

Since we know that $\mathrm{Im}\, d_{j+1}^\ell$ is equal to $\ker d_j^\ell$, in order to prove that the above sequence is exact at D_j^ℓ it would suffice to show that

$$\ker f_j^\ell = [U_{\Sigma_\ell}(\mathcal{G}) \otimes_{U(\bar P)} (\ker d_j^\ell)_\theta]_\zeta.$$

It is clear from the definition of f_j^ℓ that

$$\ker f_j^\ell \supset [U_{\Sigma_\ell}(\mathcal{G}) \otimes_{U(\bar P)} (\ker d_j^\ell)_\theta]_\zeta.$$

Then we have

$$\begin{aligned}D_j^\ell/\ker f_j^\ell \simeq \mathrm{Im}\, f_j^\ell &\simeq [U_{\Sigma_\ell}(\mathcal{G}) \otimes_{U(\bar P)} (\mathrm{Im}(d_j^\ell))_\theta]_\zeta \\ &\simeq [U_{\Sigma_\ell}(\mathcal{G}) \otimes_{U(\bar P)} (C_j^\ell/\ker d_j^\ell)_\theta]_\zeta \\ &\simeq [U_{\Sigma_\ell}(\mathcal{G}) \otimes_{U(\bar P)} (C_j^\ell)_\theta]_\zeta/[U_{\Sigma_\ell}(\mathcal{G}) \otimes_{U(\bar P)} (\ker d_j^\ell)_\theta]_\zeta \\ &\simeq D_j^\ell/[U_{\Sigma_\ell}(\mathcal{G}) \otimes_{U(\bar P)} (\ker d_j^\ell)_\theta]_\zeta\end{aligned}$$

We conclude that

$$\ker f_{j-1}^\ell = [U_{\Sigma_\ell}(\mathcal{G}) \otimes_{U(\bar P)} (\ker d_{j-1}^\ell)_\theta]_\zeta$$

which completes the proof. □

If we denote by char V the formal character of a weight module V then an immediate corollary of Theorem 4.1 is

COROLLARY 4.1. char $L(\ell, \mu, \zeta) = \sum_{w \in \overline{W}^\ell} (-1)^{\ell(w)}$ char $M(\ell, w \cdot \mu, \zeta))$.

Recall that the character formula of the generalized Verma Modules was described in section 3 in terms of partition functions.

REMARK 1. In general the resolution given in Theorem 4.1 does not apply to the cases of integral weights μ. Consider, for example, the case where $r = 2, \ell = 1$ and $\mu = -2\omega_1 + \omega_2$. In this case Mathieu's results tell us that the multiplicities of all weight spaces in any simple torsion free module $L(1, -2\omega_1 + \omega_2, \zeta)$ are equal to one. Note in this case ζ is any nonintegral multiple of α_1. Following the notation above we have that $\bar{u} = \mathbb{C}E_{23}$ and $\overline{W}^1 = \overline{W} = \{\mathrm{id}, r_{\alpha_2}\}$ which implies that the modules D_0^1 and D_1^1 are given by $M(1, -2\omega_1 + \omega_2, \zeta)$ and $M(1, r_{\alpha_2} \cdot (-2\omega_1 + \omega_2), \zeta)$ respectively. If the resolution given in Theorem 4.1 were exact in this case we should have

$$D_0^1/D_1^1 \simeq L(1, -2\omega_1 + \omega_2, \zeta).$$

By direct computation, however, we can show that the $-2\omega_1 + \omega_2 + \zeta - \alpha_2$ weight space of this quotient is two-dimensional.

References

1. G. Benkart, D. Britten and F. Lemire, *Modules with bounded weight multiplicities for simple Lie algebras*, Math. Z. **225** (1997), 333–353.
2. I. N. Bernšteĭn, I. M. Gel'fand and S. I. Gel'fand, *Differential operators on the base affine space and a study of $\mathcal{G}$-modules*, Lie Groups and Their Representations (Budapest, 1971), Halstead Press, New York, 1975, pp. 21-64.
3. D. Britten, V. Futorny and F. Lemire, *Submodule lattice of generalized Verma modules*, J. Algebra (to appear).
4. S. Fernando, *Lie algebra modules with finite-dimensional weight spaces. I*, Trans. Amer. Math. Soc. **322** (1990), 757–781.
5. V. Futorny and V. Mazorchuk, *BGG-resolution for α-stratified modules over simply-laced finite-dimensional Lie algebras*, J. Math. Kyoto Univ. **38** (1998), 229–240.
6. J. Lepowsky, *A generalization of the Bernstein–Gelfand–Gelfand resolution*, J. Algebra *49* (1977), 496-511.
7. O. Mathieu, *Classification of irreducible weight modules*, Ann. Inst. Fourier (Grenoble) **50** (2000), 537–592.
8. V. Mazorchuk, *The structure of generalized Verma modules*, Ph.D. thesis, National Taras Shevchenko University of Kyiv, 1996.

DEPARTMENT OF MATHEMATICS, UNIVERSITY OF WINDSOR, WINDSOR, ON N9B 3P4, CANADA
E-mail address, D. Britten: britten@uwindsor.ca
E-mail address, F. Lemire: lemire@uwindsor.ca

Fusion Rules and the Patera–Sharp Generating-Function Method

L. Bégin, C. Cummins, *P. Mathieu*, and M. A. Walton

ABSTRACT. We review some contributions on fusion rules that were inspired by the work of Sharp, in particular, the generating-function method for tensor-product coefficients that he developed with Patera. We also review the Kac–Walton formula, the concepts of threshold level, fusion elementary couplings, fusion generating functions and fusion bases. We try to keep the presentation elementary and exemplify each concept with the simple $\widehat{su}(2)_k$ case.

1. Introduction

The Patera–Sharp generating function for su(2) tensor products reads [23, 24]:

$$(1.1) \qquad G^{\text{su}(2)}(L, M, N) = \frac{1}{(1 - LM)(1 - LN)(1 - MN)},$$

where the multiplicity of the representation (n) in the tensor product $(\ell) \otimes (m)$ is the coefficient of $L^\ell M^m N^n$ in the series expansion of (1.1). The derivation of this expression is based on manipulations of the character generating functions [8, 10, 15, 16, 18, 22, 25].

Key concepts on fusion rules have been obtained by looking for the affine-fusion extension of this simple-looking expression and its simplest higher-rank relatives. These are: the threshold level, fusion elementary couplings and fusion bases. Before reviewing these results, we briefly discuss the Kac–Walton formula. This last result can also be linked, albeit loosely, to Bob Sharp. Indeed, it is an affine extension of the Racah–Speiser algorithm for computing tensor-product coefficients, one of Sharp's favorite techniques. He presented it in his course on group theory, where two of the authors (PM and MW) learned the fundamentals of this subject.

2. Fusion rules: the set up

Fusion rules give the number of independent couplings between three given primary fields in conformal field theories. We are interested in those conformal field

---

2000 *Mathematics Subject Classification.* Primary 81R10, 81T40, 17B65; Secondary 05A15, 17B67, 17B10.

This work was supported in part by NSERC.

This is the final form of the paper.

theories having a Lie algebra symmetry. These are the Wess–Zumino–Witten models [17, 21], whose spectrum-generating algebra is an affine Lie algebra at integer level. Their primary fields are in 1–1 correspondence with the integrable representations of the appropriate affine Lie algebra at level k. Denote this set by $P_+^{(k)}$ and a primary field by the corresponding affine weight $\hat{\lambda}$. Fusion coefficients $\mathcal{N}_{\hat{\lambda}\hat{\mu}}^{(k)\,\hat{\nu}}$ are defined by the fusion product

$$(2.1) \qquad \hat{\lambda} \times \hat{\mu} = \sum_{\hat{\nu} \in P_+^{(k)}} \mathcal{N}_{\hat{\lambda}\hat{\mu}}^{(k)\,\hat{\nu}} \hat{\nu}.$$

To simplify the presentation, we consider only the algebra $\widehat{\mathrm{su}}(N)$.

An affine weight may be written as

$$(2.2) \qquad \hat{\lambda} = \sum_{i=0}^{N-1} \lambda_i \widehat{\omega}_i = [\lambda_0, \lambda_1, \ldots, \lambda_{N-1}]$$

where $\widehat{\omega}_i$ denote the fundamental weights of $\widehat{\mathrm{su}}(N)$. If the Dynkin labels λ_i are nonnegative integers, the weight $\hat{\lambda}$ is the highest weight of an integrable representation of $\widehat{\mathrm{su}}(N)$ at level k, with k defined by $k = \sum_{i=0}^{N-1} \lambda_i$. To the affine weight $\hat{\lambda}$, we associate a finite weight λ of the finite algebra $\mathrm{su}(N)$:

$$(2.3) \qquad \lambda = \sum_{i=1}^{N-1} \lambda_i \omega_i = (\lambda_1, \ldots, \lambda_{N-1})$$

where ω_i are the fundamental weights of $\mathrm{su}(N)$. Thus $\hat{\lambda}$ is uniquely fixed by λ and k.

3. Fusion rules and tensor products: the Kac–Walton formula

The fusion coefficient $\mathcal{N}_{\hat{\lambda}\hat{\mu}}^{(k)\,\hat{\nu}}$ is fixed to a large extent by the tensor-product coefficient pertaining to the product of the corresponding finite representations. We denote by $\mathcal{N}_{\lambda\mu}^{\,\nu}$ the multiplicity of the representation ν in the tensor product $\lambda \otimes \mu$:

$$(3.1) \qquad \lambda \otimes \mu = \sum_{\nu \in P_+} \mathcal{N}_{\lambda\mu}^{\,\nu} \nu$$

By abuse of notation, we use the same symbol for the highest weight and the highest-weight representation. P_+ represents the set of integrable finite weights. The precise relation between tensor-product and fusion-rule coefficients is given by the Kac–Walton formula [13, 14, 19, 32, 33]:

$$(3.2) \qquad \mathcal{N}_{\hat{\lambda}\hat{\mu}}^{(k)\,\hat{\nu}} = \sum_{\substack{\xi \in P_+;\, w \in \widehat{W} \\ w \cdot \xi = \hat{\nu} \in P_+^{(k)}}} \mathcal{N}_{\lambda\mu}^{\,\xi} \epsilon(w)$$

w is an element of the affine Weyl group $\widehat{W}$, of sign $\epsilon(w)$, and the dot indicates the shifted action: $w \cdot \hat{\lambda} = w(\hat{\lambda} + \hat{\rho}) - \hat{\rho}$, where $\hat{\rho}$ stands for the affine Weyl vector $\hat{\rho} = \sum_{i=0}^{N-1} \widehat{\omega}_i$.

The Kac–Walton formula can be transcribed into a simple algorithm: one first calculates the tensor product of the corresponding finite weights and then extends every weight to its affine version at the appropriate level k. Weights with negative zeroth Dynkin label are then shift-reflected to the integrable affine sector. Weights that cannot be shift-reflected to the integrable sector are ignored (this is the case, for example, for those with zeroth Dynkin label equal to -1).

Here is a simple example: consider the su(2) tensor-product

(3.3) $$(2) \otimes (4) = (2) \oplus (4) \oplus (6)$$

and its affine extension at level 4:

(3.4) $$[2,2] \times [0,4] = [2,2] + [0,4] + [-2,6]$$

The last weight must be reflected since it is not integrable: the shifted action of s_0, the reflection with respect to the zeroth affine root, is

(3.5) $$s_0 \cdot [-2,6] = s_0([-2,6] + [1,1]) - [1,1] = [0,4]$$

and this contributes with a minus sign ($\epsilon(s_0) = -1$), canceling then the other $[0,4]$ representation; we thus find:

(3.6) $$[2,2] \times [0,4] = [2,2]$$

The relation between tensor products and fusion was further explored in [34], published in a special volume of the Canadian Journal of Physics dedicated to Prof. R. T. Sharp.

4. The idea of threshold level

The result of the above computation is manifestly level dependent. Let us reconsider the same product, but at level 5. The affine extension of the product becomes

(4.1) $$[3,2] \times [1,4] = [3,2] + [1,4] + [-1,6]$$

The last weight is thus ignored and the final result is $[3,2] \times [1,4] = [3,2] + [1,4]$. For $k > 5$ it is clear that there are no truncations, hence no difference between the fusion coefficients and the tensor products. Moreover, we see that the representation (4) occurs at level 5 and higher. We then say that its *threshold level*, denoted by k_0, is 5. The threshold level is thus the smallest value of k such that the fusion coefficient $\mathcal{N}_{\hat{\lambda}\hat{\mu}}^{(k)\,\hat{\nu}}$ is non-zero, when $\mathcal{N}_{\hat{\lambda}\hat{\mu}}^{(k)\,\hat{\nu}} \in \{0,1\}$, for all levels k.[1] If we indicate the threshold level by a subscript, we can write

(4.2) $$(2) \otimes (4) = (2)_4 \oplus (4)_5 \oplus (6)_6.$$

To read off a fusion at fixed level k, we only keep terms with index not greater than k. This implies directly the inequality

(4.3) $$\mathcal{N}_{\hat{\lambda}\hat{\mu}}^{(k)\,\hat{\nu}} \leq \mathcal{N}_{\hat{\lambda}\hat{\mu}}^{(k+1)\,\hat{\nu}}$$

which in turn yields

(4.4) $$\lim_{k \to \infty} \mathcal{N}_{\hat{\lambda}\hat{\mu}}^{(k)\,\hat{\nu}} = \mathcal{N}_{\lambda\mu}^{\,\nu}$$

The concept of threshold level was first introduced in [12]. Its origin is directly rooted in the generating-function method applied to fusion rules. This is reviewed in the next section, focusing again on the simple $\widehat{su}(2)$ case.

---

[1]More generally, we say there are $\mathcal{N}_{\lambda\mu}^{\,\nu}$ couplings, each having its own threshold level k_0. For fixed $\{\lambda, \mu; \nu\}$ then, one gets a multi-set of threshold levels. A simple example: su(3) with $\lambda = \mu = \nu$ labeling the adjoint representation has threshold levels $\{2,3\}$.

5. Fusion generating functions

The result (1.1) on the su(2) tensor-product generating function can be understood as follows: all couplings can be described by appropriate products of three *tensor-product elementary couplings*:

(5.1) $$E_1 = LM, \quad E_2 = LN, \quad E_3 = MN.$$

$G^{\text{su}(2)}$ can thus be written compactly as

(5.2) $$G^{\text{su}(2)}(L,M,N) = \prod_{i=1}^{3} \frac{1}{(1-E_i)}$$

How can we construct the affine extension of this generating function? One certainly needs to introduce a further dummy variable, say d, in order to keep track of the extra variable k. Then one could try to introduce factors of d appropriately. A natural guess is

(5.3) $$G^{\widehat{\text{su}}(2)} = \frac{1}{(1-d)(1-dLM)(1-dLN)(1-dMN)}$$

That turns out to be the right answer: this reproduces the $\widehat{\text{su}}(2)_k$ fusion rules. The prefactor is justified as follows: the fusion of the "vacuum" with itself, $[k,0] \times [k,0] = [k,0]$, exists at every level and this is precisely taken into account by the factor $1/(1-d)$. With the threshold level insight, we can also naturally justify the factors of d multiplying the three elementary couplings: their power yields their threshold level. This expression was first proved in [12]. From this example and that of the $\widehat{\text{su}}(3)$ generating function, we conjectured that any fusion coupling is characterized by a threshold level. This was subsequently checked with the $\widehat{\text{so}}(5)$ case [5]. This is now understood to be a consequence of the Gepner–Witten depth rule [17, 21], as shown in [20].

From the above expression, we also infer the existence of *fusion elementary couplings*. For $\widehat{\text{su}}(2)$, there are four of them

(5.4) $$\begin{aligned}&\widehat{E}_0 : d : \quad (0) \otimes (0) \supset (0)_1 \qquad &\widehat{E}_1 : dLM : (1) \otimes (1) \supset (0)_1, \\ &\widehat{E}_2 : dLN : (1) \otimes (0) \supset (1)_1, \qquad &\widehat{E}_3 : dMN : (0) \otimes (1) \supset (1)_1.\end{aligned}$$

As explained above, subscripts indicate the threshold level.

A re-derivation of (5.3) was presented in [3]. The method used there was amenable to generalization, unlike the original proof in [12]. Consequently, [3] displays further examples of fusion generating functions.

6. Tensor products, linear inequalities and elementary couplings

There are simple combinatorial methods that can be used for calculating su(N) tensor products, for instance, the Littlewood–Richardson (LR) rule. It is thus natural to ask whether we can read off the threshold level of a coupling from its LR tableau.

Integrable weights in su(N) can be represented by tableaux: the weight $(\lambda_1, \lambda_2, \ldots, \lambda_{N-1})$ is associated to a left justified tableau of $N-1$ rows with $\lambda_1 + \lambda_2 + \cdots + \lambda_{N-1}$ boxes in the first row, $\lambda_2 + \cdots + \lambda_{N-1}$ boxes in the second row, etc. Equivalently, the tableau has λ_1 columns of 1 box, λ_2 columns of 2 boxes, etc. The scalar representation has no boxes, or equivalently, any number of columns of N boxes. For instance:

(6.1) $\quad$ su(3) : (1, 1) $\longleftrightarrow$ [tableau] $\quad$ su(4) : (2, 3, 0) $\longleftrightarrow$ [tableau]

The Littlewood–Richardson rule is a simple combinatorial algorithm that calculates the decomposition of the tensor product of two su(N) representations $\lambda \otimes \mu$. The second tableau (μ) is filled with numbers as follows: the first row with 1's, the second row with 2's, etc. All the boxes with a 1 are then added to the first tableau according to following restrictions: (1) the resulting tableau must be regular: the number of boxes in a given row must be smaller or equal to the number of boxes in the row just above; (2) the resulting tableau must not contain two boxes marked by 1 in the same column. All the boxes marked by a 2 are then added to the resulting tableaux according to the above two rules (with 1 replaced by 2) and the further restriction: (3) in counting from right to left and top to bottom, the number of 1's must always be greater than or equal to the number of 2's. The process is repeated with the boxes marked by a $3, 4, \ldots, N-1$, with the additional rule that the number of i's must always be greater than or equal to the number of $i+1$'s when counted from right to left and top to bottom. The resulting Littlewood–Richardson (LR) tableaux are the Young tableaux of the irreducible representations occurring in the decomposition.

Here is a simple su(3) example: $(1,1) \otimes (1,1) \supset 2(1,1)$ since we can draw two LR tableaux with shape $(1,1)$ and an extra column of three boxes (the total number of boxes being preserved, the resulting LR tableau must have 6 boxes):

(6.2) [tableaux shown]

These rules can be rephrased in an algebraic way as follows [9]. Define n_{ij} to be the number of boxes i that appear in the LR tableau in the row j. The LR conditions read:

(6.3) $\quad \lambda_{j-1} + \sum_{i=1}^{k-1} n_{i\,j-1} - \sum_{i=1}^{k} n_{ij} \geq 0 \quad 1 \leq k < j \leq N,\ j \neq 1$

and

(6.4) $\quad \sum_{j=i}^{k} n_{i-1\,j-1} - \sum_{j=i}^{k} n_{ij} \geq 0, \quad 2 \leq i \leq k \leq N \text{ and } i \leq N-1.$

The weight μ of the second tableau and the weight ν of the resulting LR tableau are easily recovered from these data.

The combined equations (6.3) and (6.4) constitute a set of linear and homogeneous inequalities. We call this the LR (or tensor-product) basis. As described in [**30, 31**], the Hilbert basis theorem guarantees that every solution can be expanded in terms of the elementary solutions of these inequalities. This is a key concept for the following (see [**2**] for an extensive discussion of these methods). A sum of two solutions translates into the product of the corresponding couplings, more precisely, to the *stretched product* (denoted by ·) of the corresponding two LR tableaux. This is defined as follows. Denote the void boxes of a LR tableau by a 0, so that $n_{0j} = \sum_{i=j}^{N-1} \lambda_i$. A tableau is thus completely characterized by the data

$\{n_{ij}\}$ where now $i \geq 0$. Then, the tableau obtained by the stretched product of the tableaux $\{n_{ij}\}$ and $\{n'_{ij}\}$ is simply described by the numbers $\{n_{ij} + n'_{ij}\}$, e.g.,

(6.5)

Let us now turn to the su(2) case. The complete set of inequalities for su(2) variables $\{\lambda_1, n_{11}, n_{12}\}$ is simply

(6.6) $$\lambda_1 \geq n_{12}, \quad n_{11} \geq 0, \quad n_{12} \geq 0.$$

The first one expresses the fact that two boxes marked by a 1 cannot be in the same column while the other two are obvious. The other weights are fixed by the relation $\mu_1 = n_{11} + n_{12}$ and $\nu_1 = \lambda_1 + n_{11} - n_{12}$. Any solution of these inequalities describes a coupling. By inspection, the elementary solutions of this set of inequalities are

(6.7) $$(\lambda_1, n_{11}, n_{12}) = (1, 0, 1), \; (1, 0, 0), \; (0, 1, 0).$$

(For more complicated cases, we point out that powerful methods to find the elementary solutions are described in [**3**].) These correspond to the following LR tableaux, denoted respectively E_1, E_2, E_3:

(6.8) $E_1 : \boxed{} \atop \boxed{1}$, $\quad E_2 : \boxed{}$, $\quad E_3 : \boxed{1}$.

It is also manifest that there are no linear relations between these couplings. Any stretched product of these elementary tableaux is an allowed su(2) coupling. Because there are no relations between the elementary couplings, this decomposition is unique. We thus see that the description of the elementary couplings captures, in a rather economical way, the whole set of solutions of (6.6), that is, the whole set of su(2) couplings.

7. Reformulating the fusion rules in terms of linear inequalities

Consider now the affine-fusion extension of the reformulation of the su(2) tensor products in terms of linear inequalities. The elementary couplings have a natural affine extension, denoted by a hat, and their threshold levels are easily computed from the Kac–Walton formula.[2] The result is: $k_0(\widehat{E}_i) = 1$ for $i = 1, 2, 3$. We observe that these values of k_0 are the same as the number of columns. Since the product of fusion elementary couplings is also a fusion and because this decomposition is unique, we can read off the threshold level of any coupling, hence of any LR tableau, simply from the number of its columns:

(7.1) $$k_0 = \#\text{columns} = \lambda_1 + n_{11}.$$

And since k is necessarily greater that k_0, we have obtained the extra inequality:

(7.2) $$k \geq \lambda_1 + n_{11}.$$

This together with (6.6) yield a set of inequalities describing completely the fusion rules. This is what we call a *fusion basis*, here the fusion basis of $\widehat{su}(2)$. As in the finite case, the fusion couplings can be described in terms of elementary fusions.

---

[2]Let us mention here that for the classical simple Lie algebras, the tableau methods for tensor products have been modified to implement the Kac–Walton formula for fusions—see [**11**] in the Sharp volume of the Canadian Journal of Physics.

These correspond to the elementary solutions of the four inequalities, which are easily found to be

(7.3) $\quad (k, \lambda_1, n_{11}, n_{12}) = (1,0,0,0),\ (1,1,0,1),\ (1,1,0,0),\ (1,0,1,0).$

They correspond respectively to the coupling

(7.4)
$$\widehat{E}_0 : [1,0] \times [1,0] \supset [1,0], \qquad \widehat{E}_2 : [0,1] \times [1,0] \supset [0,1],$$
$$\widehat{E}_1 : [0,1] \times [0,1] \supset [1,0], \qquad \widehat{E}_3 : [1,0] \times [0,1] \supset [0,1].$$

Any fusion has an unique decomposition in terms of these elementary couplings. For instance

(7.5) $\quad [3,2] \times [1,4] \supset [1,4] \longleftrightarrow \begin{array}{|c|c|c|c|} \hline & 1 & 1 & 1 \\ \hline 1 & & & \\ \hline \end{array} \longleftrightarrow \widehat{E}_1 \widehat{E}_2 \widehat{E}_3^3 : \quad k_0 = 5.$

8. Constructing the fusion basis: Farkas' lemma

For algebras other than $\widehat{su}(2)$, the threshold level is not simply the number of columns. So the question is: how can we derive the fusion basis? The following strategy was developed in [3]:

(1) Write the LR inequalities;
(2) from these, find the tensor-product elementary couplings;
(3) from these, find fusion elementary couplings;
(4) from these, reconstruct the fusion basis.

To go from step 2 to step 3, we need some tools; we describe below a method based on the outer automorphism group. Similarly to go from 3 to 4, we need a further ingredient: this is Farkas' lemma. We discuss these techniques in turn.

Let us start from the set of tensor-product elementary couplings $\{E_i, i \in I\}$ for some set I fixed by the particular $su(N)$ algebra under study. For each E_i, we calculate the threshold level $k_0(E_i)$ and this datum specifies the affine extension of E_i, denoted $\widehat{E}_i$. We have then a partial set of fusion elementary couplings with the set $\{\widehat{E}_i, i \in I\}$. Our conjecture is that the missing fusion elementary couplings can all be generated by the action of the outer-automorphism group. For $\widehat{su}(N)$, this group is simply $\{a^n \mid n = 0, \ldots, N-1\}$, with

(8.1) $\quad a[\lambda_0, \lambda_1, \ldots, \lambda_{N-1}] = [\lambda_{N-1}, \lambda_0, \ldots, \lambda_{N-2}].$

The conjecture is based on the invariance relation

(8.2) $\quad \mathcal{N}^{(k)\ a^{n+m}\hat{\nu}}_{a^n\hat{\lambda}, a^m\hat{\mu}} = \mathcal{N}^{(k)\ \hat{\nu}}_{\hat{\lambda}\hat{\mu}}.$

It amounts to supposing that the full set is contained in $\{\mathcal{A}\widehat{E}_i \mid i \in I\}$. Here $\mathcal{A}\widehat{E}_i$ indicates a coupling of weights

(8.3) $\quad \mathcal{A}\{\hat{\lambda}, \hat{\mu}; \hat{\nu}\} = \{a^n\hat{\lambda}, a^m\hat{\mu}; a^{n+m}\hat{\nu}\},$

n, m being arbitrary integers defined modulo N, if $\widehat{E}_i$ has weights $\{\hat{\lambda}, \hat{\mu}; \hat{\nu}\}$. The conjectured completeness requires the consideration of all possible pairs (n, m).[3]

Let us illustrate this with the $\widehat{su}(2)$ case. Start with the elementary coupling $E_1 : (1) \otimes (1) \supset (0)$, which, as already indicated, arises at level 1: $k_0(E_1) = 1$. The

---

[3] Note that we do not suppose that the action of $\mathcal{A}$ on an elementary coupling will necessarily produce another elementary coupling. Indeed, the resulting coupling could be a product of elementary couplings. What is conjectured here is that all fusion elementary couplings can be generated in this way.

corresponding fusion is thus $[0,1] \times [0,1] \supset [1,0]$, denoted as $\widehat{E}_1$. We now consider all possible actions of the outer-automorphisms group on it. Since this group is of order 2, there are 4 possible choices for the doublet (n,m):

(8.4) $$(a^n, a^m) \in \{(a,a), (1,1), (1,a), (a,1)\}$$

with $a[\lambda_0, \lambda_1] = [\lambda_1, \lambda_0]$. This generates the set of four elementary couplings found previously, in the respective order $\widehat{E}_0$, $\widehat{E}_1$, $\widehat{E}_2$, $\widehat{E}_3$. Thus, from one tensor-product elementary coupling, all four fusion elementary couplings are deduced.

We now turn to Farkas' lemma. For its presentation, it is convenient to use an exponential description of the couplings, that is,

(8.5) $$(k, \lambda_i, n_{ij}) \to d^k L_i^{k_i} N_{ij}^{n_{ij}}$$

d, L_i, N_{ij} being dummy variables. For instance $\widehat{E}_1$ is represented by $dL_1 N_{12}$. If we collectively describe a coupling by the complete set of variables $\{x_i\}$, we have

(8.6) $$\{x_i\} \to \{X_i^{x_i}\}$$

A particular coupling is thus described by a given product $\prod_i X_i^{x_i}$ with fixed x_i. Its decomposition in terms of elementary couplings takes the form $\prod_i \widehat{E}_i^{a_i}$. Now, since $\widehat{E}_i$ can be decomposed in terms of the X_j as

(8.7) $$\widehat{E}_i = \prod_j X_j^{\epsilon_{ij}}$$

it means that reading off a particular coupling means that we are interested in a specific choice set of positive integers $\{x_i\}$ fixed by

(8.8) $$\sum_i a_i \epsilon_{ij} = x_j$$

in terms of some positive integers a_i. We are thus looking for the existence conditions for such a coupling, i.e., the underlying set of linear and homogeneous inequalities. This is exactly what Farkas' lemma [**29**] gives us: given the knowledge of the elementary couplings, it allows us to recover the underlying set of inequalities. For tensor products, this is of no interest since we know the corresponding set of inequalities and our elementary couplings have been extracted from them. But the situation is quite different in the fusion case, where the *fusion basis* is unknown.

For our application we need the following modification of the lemma, proved in [**3**]:

LEMMA. *Let A be an $r \times m$ integer matrix and let ϵ_j ($j = 1, \ldots, n$) be a set of fundamental solutions to*

(8.9) $$Ax \geq 0, \quad x \in \mathbb{N}^m.$$

Let V be the $m \times n$ matrix with entries $V_{ij} = (\epsilon_j)_i$ (for $i = 1, \ldots, m$, $j = 1, \ldots, n$), i.e., the columns of V are a set of fundamental solutions to (8.9). *Let e_w ($w = 1, \ldots, \ell$) be a fundamental system of solutions of $u^\top V \geq 0$, (not necessarily positive) $u \in \mathbb{Z}^m$, and let E be the $\ell \times m$ matrix with entries $E_{wi} = (e_w)_i$, i.e., the rows of E are the fundamental solutions e_w ($w = 1, \ldots, \ell$). Then the solution set of the system*

(8.10) $$Ex \geq 0, \quad x \in \mathbb{N}^m$$

is the same as the solution set of (8.9).

To link the lemma to the situation presented above, we note that the entries V_{ij} of the matrix V are given here by the numbers ϵ_{ji} appearing in (8.7). The relation $Va = x$ describes a generic coupling $\prod_i \widehat{E}_i^{a_i}$, and our goal is to find the defining system of inequalities for x that underly the existence of this coupling.

Take a simple example, the $\widehat{su}(2)$ case. The elementary couplings and the corresponding vectors ϵ_i are

(8.11)
$$\widehat{E}_0 : d : \quad \epsilon_0 = (1,0,0,0), \quad \widehat{E}_2 : dL_1 : \quad \epsilon_2 = (1,1,0,0)$$
$$\widehat{E}_1 : dL_1 N_{12} : \quad \epsilon_1 = (1,1,0,1), \quad \widehat{E}_3 : dN_{11} : \quad \epsilon_3 = (1,0,1,0)$$

From the vectors ϵ_i with components ϵ_{ij}, we form the matrix V with entries $V_{ij} = \epsilon_{ji}$:

(8.12)
$$V = \begin{pmatrix} 1 & 1 & 1 & 1 \\ 0 & 1 & 1 & 0 \\ 0 & 0 & 0 & 1 \\ 0 & 1 & 0 & 0 \end{pmatrix}$$

With a and x denoting the column matrices of entries a_i and x_i respectively, we have the matrix equation

(8.13) $$Va = x$$

Again, this equation describes a general fusion coupling $\prod_i \widehat{E}_i^{a_i}$. We now want to unravel the underlying system of inequalities. For this, we consider the fundamental solutions of

(8.14) $$u^\top V \geq 0$$

where u is the vector of entries u_i. These inequalities read

(8.15) $\quad u_0 \geq 0, \quad u_0 + u_1 + u_3 \geq 0, \quad u_0 + u_1 \geq 0, \quad u_0 + u_2 \geq 0.$

In this simple case, the elementary couplings can be found by inspection and these are:

(8.16) $e_0 = (1, -1, -1, 0), \quad e_1 = (0, 0, 0, 1), \quad e_2 = (0, 1, 0, -1), \quad e_3 = (0, 0, 1, 0).$

Finally, we consider the conditions $e_i\, x \geq 0$, with $(x_0, x_1, x_2, x_3) = (k, \lambda_1, n_{11}, n_{12})$. They read, in order,

(8.17) $\quad k \geq \lambda_1 + n_{11}, \quad n_{12} \geq 0, \quad \lambda_1 \geq n_{12}, \quad n_{11} \geq 0.$

The last three conditions define the LR basis. The first one is the additional fusion constraint. Together, they form the $\widehat{su}(2)$ fusion basis.

9. Constructing the fusion basis: polytope techniques

In the previous section, Farkas' lemma has been used to construct the fusion basis out of the set of fusion elementary couplings. There are alternative approaches, however. Another one is based on the reinterpretation of the fusion-rule computations in terms of counting points inside a polytope. A polytope can be described by its vertices or its facets. In our context, the *vertices* are represented by the fusion elementary couplings and the *facets* are the inequalities for which the elementary couplings are the elementary solutions. The reconstruction of the facets of a polytope from its vertices is thus another way to generate the fusion basis. This method is described in [1].

In the special case of su(N), Berenstein–Zelevinsky triangles can also be used to derive the polytope description of a fusion basis [27, 28], by considering so-called virtual triangles. Multiple sum formulas can then be written for fusion coefficients of various types. However, as are all methods to date, this one is difficult to extend to higher rank. Assigning a threshold level to a Berenstein-Zelevinsky triangle becomes very rapidly more difficult with increasing rank (see [4, 6]).

10. Conclusion

The fusion bases have been constructed for $\widehat{su}(3)$, $\widehat{su}(4)$ and $\widehat{sp}(4)$. Note that for algebras other that $\widehat{su}(N)$, we replace the LR basis by the Berenstein and Zelevinsky basis [7]. This leads to explicit expressions for the threshold levels, hence for the fusion coefficients.[4]

We stress that the reformulation of the problem of computing fusion rules in terms of a fusion basis solves, in principle, the quest for a combinatorial method, since it reduces a fusion computation to solving inequalities. But it is probably not the optimal solution to the quest for an efficient combinatorial description.

The main open problem concerning fusion bases is to find a fundamental and Lie algebraic way of deriving the basis, analogous in spirit to the Berenstein–Zelevinsky conjectures for generic Lie algebras in [7].

The methods described in Section 8, involving Farkas' lemma, are general and powerful, but they may not be Lie algebraic enough. Perhaps one should step back and look at a first principles description of the tensor product couplings, and its adaptation to fusion. This was done in [26]. Three-point functions were calculated that can be regarded as generating functions for tensor product couplings, and a very simple method was found for adapting the results to fusion couplings. In principle, the procedure works for any semi-simple Lie algebra. Unfortunately, these more Lie algebraic methods are inevitably more involved. It gives more information (such as operator product coefficients instead of just fusion coefficients), but it is not clear that it can be implemented effectively on higher rank algebras.

References

1. L. Bégin, C. Cummins, L. Lapointe, and P. Mathieu, *Fusion bases as facets of polytope*, J. Math. Phys. **43** (2002), no. 8, 4180–4201.
2. L. Bégin, C. Cummins, and P. Mathieu, *Generating-function method for tensor products*, J. Math. Phys. **41** (2000), no. 11, 7611–7639.
3. _____, *Generating-function method for fusion rules*, J. Math. Phys. **41** (2000), no. 11, 7640–7674.
4. L. Bégin, A. N. Kirillov, P. Mathieu, and M. Walton, *Berenstein–Zelevinsky triangles, elementary couplings, and fusion rules*, Lett. Math. Phys. **28** (1993), no. 4, 257–268.
5. L. Bégin, P. Mathieu, and M. A. Walton, *New example of a generating function for WZNW fusion rules*, J. Phys. A **25** (1992), no. 1, 135–148.
6. _____, $\widehat{su}(3)_k$ *fusion coefficients*, in Modern Phys. Lett. A **7** (1992), no. 35, 3255–3265.
7. A. D. Berenstein and A. V. Zelevinsky, *Tensor product multiplicities and convex polytopes in partition space*, J. Geom. Phys. **5** (1989), no. 3, 453–472.
8. C. Bodine and R. T. Gaskell, *Generating functions for IR multiplicity*, J. Math. Phys. **23** (1982), no. 12, 2217–2220.
9. M. Couture, C. J. Cummins, and R. T. Sharp, *Generating functions and elementary Young tableaux*, J. Phys. A **23** (1990), no. 11, 1929–1957.
10. M. Couture and R. T. Sharp, *Reduction of enveloping algebras of low-rank groups*, J. Phys. A **13** (1980), no. 6, 1925–1945.

---

[4]Related methods for obtaining the threshold level are presented in [4, 6].

11. C. J. Cummins and R. C. King, *WZW fusion rules for the classical Lie algebras*, Canad. J. Phys. **72** (1994), no. 7-8, 342–344.
12. C. J. Cummins, P. Mathieu, and M. A. Walton, *Generating functions for WZNW fusion rules*, Phys. Lett. B **254** (1991), no. 3-4, 386–390.
13. J. Fuchs and P. van Driel, *WZW fusion rules, quantum groups, and the modular matrix S*, Nuclear Phys. B **346** (1990), no. 2-3, 632–648.
14. P. Furlan, A. Ganchev, and V. B. Petkova, *Quantum groups and fusion rules multiplicities*, Nuclear Phys. B **343** (1990), no. 1, 205–227.
15. R. Gaskell, A. Peccia, and R. T. Sharp, *Generating functions for polynomial irreducible tensors*, J. Mathematical Phys. **19** (1978), no. 4, 727–733.
16. R. Gaskell and R. T. Sharp, *Generating functions for G_2 characters and subgroup branching rules*, J. Math. Phys. **22** (1981), no. 12, 2736–2739.
17. D. Gepner and E. Witten, *String theory on group manifolds*, Nuclear Phys. B **278** (1986), no. 3, 493–549.
18. Y. Giroux, M. Couture and R. T. Sharp, *Degenerate enveloping algebras of* SU(3), SO(5), G_2 *and* SU(4), J. Phys. A **17** (1984), no. 4, 715–725.
19. V. G. Kac, *Infinite-dimensional Lie algebras*, 3rd edition, Cambridge Univ. Press, Cambridge, 1990, Exercise 13.35.
20. A. N. Kirillov, P. Mathieu, D. Sénéchal and M. A. Walton, *Can fusion coefficients be calculated from the depth rule?*, Nuclear Phys. B **391** (1993), no. 3, 651–674.
21. V. G. Knizhnik and A. B. Zamolodchikov, *Current algebra and Wess–Zumino model in two dimensions*, Nuclear Phys. B **247** (1984) no. 1, 83–103.
22. R. V. Moody, J. Patera and R. T. Sharp, *Character generators for elements of finite order in simple Lie groups A_1, A_2, A_3, B_2, and G_2*, J. Math. Phys. **24** (1983), no. 10, 2387–2396.
23. J. Patera and R. T. Sharp, *Generating function techniques pertinent to spectroscopy and crystal physics*, Recent Advances in Group Theory and Their Application to Spectroscopy (Antigonish, NS, 1978), NATO Adv. Sci. Inst. Ser. B Phys., vol. 43, Plenum, New York–London, 1979, pp. 219–248.
24. ———, *Generating functions for characters of group representations and their applications*, Group Theoretical Methods in Physics (Austin, TX, 1978), Lecture Notes in Phys., vol. 94, Springer-Verlag, Berlin–New York, 1979, pp. 175–183.
25. ———, *Generating functions for plethysms of finite and continuous groups*, J. Phys. A **13** (1980), no. 2, 397–416.
26. J. Rasmussen and M. A. Walton, *On the level-dependence of Wess–Zumino–Witten three-point functions*, Nuclear Phys. B **616** (2001), no. 3, 517–536.
27. ———, *Fusion multiplicities as polytope volumes: $\mathcal{N}$-point and higher-genus* su(2) *fusion*, Nuclear Phys. B **620** (2002), no. 3, 537–550.
28. ———, *Affine* su(3) *and* su(4) *fusion multiplicities as polytope volumes*, J. Phys. A **35** (2002), no. 32, 6939–6952.
29. A. Schrijver, *Theory of linear and integer programming*, Wiley-Intersci. Ser. Discrete Math., Wiley, Chchester, 1986.
30. R. P. Stanley, *Linear homogeneous Diophantine equations and magic labelings of graphs*, Duke Math. J. **40** (1973) 607–632.
31. ———, *Combinatorics and commutative algebra*, Progr. Math., vol. 41, Birkhäuser, Boston, 1983.
32. M. A. Walton, *Fusion rules in Wess–Zumino–Witten models*, Nuclear Phys. B **340** (1990), no. 2-3, 777–790.
33. ———, *Algorithm for WZW fusion rules: a proof*, Phys. Lett. B **241** (1990), no. 3, 365–368.
34. ———, *Tensor products and fusion rules*, Canad. J. Phys. **72** (1994), no. 7-8, 527–536.

Secteur des Sciences, Campus d'Edmundston, Université de Moncton, 165, boulevard Hébert, Edmundston, NB E3V 2S8, Canada
E-mail address: `lbegin@umce.ca`

Department of Mathematics and Statistics, Concordia University, Montréal, QC H3G 1M8, Canada
E-mail address: `cummins@mathstat.concordia.ca`

Département de physique, Université Laval, Pavillon Alexandre-Vachon, Québec, QC G1K 7P4, Canada

E-mail address: `pmathieu@phy.ulaval.ca`

Department of Physics, University of Lethbridge, 4401 University Drive, Lethbridge, AB T1K 3M4, Canada

E-mail address: `walton@uleth.ca`

Transient Effects in Wigner Distribution Phase Space of a Scattering Problem

Marcos Moshinsky

ABSTRACT. At the Group24 Conference in July 2002 in Paris several participants were interested in my analysis in Wigner distribution space of my work on diffraction in time. It was suggested that I should extend it to the case where a potential is present. The simplest case is one-dimensional with a delta function at the origin, and the particles coming from the left with a given momentum. In this paper we carry out the analysis of the transient effects in this problem, using the Laplace transform. We discuss the similarities and differences of our problem in Wigner distribution phase space when compared with the case of diffraction in time.

Introduction

At a recent conference on Group Theoretical Methods in Physics (Group24, Paris, July 2002) the question was raised of what would be the description in the Wigner distribution phase space of a scattering problem evolving in time.

The question caught my attention on the possibility of finding an example simple enough to allow a full answer.

Such an example is the one-dimensional problem with a a Dirac delta function potential of strength b at the origin $x = 0$. We will use units where $\hbar$ as well as the mass m and the charge e are 1. The charge e will not appear in the equation but the units allow us to make everything dimensionless and, in the case where the particle is an electron, the unit of length is $(\hbar^2/me^2) = 0.5 \times 10^{-8}$ cm

We consider first the case when $b = \infty$ and particles come from the left with momentum k. The stationary solution of the Schrödinger equation should then be $\sin kx$. At $t = 0$ the strength of the delta barrier is lowered to a value b and thus our time dependent quantum mechanical problem will be

$$(0.1) \qquad \left[-\frac{1}{2}\frac{\partial^2}{\partial x^2} + b\delta(x)\right]\psi(x,t) = i\frac{\partial \psi}{\partial t},$$

with the initial value

$$(0.2) \qquad \psi(x,0) = (\sin kx)\theta(-x),$$

2000 *Mathematics Subject Classification.* 34L40.
The author is member of El Colegio Nacional and of the Sistema Nacional de Investigadores.
This is the final form of the paper.

©2004 American Mathematical Society

where θ is a step function

$$\theta(x) = \begin{cases} 1, & \text{if } x > 0, \\ 0, & \text{if } x < 0. \end{cases} \tag{0.3}$$

Our first objective is then to determine $\psi(x, k, t)$, where the parameter k in the wave function takes into account the initial condition (0.2).

1. Solution of the quantum mechanical problem

A usual, the Laplace transform of the function $\psi(x, , k, t)$ is defined as

$$\bar{\psi}(x, k, s) = \int_0^\infty \psi(x, k, t) e^{-st}\, dt. \tag{1.1}$$

The Laplace transform of (0.1) is

$$\left[-\frac{1}{2}\frac{d^2}{dx^2} + b\delta(x)\right]\bar{\psi}(x, k, s) = -i(\sin kx)\theta(-x) + is\bar{\psi}(x, k, s). \tag{1.2}$$

This is an ordinary differential equation and replacing s by a variable κ through the relations

$$s = -i\frac{\kappa^2}{2} \quad \text{or} \quad \kappa = \sqrt{2is} \tag{1.3}$$

it takes the form

$$-\frac{1}{2}\frac{d^2}{dx^2}\bar{\psi}(x, k, \kappa) = -i\sin kx + \frac{\kappa^2}{2}\bar{\psi}(x, k, \kappa) \qquad \text{if } x < 0, \tag{1.4}$$

$$-\frac{1}{2}\frac{d^2}{dx^2}\bar{\psi}(x, k, \kappa) = \frac{\kappa^2}{2}\bar{\psi}(x, k, \kappa) \qquad \text{if } x > 0. \tag{1.5}$$

Besides by integrating both sides of (1.2) with respect to x from $-\epsilon$ to ϵ, we obtain, when $\epsilon \to 0$,

$$-\frac{1}{2}\left[\frac{d\bar{\psi}}{dx}(x, k, \kappa)\right]_{-0}^{+0} + b\bar{\psi}(0, k, \kappa) = 0. \tag{1.6}$$

Moreover, the continuitycondition yields

$$\bar{\psi}(+0, k, \kappa) = \bar{\psi}(-0, k, \kappa). \tag{1.7}$$

From (1.4), (1.5) we have that

$$\bar{\psi}(x, k, \kappa) = C\sin kx + Be^{-i\kappa x} \qquad \text{if } x < 0, \tag{1.8}$$

$$\bar{\psi}(x, k, \kappa) = De^{i\kappa x} \qquad \text{if } \quad x > 0, \tag{1.9}$$

where the sign of κ guarantee the boundlessness of the solutions at $x = -\infty$ in (1.8) and $+\infty$ in (1.9).

A solution of the inhomogeneous equation (1.4) is given by

$$C = \frac{2i}{\kappa^2 - k^2} \tag{1.10}$$

and from (1.7) and (1.8) we get

$$D = B \tag{1.11}$$

$$-\tfrac{1}{2}(iD\kappa - Ck + i\kappa B) = 0 \tag{1.12}$$

and from (1.10)–(1.12) we finally obtain

(1.13) $$D = [2(\kappa + ib)(\kappa^2 - k^2)]^{-1}.$$

Now the question is how to get $\psi(x, k, t)$ from $\bar\psi(x, k, \kappa)$. The inverse of the Laplace transform when we use the variable s of (1.3) instead of κ is given by

(1.14) $$\psi(x, k, t) = \frac{1}{2\pi i} \int_{c-i\infty}^{c+i\infty} \bar\psi(x, k, s) e^{st}\, ds,$$

where the real constant c is chosen in such a way that the line from $c - i\infty$ is to $c + i\infty$ is above all the poles of the function $\bar\psi(x, k, s)$. If we go from s to $\kappa = \sqrt{2is}$ then (1.14) becomes [1]

(1.15) $$\psi(x, k, t) = \frac{1}{2\pi} \int_{-\infty}^{\infty} \kappa \bar\psi(x, k, \kappa) \exp[-i\kappa^2 t/2]\, d\kappa.$$

From (1.9) and (1.15) we have that

(1.16) $$\kappa \bar\psi(x, k, \kappa)$$
$$= \frac{\kappa \exp(i\kappa x)}{2(\kappa + ib)(\kappa - k)(\kappa + k)}$$
$$= \left[\frac{1}{2(k + ib)} \frac{1}{(\kappa - k)} - \frac{1}{2(-k + ib)} \frac{1}{(\kappa + k)} - \frac{2ib}{(k^2 + b^2)} \frac{1}{(\kappa + ib)} \right] e^{i\kappa x}.$$

To obtain now $\psi(x, k, t)$ we need first to recall that

(1.17) $$\frac{1}{2\pi} \int_{-\infty}^{\infty} \frac{\exp[i(\kappa x - \frac{1}{2}\kappa^2 t)]}{\kappa - \lambda}\, d\kappa = -i M(x, \lambda, t),$$

where

(1.18) $$M(x, \lambda, t) = \frac{1}{2} \exp\left[i\left(\lambda x - \frac{1}{2}\lambda^2 t\right)\right] \operatorname{erfc}\left(e^{-i\pi/4} \frac{x - \lambda t}{\sqrt{2t}} \right)$$

if the imaginary part of λ is negative.

We obtain that

(1.19) $$\psi(x, k, t)$$
$$= \left[\frac{i}{2(k+ib)} M(x, k, t) - \frac{i}{2(-k+ib)} M(x, -k, t) + \frac{2b}{(k^2+b^2)} M(x, -ib, t) \right] k.$$

The result in (1.19) is valid for $x > 0$, while for $x < 0$ we have that

(1.20) $$\kappa \bar\psi(x, k, \kappa) = \kappa C \sin kx + \kappa D e^{-i\kappa x},$$

where we used (1.9) and the fact that $B = D$ as inindicated in (1.12).

To pass now to $\psi(x.k,t)$ with $x < 0$ we consider the two terms in (1.20) separately. From (1.14) we have for the first term

(1.21) $$2i \sin kx \int_{-\infty}^{\infty} \frac{\kappa}{\kappa^2 - k^2} \exp(-i\kappa^2 t/2)\, d\kappa = 0,$$

which vanishes because if we change $\kappa \to -\kappa$ we get the same integral with a minus sign.

For the second term we get the result (1.19) but with x replaced by $-x$. Thus for the whole interval $-\infty \leq x \leq \infty$ we can write

$$(1.22) \quad \psi(x,k,t) = \left[\frac{i}{2(k+ib)}M(|x|,k,t) - \frac{i}{2(-k+ib)}M(|x|,-k,t) + \frac{2b}{k^2+b^2}M(|x|,-ib,t)\right],$$

with $|x|$ being the absolute value of x.

2. The Wigner distribution function of the problem

If we have the quantum mechanical configuration space function $\psi(x)$ of the problem, the corresponding Wigner distribution function $W(x,p)$ is given by [3]

$$(2.1) \quad W(x,p) = \frac{1}{\pi}\int_{-\infty}^{\infty} \psi^*(x+y)\psi(x-y)\exp(i2py)\,dy.$$

The $\psi(x,k,t)$ of (1.19) can be expressed as

$$(2.2) \quad \psi(x,k,t) = \sum_{j=1}^{3} A_j M(x,\lambda_j,t),$$

where

$$(2.3) \quad \lambda_1 = k, \quad \lambda_2 = -k, \quad \lambda_3 = -ib$$

and

$$(2.4) \quad A_1 = \frac{i}{2(k+ib)}, \quad A_2 = \frac{i}{2(k-ib)}, \quad A_3 = \frac{2b}{(k^2+b^2)}.$$

Thus the Wigner function is given by

$$(2.5) \quad W(x,p;k,t,b) = \sum_{i,j} A_i A_j^* W(x,p,k,\lambda_i,\lambda_j,t),$$

where

$$(2.6) \quad W(x,p,k,\lambda_i,\lambda_j,t) = \frac{1}{4\pi^3}\int_{-\infty}^{\infty}\int_{-\infty}^{\infty}\int_{-\infty}^{\infty} \frac{\exp\{-i[\kappa|x+y|-\kappa^2 t/2]\}\exp\{i[\kappa'|x-y|-\kappa'^2 t/2]\}}{(\kappa-\lambda_i^*)(\kappa'-\lambda_j)} \times e^{i2py}\,dy\,d\kappa\,d\kappa'.$$

The integral (2.6) is difficult to evaluate because we have to separate a part in which $y < x$ from the part that $y > x$. We note though that we are dealing with a scattering problem in which we are only interested in the wave function in the right-hand side, i.e., $x > 0$ and large distances from the potential that, in our units, implies $x \gg 1$. In this case we can essentially suppress the absolute values in (2.6) and are left with

(2.7) $W(x,p,k,\lambda_i,\lambda_j,t)$
$$\approx \frac{1}{4\pi^3}\int_{-\infty}^{\infty}\int_{-\infty}^{\infty}\int_{-\infty}^{\infty}\frac{\exp\{-i[\kappa(x+y)-\kappa^2 t/2]\}\exp\{i[\kappa'(x-y)-\kappa'^2 t/2]\}}{(\kappa-\lambda_i^*)(\kappa'-\lambda_j)}$$
$$\times e^{i2py}\,dy\,d\kappa\,d\kappa'$$
$$=\frac{1}{4\pi^3}\int_{-\infty}^{\infty}\int_{-\infty}^{\infty}\left\{\int_{-\infty}^{\infty}\exp\left[2iy\left(p-\frac{\kappa+\kappa'}{2}\right)\right]dy\right\}$$
$$\times\left\{\frac{\exp[-i(\kappa x-\frac{1}{2}\kappa^2 t)]\exp[-i(\kappa' x-\kappa'^2 t)]}{(\kappa-\lambda_i^*)(\kappa'-\lambda_j)}\right\}d\kappa\,d\kappa'.$$

The first integral obviously gives

(2.8) $$2\pi\delta[2p-\kappa-\kappa'],$$

and so, introducing it in (2.7) and integrating it with respect to κ', we get

(2.9) $W(x,p,k,\lambda_i,\lambda_j,t) = -\dfrac{1}{2\pi^2}\exp[2ip(x-pt)]\displaystyle\int_{-\infty}^{\infty}\dfrac{\exp[-2i\kappa(x-pt)]}{(\kappa-\lambda_i^*)(\kappa+\lambda_j-2p)}\,d\kappa.$

The integral in (2.9) can be evaluated by the method of residues if we take into account that it can be closed from above by a circle of radius going to ∞ if $x-pt<0$ and from below if the $(x-pt)>0$. We shall proceed to prove that all the zero of the denominator, i.e., the poles of the integral are in the upper half of the κ plane.

We begin by noting that (1.18) for $M(x,\lambda,t)$ is valid [**1**] only if the imaginary part of λ is negative. As k and $-k$ corresponding to λ_1 and λ_2 in (2.7) are real, we can add to them a small negative imaginary part $-\epsilon$ (we will later make $\epsilon\to 0$) and so define now $\lambda_1,\lambda_2,\lambda_3$ as

(2.10) $\qquad\qquad\lambda_1=k-i\epsilon,\quad \lambda_2=-k-i\epsilon,\quad \lambda_3=-ib,$

which implies

(2.11) $\qquad\qquad\lambda_1^*=k+i\epsilon,\quad \lambda_2^*=-k+i\epsilon,\quad \lambda_3^*=ib,$

(2.12) $-\lambda_1+2p=-k+2p+i\epsilon,\ -\lambda_2+2p=k+2p+i\epsilon,\ -\lambda_3+2p=ib+2p.$

It is clear then that all the poles of the integral (2.9) are in the upper half κ plane and thus there is a contribution in (2.9) only if $x-pt<0$ or $x<pt$.

Thus the integral (2.9) contains a factor $\theta(pt-x)$ and evaluating it by the method of residues we get

(2.13) $W(x,p,\lambda_i,\lambda_j,t)$
$$=-\frac{i}{\pi}(2p-\lambda_i^*-\lambda_j)^{-1}\left\{\exp[2i(p-\lambda_i^*)(x-pt)]-\exp[2i(-p+\lambda_j)(x-pt)]\right\}\theta(pt-x).$$

The above Wigner function has a complex value but it can easily be shown that the combination in (2.5) giving $W(x,p,k,t,b)$ is a real number but not necessarily a positive one, which is one of the problems of the physical interpretation of the Wigner distribution function.

It is interesting to note that the solutions of our problem in configuration space involves error integral functions or, equivalently, Fresnel integrals, while the solution in Wigner distribution phase space is given in terms of purely elementary functions

It is usually interesting to compare our $W(x,p,k,t,b)$ with the case where there is no δ potential at the origin and we had purely incoming waves initially, i.e., $\exp(ikx)$ instead of $\sin(kx)$. This can be obtained when the strength b of the δ function vanishes and we disregard the pole at $-k$. We then obtain

$$W(x,p,k,t,0) = \frac{\theta(pt-x)}{2\pi i(k-p)}\left\{\exp[2i(k-p)(x-pt)] - \exp[-2i(k-p)(x-pt)]\right\}, \tag{2.14}$$

which agrees with the result obtained previously for the Wigner distribution function for diffraction in time [1].

We note that in (2.14) we can pass from units in which $m = \hbar = 1$ to cgs units and then $\hbar$ and m will appear in the expression. In the classical limit ($\hbar \to 0$), we obtain [2]

$$W_{\text{cl}}(x,p;k,t) = \delta(k-p)\theta\left(\frac{kt}{m} - x\right) \tag{2.15}$$

which is the classical probability in phase space of finding the particle at the point (x,p) at the time t.

References

1. G. García-Calderón, G. Loyola, and M. Moshinsky, *The decay process: an excatly soluble example and its implications*, Symetries in Physics (A. Frank and B. Wolf, eds.), Springer-Verlag, Berlin, 1992, pp. 273–292.
2. V. Man'ko, M. Moshinsky, and A. Sharma, *Diffraction in time in terms of Wigner distributions and tomographic probabilities*, Phys. Rev. A. **59** (1999), 1809–1815.
3. E. P. Wigner, *On the quantum correction for thermodynamic equilibrium*, Phys. Rev. **40** (1932), 749–759.

INSTITUTO DE FÍSICA, UNIVERSIDAD NACIONAL AUTÓNOMA DE MÉXICO, APARTADO POSTAL 20-364, CIUDAD UNIVERSITARIA, 01000 MÉXICO, D.F., MÉXICO
E-mail address: `moshi@fisica.unam.mx`

R. T. Sharp and Generating Functions in Group Theory

Jiří Patera

ABSTRACT. Generating functions in group representation theory were by far the topic closest to Bob Sharp's heart during the last 25 years of his professional life. Milestones of his work in this direction are recalled, two special topics are underlined, and five new generating·functions are shown.

In a way it is all Bill Miller's fault... I was only an intermediary at the very beginning.

During our work on the paper [13], Willard Miller, one of its coauthors, suggested to use the first example of the generating function in Lie theory which I ever saw. On the first occasion I explained that example to Bob who got quickly very interested in the technique. His first paper on the subject [24] followed soon after.

In subsequent years he has been extending the applicability of the generating functions technique to ever new problems in group representation theory. But that is only one side of his contribution. It well may be that in the long run even more important will turn out to be efficient method of deriving the generarting functions he invented. They are scattered through his papers and would badly require a comprehensive survey.

Next I want to point out the papers where important new steps were made: [1–12, 14–23, 25].

In the rest of my contribution I would like to bring three very different topics involving generating functions. All are closely related to R. T. Sharp's work. The first topic is the character generators, undoubtedly the most basic objects in representation theory of semisimple Lie groups.

The second topic consists of five generating functions for various, what we like to call, polynomial tensors. That is tensors with well defined group transformation properties and with polynomial components. The five functions were left in my files, written in Bob's handwriting. None of them was published, as far as I know.

The last topic is a functional equation like no others involving representations of $SU(2)$. Its solution is an unsolved challenging problem. In my opinion, any new property of the equation and of its solution as well as generalizations to other groups would be rather interesting.

2000 *Mathematics Subject Classification.* 33D90, 33D52.
This is the final form of the paper.

1. Character generators

Shortly after the generating functions for characters, or simply character generators, were invented in [**17**], the subject was continued also by Richard Stanley, Ronald King, Kenneth Baclawski. New combinatorial methods of derivation of the generators were invented for groups of certain types. Even today the largest character generator in the literature is that of G_2 in [**9**].

Considered as power series, a character generator provides the characters of all finite dimensional irreducible representations of a given simple Lie group. In the following examples, the variables denoted by capital letters carry, in the power series, exponents which are the components of the highest weight of the representation given relative to the basis of dominant weights. In case of $sl(2,\mathbb{C})$, the exponent of A has the value of twice the "angular momentum" of the representation. Lower case Greek letters are the variables of the characters. Their exponents are the weights of the representations and their coefficients are the multiplicities of weights.

(1.1) $sl(2,\mathbb{C})$:
$$X_A(\alpha) = \frac{1}{(1-A\alpha)(1-A\alpha^{-1})} = 1 + A(\alpha + \alpha^{-1}) + \cdots$$

(1.2) $sl(3,\mathbb{C})$:
$$X_{PQ}(\alpha,\beta) = \left(\frac{1}{1-P\beta^{-1}} + \frac{Q\beta}{1-Q\beta}\right)$$
$$\times \frac{1}{(1-P\alpha)(1-P\alpha^{-1}\beta)(1-Q\alpha\beta^{-1})(1-Q\alpha^{-1})}$$
$$= 1 + P(\alpha + \alpha^{-1}\beta + \beta^{-1}) + Q(\beta + \alpha\beta^{-1} + \alpha^{-1})$$
$$+ PQ(\alpha\beta + \alpha^2\beta^{-1} + \alpha^{-1}\beta^2 + 2 + \alpha^{-2}\beta + \alpha\beta^{-2} + \alpha^{-1}\beta^{-1})$$
$$+ \cdots$$

See [**17**] for a general method of deriving character generators, as well as for their transformations into generation functions for other problems. In principle, character generator is the starting point for derivation virtually all other generating problems involving the initial Lie group. Often there are simpler ways to calculate a generating function then to start from the character generator.

Some of the transformations may be rather simple. Thus putting $\alpha = \beta = 1$ in (1.2) and making obvious simplifications, one gets the generating function for the dimensions of the representations. More generally, one can fix the character variables in a way that their values equal to various element of finite order in the Lie group, not only the identity element as before. A drastic simplification of the generating function then follows. Resulting function provides character values of the corresponding element of finite order on irreducible representations of the Lie group [**16**]. Transformations of other types would involve a substitution of variables of the characters, elimination of all by the highest weight terms, substitution of one generating function into another, a fusion of two functions while retaining only variables of certain kind in both, etc.

2. Some new generating functions

Bob Sharp was in a habit of calculating weird generating functions when he was bored doing other less amusing things. His skills in avoiding mistakes, while writing on casual pieces of paper, were unsurpassed. Some of the functions were left waiting to be put as a suitable example into some future papers. I know about the following ones. It is likely that there are other ones somewhere.

- Consider the 16-dimensional irreducible representation $(0\,0\,0\,1)$ of the simple Lie algebra B_4. The representation defines 16 variables-coordinates of a generic vector of the space transforming irreducibly under it. Transformation properties of polynomials of all degrees in the variables are described by the following generating function,

$$X(U,A,B,C,D) = \frac{1}{(1-U^2)(1-UD)(1-U^2A)} = \cdots + mU^u A^a D^d + \cdots.$$

Here u is the degree of polynomials, a and d indicate an irreducible representation with the highest weight $(a\,0\,0\,d)$, and m is the multiplicity of the representation among the polynomials of degree u. Because the generating function does not depend on the variables B and C, transformations of the polynomials involve only the B_4-representations with the highest weights $(a\,0\,0\,d)$, a, $d \geq 0$, and no others.

- Next let $(0\,0\,0\,1)$ denote the 26-dimensional irreducible representation of F_4. Similarly as in the previous case, F_4-transformation properties of the polynomials in 26 variables are described by the generating function

$$X(U,A,B,C,D) = \frac{1}{(1-U^2)(1-U^3)(1-UD)(1-U^2D)(1-U^3C)}.$$

Again from the absence of the variable B, one infers that the irreducible representations appearing in the symmetric powers of the initial representation (equivalently polynomials) decompose into direct sum of F_4-representations $(a\,0\,c\,d)$ only.

- Analogous interpretation is given to the following generating function based on the 27-dimensional representation $(0\,0\,0\,0\,1\,0)$ of E_6,

$$X(U,A,B,C,D,E,F) = \frac{1}{(1-U^3)(1-UE)(1-U^2A)}.$$

- Similarly the polynomial E_7-generating function based on $(0\,0\,0\,0\,0\,1\,0)$ of dimension 56 is the following,

$$X(U,A,B,C,D,E,F,G) = \frac{1}{(1-U^4)(1-UF)(1-U^2A)(1-U^3F)(1-U^4F)}.$$

- The last case is the polynomial B_5-generating function based on $(0\,0\,0\,1)$ of dimension 32,

$$X(U,A,B,C,D)$$
$$= \frac{1}{(1-U^4)(1-UE)(1-U^2A)(1-U^2B)(1-U^3E)(1-U^4C)(1-U^4D)}.$$

The structure of the generating funcrions, before they are expanded into power series, provide invaluable information about the structure of the corresponding polynomial rings. For that however, the generating function has to be written in, what we call, positive form. That is, contribution to the expansions from various parts of the function, must all be with the same sign, no cancellations should take place. Then only each bracket of the denominator indicate the existence of a basic polynomial (element of the integrity basis of the polynomial ring), its degree and transformation properties. Furthermore from the numerators in the five cases above, one concludes that the polynomial rings in those cases are free; i.e., there are no syzygies involving the basic polynomials.

3. An SU(2) functional equation

Consider an irreducible representation of SU(2) of dimension $j+1$ and the symmetric part of its tensor power of degree u. Denote it by $\mathrm{Sym}[(j)^u]$. Such a product is fully reducible, i.e., it can be decomposed into the direct sum of the irreducible representations of the general form

$$\mathrm{Sym}[(j)^u] = \bigoplus_s m_s^{(j,u)}(s). \tag{3.1}$$

Here (s) stands for the irreducible representations of SU(2) of dimension $s+1$ appearing in the reduction, and $m_s^{(j,u)}$ denotes the multiplicity of (s) in the direct sum. Of course, the highest term in the sum is the representation (uj), its multiplicity is 1.

Note that there are three integers involved in all the problems of this type, namely j, u, and s. They take the values between 0 and ∞. Only s has its upper limit in these problems, the values of j and u are independent.

Suppose that there exists the generating function

$$\mathbf{F}(U,J,S) = \sum_{j,u,s}^{\infty} m_s^{(j,u)} U^u J^j S^s$$

which provides the solution to all the problems (3.1) at the same time. It was shown in [**19**] that such a function, if it exists, must satisfy the functional equation

$$1 + J + U = (1 - J^2 - U^2)\mathbf{F}(U,J,S) + \frac{UJ}{S}\mathbf{F}\left(US, JS, \frac{1}{S}\right)$$
$$- UJS\,\mathbf{F}(US, JS, S) - \frac{UJ}{S}\mathbf{F}\left(\frac{U}{S}, \frac{J}{S}, S\right).$$

The equality here is understood with the proviso that one should disregard the terms, containing negative powers of the variables, in the expansion of $\mathbf{F}(U,J,S)$ into the power seriers.

References

1. J. Bystrický, R. Gaskell, J. Patera, and R. T. Sharp, *Generalized* SU(2) *spherical harmonics*, J. Math.Phys. **23** (1982), 1560–1565.
2. M. Couture, C. J. Cummins, and R. T. Sharp, *Generating functions, elementary multiplets and Young tableaux*, J. Phys. A **23** (1990), 1929–1957.
3. M. Couture and R. T. Sharp, *Reduction of enveloping algebras of low-rank groups*, J. Phys. A **13** (1980), 1925.
4. _____, *Irreducible embedings and polynomial tensors*, J. Phys. A **22** (1989), 1525–1542.
5. P. E. Desmier, J. Patera, and R. T. Sharp, *Finite subgroup basis for compact Lie groups*, Phys. A **114** (1982), 336–340.
6. _____, *Analytic* SU(3) *states in a finite subgroup basis*, J. Math. Phys. **23** (1982), 1393–1398.
7. P. Desmier and R. T. Sharp, *Polynomial tensors for double point groups*, J. Math. Phys. **19** (1978), 2362–2376.
8. R. Gaskell, A. Peccia, and R. T. Sharp, *Generating functions for polynomial irreducible tensors*, J. Math. Phys. **19** (1978), 727–733.
9. R. Gaskell and R. T. Sharp, *Generating fuctions for* G_2 *characters and subgroup branching rules*, J. Math. Phys. **22** (1981), 2736.
10. _____, *Fixed symmetry and fixed class generating functions*, J. Math. Phys. **25** (1984), 2144–2148.
11. F. Gingras, J. Patera, and R. T. Sharp, *Orbit-orbit branching rules between simple low-rank algebras and equal-rank subalgebras*, J. Math. Phys. **33** (1992), 1618–1626.

12. N. Hambli, J. Michelson, and R. T. Sharp, *Character states and generator matrix elements for* $Sp(4) \supset SU(2) \times U(1)$, J. Math. Phys. **37** (1996), 3022–3031.
13. B. Judd, W. Miller, J. Patera, and P. Winternitz, *Complete set of cummuting operators and O(3) scalars in the enveloping algebra of* $SU(3)$, J. Math. Phys. **15** (1974), 1787–1799.
14. R. C. King, J. Patera, and R. T. Sharp, *On finite and continuous little groups of representations of semisimple Lie groups*, J. Phys. A **15** (1982), 1143–1158.
15. C. S. Lam, J. Patera, and R. T. Sharp, *Generating functions for the Coxeter group* H_4, J. Phys. A **29** (1996), 7705–7719.
16. R. V. Moody, J. Patera, and R. T. Sharp, *Character generators for elements of finite order in simple Lie groups* A_l, A_2, A_3, B_2 *and* G_2, J. Math. Phys. **24** (1983), 2387–2397.
17. J. Patera and R. T. Sharp, *Generating functions for characters of group representations and their applications*, Group Theoretical Methods in Physics (Austin, TX, 1978), Lecture Notes in Phys., vol. 94, Springer-Verlag, Berlin–New York, 1979, pp. 175–183.
18. _____, *Generating functions for plethysms of finite and continuous groups*, J. Phys. A **13** (1980), 397–414.
19. _____, *Generating functions for* $SU(2)$ *plethysms with fixed exchange symmetry*, J. Math. Phys. **22** (1981), 261–266.
20. _____, *Signatures of all finite representations of* $SU(p,q)$, J. Math. Phys. **25** (1984), 2128–2132.
21. _____, *Branching rules for representations of simple Lie algebras through Weyl group orbit reduction*, J. Phys. A **22** (1989), 2329–2340.
22. J. Patera, R. T. Sharp, and P. Winternitz, *Polynomial irreducible tensors for point groups*, J. Math. Phys. **19** (1978), 2362–2376.
23. D. Phaneuf and R. T. Sharp, *Polynomial tensors for the space groups*, J. Math. Phys. **26** (1985), 1534–1539.
24. R. T. Sharp, *Internal-labeling operators*, J. Math. Phys. **16** (1975), 2050–2053.
25. J. Van der Jeugt, B. J. W. Huges, and R. T. Sharp, *Applications of generating function techniques to Lie superalgebras*, J. Math. Phys. **26** (1985), 901–912.

CENTRE DE RECHERCHES MATHÉMATIQUES, UNIVERSITÉ DE MONTRÉAL, C.P.6128, SUCC. CENTRE-VILLE, MONTRÉAL, QC H3C3J7, CANADA
E-mail address: `patera@crm.umontreal.ca`

Quasi-Exact Solvability in Nonlinear Optics

G. Álvarez, F. Finkel, A. González-López, and M. A. Rodríguez

ABSTRACT. We derive a unified formalism for the study of a large class of models described by Hamiltonians which are polynomial in bosonic creation and annihilation operators. These models include as particular cases the effective Hamiltonians for nth harmonic generation and photon cascades in nonlinear optics. We construct a complete set of commuting integrals of motion of the Hamiltonian, describe the common eigenspaces of the integrals of motion, and show that the action of the Hamiltonian in each of these common eigenspaces can be represented both by an explicitly constructed finite-dimensional matrix (from which we infer several general properties of the spectrum) and by a quasi-exactly solvable differential operator, whose expression in terms of the usual generators of the $\mathfrak{sl}_2$ Lie algebra is also computed explicitly.

1. Introduction

In 1978 Singh, Biswas and Datta [23] found a family of sextic polynomial potentials for which a finite number of the infinite bound states of the corresponding Schrödinger equation could be calculated exactly by diagonalization of a finite-dimensional matrix. This family of potentials was later rediscovered by Flessas and Das [11, 12] and Turbiner and Ushveridze [32]. Although there are several elementary ways of constructing nontrivial potentials with exactly calculable bound states, the model discussed in [23] turned out to have an underlying algebraic structure that accounts for the phenomenon, which has been termed "quasi-exact solvability" (QES) by Turbiner [28].

In 1990, Zaslavskii [35] pointed out the relation between the exactly calculable part of the spectrum of some QES systems [15, 21, 28, 33, 36, 37] and the intrinsically finite-dimensional Dicke and Heisenberg models for spin systems. Zaslavskii used a suitable ansatz for the coefficients of the eigenvectors of each multiplet of the spin system to build a generating function for these coefficients. This generating function, in turn, was related by elementary transformations to the wavefunction of the QES system.

2000 *Mathematics Subject Classification.* Primary 81U15; Secondary 81R05.

This work was supported by Grant BFM2002-02646 from the Spanish Ministerio de Ciencia y Tecnología.

This is the final form of the paper.

Further research lead to the identification of an increasing number of physical systems whose behavior was partially or totally described by QES Hamiltonians, including electrons in an external oscillator potential [20,24,30], and two-dimensional Schrödinger [25–27], Klein–Gordon [34], and Dirac [6,16] equations for charged particles in Coulomb and magnetic fields. Very recently, Chiang and Ho [7] have given a unified treatment of most of these systems.

Simultaneously, the algebraic structure of these Hamiltonians was being studied from a more mathematical point of view. All the known examples in one dimension turned out to be quadratic polynomials in the generators of a certain realization of $\mathfrak{sl}_2$ by first-order differential operators (see (4.6)). The corresponding Hamiltonians were completely classified with due care of the normalizability of the eigenfunctions [14,17,21,22,28,31].

In 1995, Álvarez and Álvarez-Estrada [1] noted that the nonlinear optical process known as second harmonic generation (in which two photons of angular frequency ω yield a single photon of angular frequency 2ω) was another instance of a QES system, and studied the second harmonic generation effective Hamiltonian by capitalizing on its equivalence to the infinite family of QES sextic polynomial potentials found by Singh, Biswas and Datta [23]. In their method, the transition from the "discrete" photon system to the "continuous" QES systems is achieved by first transforming the effective Hamiltonian from the second quantization to the Bargmann representation. The key idea for separating variables and identifying the one-dimensional QES systems—the second step of their procedure—is to use as a new variable an appropriate quotient of powers of the Bargmann variables that describe each oscillator.

It was natural to ask whether a similar situation holds for higher-order optical processes, and later the same authors [2] used their method to study third harmonic generation (in which three photons of angular frequency ω yield a photon of angular frequency 3ω), whose effective Hamiltonian turned out to be equivalent to a cubic polynomial in the generators of $\mathfrak{sl}_2$, and pointed out that the method works in general for the nth harmonic generation. (Incidentally, third harmonic generation is physically interesting not only as a higher-order effect, but as the dominant effect in centrosymmetric nonlinear materials, in which second harmonic generation is suppressed.)

Recently, Dolya and Zaslavskii [8,9] have studied models whose Hamiltonians are even polynomials in the creation and annihilation operators (but of *a single* degree of freedom) which under certain conditions also lead to QES systems.

Finally, an important idea for our work can be traced back to the early papers on QES models by Ushveridze and Zaslavskii, and appears with more or less emphasis in [1,2,9,33,35]: the use of integrals of motion to separate variables in completely integrable Hamiltonians with more than one degree of freedom.

In this contribution we report on a large class of models with a finite number of degrees of freedom described by Hamiltonians which are polynomial in the creation and annihilation operators [3]. These models include as particular cases the effective Hamiltonians of nth harmonic generation [5,19], as well as photon cascades [18] (for example, two photons of different angular frequencies ω_1 and ω_2 yielding a single photon of angular frequency $\omega_1 + \omega_2$) which, to the best of the authors' knowledge, have not been analyzed before as QES systems.

For each model, we first find a complete set of commuting integrals of motion of the corresponding Hamiltonian. By using the Bargmann representation, we are able to characterize completely the common eigenspaces of these integrals of motion. Finally, we show that the action of the Hamiltonian in these common eigenspaces can be represented by a QES reduced Hamiltonian, whose expression in terms of the usual generators of $\mathfrak{sl}_2$ is computed explicitly.

The derivation of this explicit expression for the differential operator is not just an exercise of academic interest, but the starting point for the application of asymptotic methods to calculate the corresponding eigenvalues in the limit of a large number of photons. By way of example, we give explicitly the expressions for nth harmonic generation and N-photon cascades, which in the former case can be compared with the particular instances of second and third harmonic generation studied in [1, 2].

2. The models

We shall study Hamiltonians which describe the energy-conserving conversion of a number of photons of angular frequencies $\nu_1, \ldots, \nu_N$ into photons of angular frequencies $\mu_1, \ldots, \mu_M$, initially written as is customary in nonlinear optics in the second quantization formalism. More concretely, we will denote by a_l, b_k ($a_l^\dagger, b_k^\dagger$) the boson annihilation (creation) operators of angular frequencies ν_l, μ_k, respectively.

2.1. The Hamiltonian.
The common form of all these Hamiltonians will be

(2.1) $$H = H_0 + gH_1,$$

where the unperturbed term H_0 is the usual Hamiltonian for free modes

(2.2) $$H_0 = \sum_{l=1}^{N} \nu_l a_l^\dagger a_l + \sum_{k=1}^{M} \mu_k b_k^\dagger b_k, \quad \nu_l, \mu_k > 0 \quad N, M \in \mathbb{N},$$

and the perturbation (coupling among the modes) will be

(2.3) $$H_1 = \prod_{k=1}^{M}(b_k^\dagger)^{m_k} \cdot \prod_{l=1}^{N} a_l^{n_l} + \prod_{l=1}^{N}(a_l^\dagger)^{n_l} \cdot \prod_{k=1}^{M} b_k^{m_k} \quad n_l, m_k \in \mathbb{N},$$

with the additional requirement that the frequencies ν_l and μ_k satisfy the energy-conservation constraint

(2.4) $$\sum_{l=1}^{N} n_l \nu_l = \sum_{k=1}^{M} m_k \mu_k.$$

As we anticipated in the Introduction, the model includes as particular cases the usual effective Hamiltonians for nth harmonic generation (if $N = M = m_1 = 1$, $n_1 = n$, and consequently $n\nu_1 = \mu_1$) and for photon cascades (if $M = m_1 = n_1 = \cdots = n_N = 1$, and consequently $\nu_1 + \cdots + \nu_N = \mu_1$), which have recently received considerable attention in the literature [5, 18, 19].

2.2. Complete integrability.
Our first main observation is that these models are *completely integrable*, in the sense that there exists a set of $N + M$ (the total number of degrees of freedom) pairwise commuting, functionally independent operators including the Hamiltonian. In fact, since the unperturbed Hamiltonian

H_0 and the perturbation H_1 commute, we have found it convenient to include them separately in our complete set of commuting operators

$$(2.5) \qquad C = \{H_0, H_1, A_l, B_k \mid l = 1, \ldots, N-1; \; k = 1, \ldots, M-1\},$$

where

$$(2.6) \qquad A_l = n_{l+1} a_l^\dagger a_l - n_l a_{l+1}^\dagger a_{l+1},$$

$$(2.7) \qquad B_k = m_{k+1} b_k^\dagger b_k - m_k b_{k+1}^\dagger b_{k+1}.$$

A straightforward computation using the standard bosonic commutation relations shows that these operators commute pairwise, and we shall see below that they are indeed functionally independent.

As a consequence, there exists a basis of common eigenfunctions of all the operators in the set C. In the next section we will give an explicit description of the simultaneous eigenspaces of H_0, A_l and B_k, and subsequently, by studying the action of H_1 within these subspaces, we shall establish the exact solvability of the model (2.1)–(2.4) and derive some general properties of its spectrum.

3. Invariant subspaces and matrix representation

In this section we first generalize the method of [1] to study the spectrum of the Hamiltonian (2.1) using the Bargmann representation.

3.1. The Bargmann representation for a single boson.

The Bargmann representation of the Hilbert space for a single boson [13] is the space of entire functions of the form

$$(3.1) \qquad f(z) = \sum_{n=0}^{\infty} \frac{c_n}{\sqrt{n!}} z^n, \quad z \in \mathbb{C},$$

where the complex numbers c_n are such that

$$(3.2) \qquad \sum_{n=0}^{\infty} |c_n|^2 < \infty$$

and the scalar product is defined by

$$(3.3) \qquad (g, f) = \frac{1}{\pi} \int_{\mathbb{R} \times \mathbb{R}} d(\operatorname{Re} z) \, d(\operatorname{Im} z) \, \overline{g(z)} f(z) e^{-|z|^2}.$$

In this representation, the orthonormal harmonic oscillator eigenstates $|n\rangle$ are given by

$$(3.4) \qquad |n\rangle \to \frac{z^n}{\sqrt{n!}}, \quad n = 0, 1, \ldots$$

and the annihilation and creation operators, defined by

$$(3.5) \qquad a|n\rangle = \sqrt{n}|n-1\rangle,$$
$$(3.6) \qquad a^\dagger|n\rangle = \sqrt{n+1}|n+1\rangle,$$

are represented by derivation with respect to z and multiplication by z, respectively

$$(3.7) \qquad a \to \frac{d}{dz}, \quad a^\dagger \to z.$$

3.2. The Bargmann representation for H.

The Hilbert space $\mathcal{H}$ of the model (2.1) is the tensor product of $N + M$ copies of the single boson Hilbert space described in the previous subsection. We denote by x_l ($l = 1, \ldots, N$) and y_k ($k = 1, \ldots, M$) the complex variables associated to the "a" and "b" degrees of freedom respectively, and therefore have the following assignments:

$$(3.8) \qquad a_l \to \partial_{x_l}, \quad a_l^\dagger \to x_l, \quad b_k \to \partial_{y_k}, \quad b_k^\dagger \to y_k.$$

A basis of the Hilbert space $\mathcal{H}$ of our models is the set of (in general unnormalized) monomials

$$(3.9) \qquad x^i y^j, \quad i \in \mathbb{Z}_+^N, \; j \in \mathbb{Z}_+^M,$$

where $\mathbb{Z}_+$ denotes the set of nonnegative integers, and we have used the multiindex notation

$$(3.10) \qquad x^i \equiv \prod_{l=1}^{N} x_l^{i_l}, \quad y^j \equiv \prod_{k=1}^{M} y_k^{j_k}.$$

Finally, the corresponding expressions for the operators in the set C are

$$(3.11) \qquad H_0 = \sum_{l=1}^{N} \nu_l x_l \partial_{x_l} + \sum_{k=1}^{M} \mu_k y_k \partial_{y_k},$$

$$(3.12) \qquad H_1 = y^m \partial_x^n + x^n \partial_y^m,$$

$$(3.13) \qquad A_l = n_{l+1} x_l \partial_{x_l} - n_l x_{l+1} \partial_{x_{l+1}},$$

$$(3.14) \qquad B_k = m_{k+1} y_k \partial_{y_k} - m_k y_{k+1} \partial_{y_{k+1}},$$

where x^n and y^m are monomials as in (3.10) and

$$(3.15) \qquad \partial_x^n \equiv \prod_{l=1}^{N} \partial_{x_l}^{n_l}, \quad \partial_y^m \equiv \prod_{k=1}^{M} \partial_{y_k}^{m_k}.$$

In the rest of the paper we shall make use of the multiindex notation without further notice whenever there are no ambiguities in the interpretation of the formulas.

3.3. Invariant subspaces.

Each monomial (3.9) is a common eigenfunction of the operators H_0, A_l, B_k, $l = 1, \ldots, N-1$, $k = 1, \ldots, M-1$, with eigenvalues respectively given by

$$(3.16) \qquad E_0 = \sum_{l=1}^{N} \nu_l i_l + \sum_{k=1}^{M} \mu_k j_k,$$

$$(3.17) \qquad \alpha_l = n_{l+1} i_l - n_l i_{l+1}, \quad l = 1, \ldots, N-1,$$

$$(3.18) \qquad \beta_k = m_{k+1} j_k - m_k j_{k+1}, \quad k = 1, \ldots, M-1.$$

The spectrum of the Hamiltonian (2.1) can thus be computed by diagonalization of the perturbation Hamiltonian H_1 restricted to each of the common eigenspaces $\mathcal{S}_{E_0, \alpha_1, \ldots, \alpha_{N-1}, \beta_1, \ldots, \beta_{M-1}}$ (hereafter "the unperturbed eigenspaces") spanned by the monomials (3.9) whose exponents satisfy (3.16)–(3.18) for fixed values of $E_0, \alpha_1, \ldots, \beta_{M-1}$. Furthermore, since the unperturbed energy E_0 is a finite sum of nonnegative terms, it follows that each unperturbed subspace is finite-dimensional. Thus, the model (2.1) is exactly solvable, because its whole spectrum can be computed by diagonalization of finite matrices.

We shall now describe explicitly the unperturbed eigenspaces, which shall be hereafter denoted by $\mathcal{S}$, without explicit mention of the values $E_0, \alpha_1, \ldots, \beta_{M-1}$. Let $x^p y^q$ be a given (fixed) monomial in $\mathcal{S}$. We shall first prove that $\mathcal{S}$ is spanned by monomials of the form

(3.19) $$f_s = x^p y^q \zeta^s$$

where

(3.20) $$\zeta = \frac{x^n}{y^m}$$

and s ranges over a finite interval $s_0 \leq s \leq s_1$ of $\mathbb{Z}$. Indeed, let $x^{p'} y^{q'}$ be any other monomial in $\mathcal{S}$. The quotient $Q = x^{p'-p} y^{q'-q}$ is then a generalized monomial (i.e., a monomial with possibly negative integer exponents) and thus may not belong to $\mathcal{H}$. However, the action of the operators (3.11)–(3.14) on such generalized monomials is well-defined, and a simple calculation shows that Q satisfies the system of first-order partial differential equations

(3.21) $$H_0 Q = A_1 Q = \cdots = A_{N-1} Q = B_1 Q = \cdots = B_{M-1} Q = 0.$$

In other words, Q must be a joint invariant of the $M + N - 1$ commuting vector fields H_0, A_l, B_k. A second straightforward computation shows that the function ζ defined in (3.20) is also a joint invariant of this set of vector fields. Since the number of independent variables is $N + M$, the general solution of the system (3.21) is an arbitrary smooth function of ζ, and since Q is a generalized monomial, it must be a power of ζ, which proves the first part of our claim.

Since $\mathcal{S}$ is finite-dimensional, the values of s for which f_s lies in $\mathcal{S}$ must be bounded above and below. If

(3.22) $$s_0 = \min\{s \in \mathbb{Z} : f_s \in \mathcal{S}\}, \quad s_1 = \max\{s \in \mathbb{Z} : f_s \in \mathcal{S}\}$$

we shall next prove that

(3.23) $$\mathcal{S} = \langle f_s \mid s \in \mathbb{Z}, \ s_0 \leq s \leq s_1 \rangle$$

where $\langle \cdot \rangle$ denotes the linear span. Indeed, since clearly

(3.24) $$H_0 f_s = E_0 f_s, \quad A_l f_s = \alpha_l f_s, \quad B_k f_s = \beta_k f_s$$

for all integer values of s, it suffices to verify that for all $s \in [s_0, s_1] \cap \mathbb{Z}$ the exponents of f_s are all nonnegative integers. Since $s \in [s_0, s_1]$ and both f_{s_0} and f_{s_1} belong to $\mathcal{S}$, this follows from the inequalities

(3.25) $$p_l + n_l s \geq p_l + n_l s_0 \geq 0, \quad q_k - m_k s \geq q_k - m_k s_1 \geq 0.$$

This completes the proof of our claim.

Note that (3.22) and the inequalities (3.25) imply that the bounds s_0 and s_1 are given by

(3.26) $$s_0 = \max\left\{-\left[\frac{p_l}{n_l}\right] : 1 \leq l \leq N\right\}, \quad s_1 = \min\left\{\left[\frac{q_k}{m_k}\right] : 1 \leq k \leq M\right\}$$

where $[\cdot]$ denotes the integer part. Calling

(3.27) $$\mathcal{N}_l = p_l + s_0 n_l \geq 0, \quad \mathcal{M}_k = q_k - s_0 m_k \geq 0$$

the eigenspace $\mathcal{S}$ can be alternatively written as

(3.28) $$\mathcal{S} \equiv \mathcal{S}_\mathcal{M}^\mathcal{N} = x^\mathcal{N} y^\mathcal{M} \langle 1, \zeta, \ldots, \zeta^r \rangle$$

where

(3.29) $$r = \min\left\{\left[\frac{\mathcal{M}_k}{m_k}\right] : 1 \leq k \leq M\right\}$$

and the nonnegative integers $\mathcal{N}_l$ and $\mathcal{M}_k$ are subject to the single restriction

(3.30) $$\mathcal{N}_l < n_l \quad \text{for al least one } l \in \{1,\ldots,N\}$$

which is an immediate consequence of the definition of s_0. The set of $N + M$ nonnegative integers $\{\mathcal{N}_1,\ldots,\mathcal{N}_N,\mathcal{M}_1,\ldots,\mathcal{M}_M\}$ subject to the condition (3.30) define uniquely the eigenspace $\mathcal{S}$. Indeed, $\mathcal{N}_l$ is the minimum power of x_l and $\mathcal{M}_k$ is the maximum power of y_k of the monomials in $\mathcal{S}$. We have thus proved the main result of this section:

THEOREM 3.1. *The common eigenspaces of the operators H_0, A_l, B_k ($l = 1,\ldots,N-1$, $k = 1,\ldots,M-1$) are the spaces $\mathcal{S}_\mathcal{M}^\mathcal{N}$ given in (3.28)–(3.30).*

3.4. Matrix representation. Since the eigenspaces $\mathcal{S}_\mathcal{M}^\mathcal{N}$ are invariant under H_1 (because this operator commutes with H_0 and A_l, B_k for all l, k), we devote this subsection to study the corresponding action of H_1. Consider the basis $\mathcal{B}$ of $\mathcal{S}_\mathcal{M}^\mathcal{N}$ consisting of the normalized vectors

(3.31) $$e_s = \frac{1}{\sqrt{C_s}} x^{\mathcal{N}+sn} y^{\mathcal{M}-sm}, \quad s = 0, 1, \ldots, r$$

where the normalization constants

(3.32) $$C_s = \prod_{l=1}^{N}(\mathcal{N}_l + n_l s)! \cdot \prod_{k=1}^{M}(\mathcal{M}_k - m_k s)!$$

have been calculated with the help of (3.4). A straightforward computation using (3.12), (3.31) and (3.32) yields

(3.33) $$H_1 e_s = \sqrt{\frac{C_{s-1}}{C_s}} \prod_{l=1}^{N}(\mathcal{N}_l + n_l s)\cdots(\mathcal{N}_l + n_l(s-1)+1)\, e_{s-1}$$
$$+ \sqrt{\frac{C_{s+1}}{C_s}} \prod_{k=1}^{M}(\mathcal{M}_k - m_k s)\cdots(\mathcal{M}_k - m_k(s+1)+1)\, e_{s+1}.$$

Therefore, we have the following

THEOREM 3.2. *The matrix H_1 representing H_1 in the basis $\mathcal{B}$ is tridiagonal, Hermitian, and has zero diagonal entries. The only nonzero entries of H_1 are given by $(\mathsf{H}_1)_{s+1,s} = (\mathsf{H}_1)_{s,s+1}$ where*

(3.34) $$(\mathsf{H}_1)_{s+1,s} = \left(\prod_{l=1}^{N}\prod_{j_l=0}^{n_l-1}(\mathcal{N}_l + n_l s + j_l + 1) \cdot \prod_{k=1}^{M}\prod_{i_k=0}^{m_k-1}(\mathcal{M}_k - m_k s - i_k)\right)^{1/2}$$

and $s = 0, 1, \ldots, r-1$.

REMARK 3.3. Note that H_1 acts irreducibly in $\mathcal{S}_\mathcal{M}^\mathcal{N}$. This implies that H_1 is functionally independent from H_0, A_l and B_k for all k,l. Since $x^n y^m$ is annihilated by all the A_l and B_k but not by H_0, the latter operator is functionally independent from the former ones. Since all the A_l and B_k are clearly functionally independent, the preceding arguments show the functional independence of all the operators in the set C and finish the proof of the complete integrability claimed at the end of Section 2.

We also have the following

THEOREM 3.4. *The eigenvalues of* H_1 *are real, simple, and symmetrically distributed around zero.*

Indeed, let $\delta_s(E)$, $1 \leq s \leq r+1$, be the sth principal minor of the matrix $E\mathbb{1} - \mathsf{H}_1$. A straightforward computation shows that $\delta_s(E)$ satisfies the three-term recursion relation

$$(3.35) \qquad \delta_{s+1}(E) = E\delta_s(E) - h_s^2 \delta_{s-1}(E), \quad s \geq 1$$

where $h_s = (\mathsf{H}_1)_{s,s-1}$ and $\delta_1(E) = E$, $\delta_0(E) = 1$. Since $h_s^2 > 0$ for $1 \leq s \leq r$, [4, Lemma 1 (Section 1.8)] implies that the roots of the characteristic polynomial $\delta_{r+1}(E)$ are real and simple. That the eigenvalues are symmetrically distributed around zero follows from the fact that $\delta_{r+1}(E)$ contains only either even or odd powers of E on account of the form of the recursion relation (3.35).

4. Quasi-exact solvability of the reduced Hamiltonians

For small values of $r+1 = \dim \mathcal{S}_\mathcal{M}^\mathcal{N}$, the eigenvalues of the matrix H_1 representing H_1 can be easily computed in closed form, and in any case the spectrum of H_1 can be determined numerically for fixed values of the parameters $\mathcal{N}_l$, $\mathcal{M}_k$ labelling the space $\mathcal{S}_\mathcal{M}^\mathcal{N}$.

But in many situations of physical interest the number of photons of each frequency—and therefore the parameter r, see (3.29)—can actually be very large. In these cases one is often interested in the behavior of its eigenvalues as a function of $\mathcal{N}_l$, $\mathcal{M}_k$ (and thus r).

In the case of second-harmonic generation, Álvarez and Álvarez-Estrada [1] derived asymptotic formulae for the eigenvalues of H_1 by performing a semiclassical analysis of the second-order ordinary differential equation (ODE) obtained by restricting the perturbation Hamiltonian H_1 to $\mathcal{S}_{\mathcal{M}_1}^{\mathcal{N}_1} = \mathcal{S}_r^\epsilon$, where $\epsilon = 0, 1$. It was also shown in that paper that the restriction of H_1 to $\mathcal{S}_r^\epsilon$ is in fact equivalent to a QES operator in a single variable, in agreement with previous results due to Zaslavskii [35].

In this section we show that the restriction of our general perturbation H_1 to an eigenspace $\mathcal{S}_\mathcal{M}^\mathcal{N}$ is also equivalent to a QES differential operator in a single variable.

4.1. Perturbation Hamiltonian as a differential operator.
Consider the action of H_1 on an element $x^\mathcal{N} y^\mathcal{M} P(\zeta) \in \mathcal{S}_\mathcal{M}^\mathcal{N}$

$$(4.1) \quad H_1\big(x^\mathcal{N} y^\mathcal{M} P(\zeta)\big) = \prod_{k=1}^M y_k^{m_k + \mathcal{M}_k} \cdot \prod_{l=1}^N \partial_{x_l}^{n_l}\big(x_l^{\mathcal{N}_l} P(\zeta)\big)$$
$$+ \prod_{l=1}^N x_l^{n_l + \mathcal{N}_l} \cdot \prod_{k=1}^M \partial_{y_k}^{m_k}\big(y_k^{\mathcal{M}_k} P(\zeta)\big)$$

where P is a polynomial of degree at most r. Since

$$(4.2) \qquad \partial_{x_l}^{n_l}\big(x_l^{\mathcal{N}_l} P(\zeta)\big) = x_l^{\mathcal{N}_l - n_l} \prod_{j_l=0}^{n_l-1}(\mathcal{N}_l - j_l + n_l \zeta \partial_\zeta) P(\zeta)$$

(4.3) $$\partial_{y_k}^{m_k}\left(y_k^{\mathcal{M}_k}P(\zeta)\right) = y_k^{\mathcal{M}_k-m_k}\prod_{i_k=0}^{m_k-1}(\mathcal{M}_k - i_k - m_k\zeta\partial_\zeta)P(\zeta)$$

it follows that

(4.4) $$H_1\left(x^{\mathcal{N}}y^{\mathcal{M}}P(\zeta)\right) = x^{\mathcal{N}}y^{\mathcal{M}}H_{1,\text{red}}P(\zeta)$$

where the reduced Hamiltonian is given by

(4.5) $$H_{1,\text{red}} = \frac{1}{\zeta}\prod_{l=1}^{N}\prod_{j_l=0}^{n_l-1}(\mathcal{N}_l - j_l + n_l\zeta\partial_\zeta) + \zeta\prod_{k=1}^{M}\prod_{i_k=0}^{m_k-1}(\mathcal{M}_k - i_k - m_k\zeta\partial_\zeta).$$

Thus the action of H_1 restricted to $\mathcal{S}_{\mathcal{M}}^{\mathcal{N}}$ is equivalent to that of $H_{1,\text{red}}$ on the space $\mathcal{P}_r$ of polynomials in ζ of degree at most r. Note in particular that the invariance of $\mathcal{S}_{\mathcal{M}}^{\mathcal{N}}$ under H_1 implies that $\mathcal{P}_r$ is invariant under $H_{1,\text{red}}$.

4.2. Perturbation Hamiltonian as a polynomial in the generators of $\mathfrak{sl}_2$.

Let $d = \max(\sum_{l=1}^{N}n_l, \sum_{k=1}^{M}m_k)$ be the order of the differential operator $H_{1,\text{red}}$. If $d \leq r$, a well-known theorem due to Turbiner [29] (see [10] for a simplified proof) states that the invariance of the polynomial space $\mathcal{P}_r$ under $H_{1,\text{red}}$ implies that this operator can be written as a polynomial of degree d in the generators of the realization of $\mathfrak{sl}_2$ spanned by

(4.6) $$J^+ = \zeta^2\partial_\zeta - r\zeta, \quad J^0 = \zeta\partial_\zeta - \frac{r}{2}, \quad J^- = \partial_\zeta.$$

If $d > r$, Turbiner's theorem only guarantees that the rth order part of $H_{1,\text{red}}$ is a polynomial of degree d in the operators (4.6). In fact, even if $d > r$ the reduced Hamiltonian $H_{1,\text{red}}$ can be written as a polynomial of degree d in the generators (4.6). Indeed, choose $k' \in \{1, \ldots, M\}$ and $l' \in \{1, \ldots, N\}$ such that (see (3.29) and (3.30))

(4.7) $$r = \left[\frac{\mathcal{M}_{k'}}{m_{k'}}\right] \quad \text{and} \quad \mathcal{N}_{l'} < n_{l'}.$$

Since all the factors in the products of the expression (4.5) commute, it follows that

(4.8) $$H_{1,\text{red}} = n^n J^- \cdot \prod_{\substack{j_{l'}=0,\\ j_{l'}\neq\mathcal{N}_{l'}}}^{n_{l'}-1}\left(J^0 + \frac{\mathcal{N}_{l'} - j_{l'}}{n_{l'}} + \frac{r}{2}\right) \cdot \prod_{\substack{l=1,\\ l\neq l'}}^{N}\prod_{j_l=0}^{n_l-1}\left(J^0 + \frac{\mathcal{N}_l - j_l}{n_l} + \frac{r}{2}\right)$$
$$+ (-m)^m J^+ \cdot \prod_{\substack{i_{k'}=0\\ i_{k'}\neq\mathcal{M}_{k'}\text{ mod } m_{k'}}}^{m_{k'}-1}\left(J^0 + \frac{i_{k'} - \mathcal{M}_{k'}}{m_{k'}} + \frac{r}{2}\right)$$
$$\cdot \prod_{\substack{k=1\\ k\neq k'}}^{M}\prod_{i_k=0}^{m_k-1}\left(J^0 + \frac{i_k - \mathcal{M}_k}{m_k} + \frac{r}{2}\right).$$

This general expression of the differential operator $H_{1,\text{red}}$ as a polynomial in the generators of the $\mathfrak{sl}_2$ algebra is of course not unique, and can be written in different forms using the commutation relations of the generators (4.6). We conclude this section by showing explicitly the physically most important particular cases, in which the general expression (4.8) simplifies considerably.

4.3. Examples.

EXAMPLE 4.1. Consider the problem of nth harmonic generation, in which $N = M = m_1 = 1$, $n_1 = n$ (and therefore multiindices reduce to ordinary indices). The eigenspaces of H_0 are $\mathcal{S}_r^\epsilon = x^\epsilon y^r \mathcal{P}(\zeta)$, where $r \in \mathbb{Z}_+$, $\epsilon = 0, \ldots, n-1$, and $\zeta = x^n/y$. The corresponding expression of $H_{1,\mathrm{red}}$ reads

$$(4.9) \qquad H_{1,\mathrm{red}} = n^n J^- \prod_{\substack{j=0 \\ j \neq \epsilon}}^{n-1} \left(J^0 + \frac{\epsilon - j}{n} + \frac{r}{2} \right) - J^+.$$

This expression can be easily shown to be equivalent to the expressions obtained in [1] for second harmonic generation ($n = 2$) and in [2] for third-harmonic generation ($n = 3$) and $\epsilon = 0$ (incidentally, in [2] the cases $\epsilon = 1$ and 2 were not considered).

EXAMPLE 4.2. In the case of a multiple photon cascade, $M = m_1 = n_1 = \cdots = n_N = 1$, the eigenspaces of H_0 and A_l, $l = 1, \ldots, N-1$, are $\mathcal{S}_r^{\mathcal{N}} = x^{\mathcal{N}} y^r \mathcal{P}(\zeta)$, where $\mathcal{N}_{l'} = 0$ for some $l' \in \{1, \ldots, N\}$, $r \in \mathbb{Z}_+$, and $\zeta = (x_1 \cdots x_N)/y$. The expression of $H_{1,\mathrm{red}}$ in terms of the generators (4.6) is

$$(4.10) \qquad H_{1,\mathrm{red}} = J^- \prod_{\substack{l=1 \\ l \neq l'}}^{N} \left(J^0 + \mathcal{N}_l + \frac{r}{2} \right) - J^+.$$

5. Summary

In this paper we have given a unified formalism for studying a large class of processes in nonlinear optics, including nth harmonic generation and multiple photon cascades. The distinguishing feature of these models from the physical point of view is the condition of energy conservation, which is the key mathematical condition to prove their complete integrability and the exact solvability.

By using the Bargmann representation, we have been able to provide an explicit description of the common unperturbed eigenspaces, as well as equally explicit expressions of the action of the perturbation Hamiltonian in these unperturbed eigenspaces, both as finite matrices and as a differential operators in one variable.

The explicit form of the finite-matrix representation allowed us to derive some general properties of the spectrum of the restriction of the perturbation H_1 to each unperturbed eigenspace, such as the nondegeneracy and symmetric distribution of the perturbation energies around zero.

The differential operator representation opens the possibility of using asymptotic methods to derive approximate analytic expressions for the eigenvalues valid in the limit of a large number of photons.

The link between these finite-matrix representations and the corresponding differential operators is most conveniently established by the introduction of a new "projective" coordinate, which is a quotient of powers of the Bargmann variables that describe the different oscillators (and physically carries information on the phase difference among the oscillators).

Furthermore, we have been able to give explicit expressions of the reduced QES Hamiltonians as polynomials in the generators of the standard QES realization of $\mathfrak{sl}_2$ by first-order differential operators.

References

1. G. Álvarez and R. F. Álvarez-Estrada, *Semiclassical analysis of a quasi-exactly solvable system: second harmonic generation*, J. Phys. A **28** (1995), 5767–5782.
2. _____, *Third harmonic generation as a third-order quasi-exactly solvable system*, J. Phys. A **34** (2001), 10045–100056.
3. G. Álvarez, F. Finkel, A. González-López, and M. A. Rodríguez, *Quasi-exactly solvable models in nonlinear optics*, J. Phys. A **35** (2002), 8705–8713.
4. F. M. Arscott, *Periodic differential equations*, Internat. Ser. Monogr. Pure Appl. Math., vol. 66, Pergamon, Oxford, 1964.
5. J. Bajer and A. Miranowicz, *Sub-Poissonian photon statistics of higher harmonics: quantum predictions via classical trajectories*, J. Opt. B Quantum Semiclass. Opt. **2** (2000), L10–L14.
6. Y. Brihaye and P. Kosinski, *Quasi-exactly solvable radial Dirac equations*, Modern Phys. Lett. A **13** (1998), 1445–1452.
7. C.-M. Chiang and C.-L. Ho, *Charged particles in external fields as physical examples of quasi-exactly-solvable models: a unified treatment*, Phys. Rev. A **63** (2001), 062105.
8. S. N. Dolya and O. B. Zaslavskii, *The quantum anharmonic oscillator and quasi-exactly solvable Bose systems*, J. Phys. A **33** (2000), L369–L374.
9. _____, *Quasi-exactly solvable quartic Bose Hamiltonians*, J. Phys. A **34** (2001), 5955–5968.
10. F. Finkel and N. Kamran, *The Lie algebraic structure of differential operators admitting invariant spaces of polynomials*, Adv. in Appl. Math. **20** (1998), 300–322.
11. G. P. Flessas, *Exact solutions for a doubly anharmonic oscillator*, Phys. Lett. A **72** (1979), 289–290.
12. G. P. Flessas and K. P. Das, *On the three-dimensional anharmonic oscillator*, Phys. Lett. A **78** (1980), 19–21.
13. A. Galindo and P. Pascual, *Quantum mechanics. I*, Texts Monogr. Phys., Springer-Verlag, Berlin, 1990.
14. A. González-López, N. Kamran, and P. J. Olver, *Normalizability of quasi-exactly solvable Schrödinger operators*, Comm. Math. Phys. **153** (1993), 117–146.
15. _____, *Quasi-exact solvability*, Lie Algebras, Cohomology, and New Applications to Quantum Mechanics (Springfield, MO, 1992), Contemp. Math. vol. 160, Amer. Math. Soc., Providence, RI, 1994, pp. 113–140.
16. C. L. Ho and V. R. Khaliov, *Planar Dirac electron in Coulomb and magnetic fields*, Phys. Rev. A **61** (2000), 032104.
17. N. Kamran and P. J. Olver, *Lie algebras of differential operators and Lie-algebraic potentials*, J. Math. Anal. Appl. **145** (1990), 342–356.
18. V. P. Karassiov, A. A. Gusev and S. I. Vinitsky, *Polynomial Lie algebra methods in solving the second-harmonic generation model: some exact and approximate calculations*, Phys. Lett. A **295** (2002), 247–255.
19. A. B. Klimov and L. L. Sánchez-Soto, *Method of small rotations and effective Hamiltonians in nonlinear quantum optics*, Phys. Rev. A **61** (2000), 063802.
20. A. Samanta and S. K. Ghosh, *Correlation in an exactly solvable two-particle quantum system*, Phys. Rev. A **42** (1990), 1178–1183.
21. M. A. Shifman, *New findings in quantum mechanics (partial algebraization of the spectral problem)*, Internat. J. Modern Phys. A **4** (1989), 2897–2952.
22. M. A. Shifman and A. V. Turbiner, *Quantal problems with partial algebraization of the spectrum*, Comm. Math. Phys. **126** (1989), 347–365.
23. V. Singh, S. N. Biswas and K. Datta, *Anharmonic oscillator and the analytic theory of continued fractions*, Phys. Rev. D **18** (1978), 1901–1908.
24. M. Taut, *Two electrons in an external oscillator potential: particular analytic solutions of a Coulomb correlation problem*, Phys. Rev. A **48** (1993), 3561–3566.
25. _____, *Two electrons in a homogeneous magnetic field: particular analytical solutions*, J. Phys. A **27** (1994), 1045–1055.
26. _____, *Two-dimensional hydrogen in a magnetic field: analytical solutions*, J. Phys. A **28** (1995), 2081–2085.
27. _____, *Two particles with opposite charge in a homogeneous magnetic field: particular analytical solutions of the two-dimensional Schrödinger equation*, J. Phys. A **32** (1999), 5509–5515.

28. A. V. Turbiner, *Quasi-exactly-solvable problems and* sl(2) *algebra*, Comm. Math. Phys. **118** (1988), 467–474.
29. _____, *Lie algebras and polynomials in one variable*, J. Phys. A **25** (1992), L1087–L109
30. _____, *Two electrons in an external oscillator potential: the hidden algebraic structure*, Phys. Rev. A **50** (1994), 5335–5337.
31. _____, *Lie-algebras and linear operators with invariant subspaces*, Lie Algebras, Cohomology, and New Applications to Quantum Mechanics (Springfield, MO, 1992), Contemp. Math., vol. 160, Amer. Math. Soc., Providence, RI, 1994, pp. 263–310.
32. A. V. Turbiner and A. G. Ushveridze, *Spectral singularities and quasi-exactly solvable quantal problem*, Phys. Lett. A **126** (1987), 181–183.
33. A. G. Ushveridze, *Quasi-exactly solvable models in quantum mechanics*, Institute of Physics Publishing, Bristol, 1994.
34. V. M. Villalba and R. Pino, *Analytic solution of a relativistic two-dimensional hydrogen-like atom in a constant magnetic field*, Phys. Lett. A **238** (1998), 49–53.
35. O. B. Zaslavskii, *Effective potential for spin-boson systems and quasi-exactly solvable problems*, Phys. Lett. A **149** (1990), 365–368.
36. O. B. Zaslavskii and V. V. Ulyanov, *New classes of exact solutions of the Schrödinger equation and potential-field description of spin systems*, Zh. Èksper. Teoret. Fiz. **87** (1984), 1724–1733 (Russian); English transl., Soviet Phys. JETP **60** (1984), 991–996.
37. _____, *Periodic effective potentials for spin systems and new exact solutions of the one-dimensional Schrödinger equation for energy zones*, Teor. Mat. Fiz. **71** (1987), 260–271 (Russian); English transl., Theor. Math. Phys. **71** (1987), 520–528.

DEPARTMENTO DE FÍSICA TEÓRICA II, FACULTAD DE CIENCIAS FÍSICAS, UNIVERSIDAD COMPLUTENSE, 28040 MADRID, SPAIN
E-mail address, G. Álvarez: galvarez@fis.ucm.es
E-mail address, F. Finkel: federico@ciruelo.fis.ucm.es
E-mail address, A. González-López: artemio@fis.ucm.es
E-mail address, M. A. Rodríguez: rodrigue@fis.ucm.es

Coherent States, Induced Representations, Geometric Quantization, and Their Vector Coherent State Extensions

D. J. Rowe

ABSTRACT. The theory of coherent state representations is both a theory of induced representations and of quantization. The different perspectives lead to valuable insights and new ways of unifying and extending the two theories. The coherent state approach, and its vector coherent state generalization, have the merit that their physical content is transparent and highly practical. Indeed, they have been shown in numerous applications to be applicable to a wide range of symmetry problems in physics. In particular, in the language of geometric quantization, they are able to quantize interesting physical systems with intrinsic (gauge) degrees of freedom.

1. Introduction

I will talk about some advances in coherent state representation theory which outline how it provides a simple and practical way of constructing induced representations and carrying out the prescriptions of geometric quantization. A typical problem that motivates such developments is from nuclear physics and can be expressed qualitatively as follows.

> Given a microscopic Hamiltonian on a Hilbert space for 168 nucleons, derive the low-energy vibrational and rotational states of the ^{168}Er nucleus.

The low energy-level spectrum of ^{168}Er is shown in Fig. 1. It is seen to comprise a sequence of rotational bands. For example, the lowest energy states of angular momentum $L = 0, 2, 4, \ldots$ make up the ground state band. But many other bands are based on what are presumably vibrational excitations. This is a beautiful problem whose solution invokes some sophisticated representation theory and the quantization of a non-trivial model. Of course, the theory is not restricted in application to the ^{168}Er nucleus or even to nuclear physics. But, it is useful to have a specific problem in mind.

As might be expected, a simple model of nuclear rotations and vibrations does not explain the ^{168}Er spectrum in any detail. The complete dynamics is complex and involves spin as well as many spatial degrees of freedom. However, we are able to understand, in qualitative microscopic terms, the rotational dynamics of the ^{168}Er ground-state and some vibrational bands by constructing a model of nuclear

---

2000 *Mathematics Subject Classification.* 22E47, 22E70, 20C35.
This is the final form of the paper.

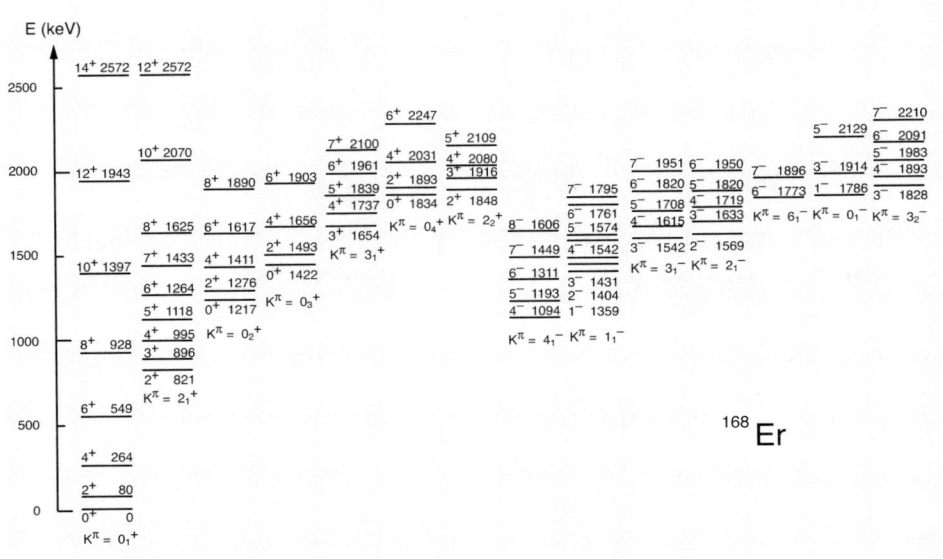

FIGURE 1. The low-energy level spectrum of the ^{168}Er nucleus shown as sequences of vibrational bands. The positive parity bands are shown on the left and the negative parity bands on the right.

rotations and vibrations, whose basic observables can be expressed in many-nucleon coordinates and momenta.

This talk will be aimed at giving a qualitative overall description of the underlying ideas and a general idea of what coherent state theory can do and the new light it sheds on induced representation theory [7–9] and the theory of geometric quantization [5, 6, 12, 13]. I will outline how coherent state theory and its vector coherent extension, incorporates the basic theories of induced representations, e.g., Mackey theory [7–9] and Harish Chandra's method [2–4] of holomorphic induction, in a way that is readily accessible to physicists and provides the explicit expression of the results that physicists need to exploit the symmetries of their problems in real calculations. For example, the coherent state methods don't just give abstract representations of Lie groups, they also give explicit algorithms for constructing matrix elements of these representations and of their infinitesimal generators. I will focus particularly on a recent development in which we have shown that coherent state theory reproduces the three fundamental representations of an algebraic model given by geometric quantization. An algebraic model is essentially a model whose basic observables can be expressed in terms of the infinitesimal generators of a Lie group, known as a dynamical group for the model. The three representations of a model given by geometric quantization are: the classical representations with their Poisson bracket, the representations of prequantization, which are unitary but reducible, and finally the irreducible unitary representations of a full quantization. I will then give an illustration of how the vector coherent state methods make fuller use of the geometrical structures available to geometric quantization and lead to quantizations with interesting non-Abelian intrinsic gauge degrees of freedom.

The developments reported here were done primarily in collaboration with S. D. Bartlett, J. Repka and G. Rosensteel.

2. A classical model with symmetry

The space of states for a classical model is a smooth symplectic manifold (phase space) $\mathbb{P}$. Such a manifold admits local systems of canonical position and momentum coordinates and supports functions which satisfy Poisson bracket relationships defined by the symplectic form for $\mathbb{P}$. A *state* of a classical model is represented by a point on $\mathbb{P}$.

An observable for a classical model, is a smooth function on the phase space having values at each point corresponding to the physical values of the observable for the corresponding state of the model. Thus, a state of the model is characterized by the values of a set of observables. We therefore consider Lie algebras of classical observables with Lie bracket given by their Poisson bracket. A given classical model can support many algebras of observables most of which are infinite-dimensional. However, when we speak of a classical model with symmetry, we suppose that there exists a minimal finite-dimensional algebra of observables having the property that the values of its elements are sufficient to uniquely define any point on the phase space. I shall refer to such a finite-dimensional algebra of observables as a *spectrum generating algebra* for the model.

The gradient of any smooth function on $\mathbb{P}$ defines a tangent vector at every point of $\mathbb{P}$, called a Hamiltonian vector field, which can be interpreted as an infinitesimal generator of displacements on $\mathbb{P}$. Thus, the elements of an algebra of observables integrate to a Lie group of transformations of $\mathbb{P}$.

A finite-dimensional Lie group G of transformations of the phase space $\mathbb{P}$ for a model is said to be a dynamical group for the model if it preserves the symplectic structure of $\mathbb{P}$, and hence Poisson brackets, and also has the property that if m is any point of $\mathbb{P}$ then any other point $m' \in \mathbb{P}$ can be reached by applying some element $g \in G$ to m; thus, we write $m' = g \cdot m$. In other words, a dynamical group should be transitive on the model's phase space. Identifying a dynamical group for a model, has the valuable property that it puts the phase space $\mathbb{P}$ in one-to-one correspondence with a factor space G/H where

(2.1) $$H = \{h \in G \mid h \cdot m = m\}.$$

The concepts of a spectrum generating and a dynamical group are clearly related. In fact, as readily shown by coherent state theory, a classical representation of the Lie algebra of a dynamical group by functions on $\mathbb{P}$ is, in fact, a spectrum generating algebra for the model.

The objective of quantization is to construct a Hilbert space for the model on which its observables act as Hermitian operators. The idea of so-called *canonical quantization*, proposed by Dirac, is to map the classical algebra of observables to an irreducible unitary (Hilbert space) representation. It is now known that this cannot be done for the infinite-dimensional algebra of all classical observables. However, it is also recognized that an appropriate quantization is given by making the Dirac map for a finite-dimensional subalgebra; hence we focus on a finite-dimensional dynamical algebra of observables as defined above.

The first step towards quantization of a given classical representation starts with the observation of the remarkable fact that the classical representation already defines a (generally projective) unitary representation of the subgroup H, defined by (2.1). This is seen in the following simple example.

An example. Consider a phase space $\mathbb{P}$ which is a two sphere of radius M. A suitable dynamical group for this phase space is then the group $G = SO(3)$ of rotations of points on the sphere and the sphere is identified with the factor space $SO(3)/SO(2)$. A corresponding dynamical algebra of observables is given by the components $(\mathcal{L}_x, \mathcal{L}_y, \mathcal{L}_z)$ of the vector from the centre to a point on the sphere. The Hamiltonian vector fields $(\hat{\mathcal{L}}_x, \hat{\mathcal{L}}_y, \hat{\mathcal{L}}_z)$ defined by these functions are then tangent to the sphere and interpreted as infinitesimal generators of rotations.

In particular, $\hat{\mathcal{L}}_z$ is an infinitesimal generator of rotations about the z axis. But now observe that the north pole of the sphere, at which the angular momenta have the values

$$(2.2) \qquad \mathcal{L}_x = \mathcal{L}_y = 0, \quad \mathcal{L}_z = M,$$

is unmoved by a rotation about the z axis. In fact, the subgroup of $SO(2)$ rotations generated by $\hat{\mathcal{L}}_z$ is the isotropy subgroup of all elements of $SO(3)$ rotations that leave the point $(0,0,M)$ fixed. The interesting observation is that the value M of $\mathcal{L}_z$ at the point $(0,0,M)$ defines a unitary representation of an $SO(2)$ rotation through angle θ by $\theta \to e^{iM\theta}$. However, unless M is an integer, this representation is a *projective* representation.

In the language of geometric quantization, we say that the classical representation of the $\mathfrak{so}(3)$ Lie algebra, given by the classical model, is quantizable if $2M$ is an integer.

> *In general, we say that a classical representation of a dynamical algebra $\mathfrak{g}$ on a phase space G/H is quantizable if the unitary representation of H defined by the classical representation of $\mathfrak{g}$ appears in the restriction of some unitary representation of G to its subgroup $H \subset G$.*

3. The coherent state perspective

Whereas geometric quantization starts from a classical model and proceeds to derive its quantizations, coherent state theory starts with an abstract realization of the dynamical group G for the model and its Lie algebra $\mathfrak{g}$ and derives the classical and quantized representations of the model.

The coherent state perspective assumes the existence of some unspecified (generally reducible) representation T of the dynamical group G on a Hilbert space $\mathbb{H}$ and from it derives both classical and quantum representations of the model. If one wishes to be specific, T might be thought of as the regular representation of G. Or, in thinking of the model as a submodel of a many-particle system, it could be a representation of G as a group of unitary transformations of a many-body Hilbert space.

3.1. Classical representations. If $|0\rangle$ is any state in $\mathbb{H}$, the set of states

$$(3.1) \qquad \mathcal{C} = \{|g\rangle = T(g^{-1})|0\rangle \mid g \in G\}$$

is called a *system of coherent states*, in accord with Perelomov's general definition of coherent states [**10**]. (The use of an inverse group element in this definition is purely for convenience; it has no significance.) The map $A \to \mathcal{A}$ from elements of $\mathfrak{g}$ to functions on G, defined by

$$(3.2) \qquad \mathcal{A}(g) = \langle g|\hat{A}|g\rangle, \quad \text{with } \hat{A} = -i\frac{d}{dt}T(e^{iAt})\Big|_{t=0},$$

associates a function $\mathcal{A}$ on the factor space $H\backslash G$ with each element $A \in \mathfrak{g}$, where

$$(3.3) \qquad H = \{h \in G \mid \mathcal{A}(hg) = \mathcal{A}(g), \forall A \in \mathfrak{g}\}.$$

Moreover, this map is a classical representation with Lie bracket given by the Poisson bracket related to the abstract Lie bracket for $\mathfrak{g}$ by

$$(3.4) \qquad \{\mathcal{A}, \mathcal{B}\}(g) = -\frac{i}{\hbar}\langle g|[\hat{A}, \hat{B}]|g\rangle.$$

The notable fact, well known in the theory of geometric quantization, is that the symplectic form defined on the factor space $H\backslash G$ by this Poisson bracket is non-degenerate, Thus, $H\backslash G$ is a phase space and the map (3.2) is a classical representation of the spectrum generating algebra.

The Lie algebra $\mathfrak{h}$ of H is the subalgebra of $\mathfrak{g}$

$$(3.5) \qquad \mathfrak{h} = \{A \in \mathfrak{g} \mid \langle 0|[\hat{A}, \hat{B}]|0\rangle = 0, \forall B \in \mathfrak{g}\}.$$

It follows that, for $A \in \mathfrak{h}$, the map

$$(3.6) \qquad A \to \mathcal{A}(e) = \langle 0|\hat{A}|0\rangle$$

defines a one-dimensional representation of $\mathfrak{h}$ (e denotes the identity element of G). The corresponding one-dimensional representation of H

$$(3.7) \qquad h \to \chi(h) = \langle 0|T(h)|0\rangle, \quad \forall h \in H,$$

is unitary but generally projective.

3.2. The reducible unitary representations of prequantization.
If a classical representation is quantizable according to the definition and provided the restriction to H of the representation T of G actually contains the unitary representation χ as a subrepresentation, then there exists some state $|0\rangle \in \mathbb{H}$ which not only has the property that $\langle 0|T(h)|0\rangle = \chi(h)$ but also satisfies the stronger condition

$$(3.8) \qquad T(h)|0\rangle = \chi(h)|0\rangle, \quad \forall h \in H.$$

Given such a state, there is a map from $\mathbb{H}$ to a space of coherent-state wave function on G in which a state $|\psi\rangle \in \mathbb{H}$ maps to a function ψ with values

$$(3.9) \qquad \psi(g) = \langle 0|T(g)|\psi\rangle, \quad g \in G,$$

and for which

$$(3.10) \qquad \psi(hg) = \chi(h)\psi(g), \quad \forall h \in H.$$

Equivalently, if $\psi(Ag)$ is defined by

$$(3.11) \qquad \psi(Ag) = i\frac{d}{dt}\psi(e^{-iAt}g)\Big|_{t=0},$$

then the constraint condition (3.10) is expressed

$$(3.12) \qquad \psi(Ag) = \mathcal{A}(e)\psi(g), \quad \forall A \in \mathfrak{h}.$$

In fact, the above construction has a natural generalization. Suppose that the representation χ of H is contained, not as a subrepresentation of the restriction of the representation T to the subgroup $H \subset G$, but as a component in a direct integral decomposition of this restriction. Then there will not exist a state $|0\rangle \in \mathbb{H}$ having the property (3.10) but there will exist a functional $\langle\varphi|$ on a dense subspace $\mathbb{H}_D$ of $\mathbb{H}$ having the property that the coherent state wave functions

$$\psi(g) = \langle\varphi|T(g)|\psi\rangle, \quad g \in G, \tag{3.13}$$

are well-defined for all $|\psi\rangle \in \mathbb{H}_D$ and satisfy the conditions (3.10) and (3.12).

The space of such coherent-state wave functions carries a (generally reducible) coherent-state representation of G given by

$$[\Gamma(g)\psi](g') = \psi(g'g). \tag{3.14}$$

This representation is known to be unitary wrt to a suitably defined inner product.

For example, if $\mathbb{H}$ is the Hilbert space of square integrable functions over G (relative to the invariant measure) and T is the regular representation then, for $|\varphi\rangle$ a normalizable state in $\mathbb{H}$, the coherent state wave functions are a subset of $\mathbb{H}$. Moreover, if $|\varphi\rangle$ is a generic state with isotropy subgroup $H \subset G$, i.e., for which $T(h)|\varphi\rangle = \chi(h)|\varphi\rangle$, then the coherent-state representation Γ of G is identical to the representation of G induced from the representation χ of the subgroup $H \subset G$.

The representation (3.14) also defines a corresponding unitary representation of the Lie algbra $\mathfrak{g}$ for which

$$[\Gamma(A)\psi](g) = i\frac{d}{dt}\psi(ge^{-iAt})\Big|_{t=0}, \quad \forall A \in \mathfrak{g}. \tag{3.15}$$

This representation can be expressed in the form

$$[\Gamma(A)\psi](g) = \langle\varphi|\hat{A}(g)T(g)|\psi\rangle = \mathcal{A}(g)\psi(g) + i\hbar[\nabla_{\mathcal{A}}\psi](g), \tag{3.16}$$

where $\hat{A}(g) = T(g)\hat{A}T(g^{-1})$, $\mathcal{A}(g) = \langle\varphi|\hat{A}(g)|\varphi\rangle$ is the value of the classical observable $\mathcal{A}$ at g, and $\nabla_{\mathcal{A}}$ is defined by

$$i\hbar[\nabla_{\mathcal{A}}\psi](g) = \langle\varphi|\hat{A}(g)T(g)|\psi\rangle - \mathcal{A}(g)\psi(g). \tag{3.17}$$

Thus, if $\{A_\alpha\}$ is a basis for the Lie algebra $\mathfrak{g}$ and we make the expansion

$$\hat{A}(g) = \sum_\alpha A^\alpha(g)\hat{A}_\alpha, \tag{3.18}$$

it follows that

$$i\hbar[\nabla_{\mathcal{A}}\psi](g) = \sum_\alpha A^\alpha(g)(i\hbar\partial_\alpha - \theta_\alpha)\psi(g), \tag{3.19}$$

where

$$i\hbar\partial_\alpha\psi(g) = i\hbar\frac{d}{dt}\psi(e^{-\frac{i}{\hbar}A_\alpha t}g)\Big|_{t=0}, \tag{3.20}$$

and

$$\theta_\alpha = \langle\varphi|\hat{A}_\alpha|\varphi\rangle = \mathcal{A}_\alpha(e). \tag{3.21}$$

As shown explicitly in [1], $\nabla_{\mathcal{A}}$ is the covariant derivative for the function $\mathcal{A}$ and the coherent state representation

$$\Gamma(A) = \mathcal{A} + i\hbar\nabla_{\mathcal{A}}, \tag{3.22}$$

is identical to the corresponding representation of prequantization.

Note, however, that the coherent state construction also gives representations on other spaces. Moreover, by choosing a functional $\langle\varphi|$ with special properties, it is possible to induce irreducible coherent state representations.

3.3. The irreducible unitary representations of a full quantization.
For a full quantization, the coherent state representation should be irreducible as well as unitary. It will be irreducible if $\langle\varphi|$ can be chosen such that only the component of a state $|\psi\rangle \in \mathbb{H}_D$ belonging to an irrep gives a non-vanishing contribution to the wave function

$$\psi(g) = \langle\varphi|T(g)|\psi\rangle, \quad g \in G. \tag{3.23}$$

Such a functional can often be defined, for example, by extending the condition (3.12) to a suitable subalgebra $\mathfrak{p}$ in the chain $\mathfrak{h} \subset \mathfrak{p} \subset \mathfrak{g}^c$, where $\mathfrak{g}^c$ is the complex extension of $\mathfrak{g}$. Let $\tilde{\chi}$ denote a one-dimensional irrep of $\mathfrak{p} \subset \mathfrak{g}^c$ for which

$$\tilde{\chi}(A) = \mathcal{A}(e), \quad \forall A \in \mathfrak{h}, \tag{3.24}$$

and let $\langle\varphi|$ be a functional on a dense subspace $\mathbb{H}_D \subseteq \mathbb{H}$ such that

$$\langle\varphi|XT(g)|\psi\rangle = \tilde{\chi}(X)\psi(g), \quad \forall X \in \mathfrak{p}. \tag{3.25}$$

This condition includes the conditions expressed by (3.12) and, for a suitable choice of $\mathfrak{p}$, it may be sufficient to ensure that the coherent state representation is irreducible.

For example, if $\mathfrak{g}$ is semisimple and $\mathfrak{h}$ is a Cartan subalgebra, $|\varphi\rangle$ could be a highest or a lowest weight state for an irrep. An appropriate choice of $\mathfrak{p}$ would then be the parabolic subalgebra of $\mathfrak{g}^c$ containing $\mathfrak{h}$ and a set of raising (or lowering) operators. A suitable subalgebra $\mathfrak{p} \subset \mathfrak{g}^c$ is known in the language of geometric quantization as a *polarization*.

3.4. An example.
Let $\mathfrak{g}$ be the Heisenberg–Weyl algebra; it is spanned by the position x and momentum p observables plus the identity 1 with Poisson brackets

$$\{x,p\} = 1, \quad \{1,x\} = \{1,p\} = 0. \tag{3.26}$$

Let $x \to \hat{x}$, $p \to \hat{p}$, $1 \to \hat{I}$ denote a generic representation of this algebra such that

$$[\hat{x},\hat{p}] = i\hbar\hat{I}, \quad [\hat{I},\hat{x}] = [\hat{I},\hat{p}] = 0. \tag{3.27}$$

For any normalized state $|0\rangle$ in the representation space, define the coherent states

$$\{|x,p\rangle = e^{-(i/\hbar)x\hat{p}}e^{(i/\hbar)p\hat{x}}|0\rangle \mid x,p \in \mathbb{R}\}. \tag{3.28}$$

A classical representation of the HW algebra is then given by

$$\begin{aligned}\hat{x} &\to \mathcal{X}(x,p) = \langle x,p|\hat{x}|x,p\rangle = x, \\ \hat{p} &\to \mathcal{P}(x,p) = \langle x,p|\hat{p}|x,p\rangle = p, \\ \hat{I} &\to \mathcal{I}(x,p) = 1,\end{aligned} \tag{3.29}$$

with Poisson brackets given, for example, by

$$\{x,p\} = -\frac{i}{\hbar}\langle x,p|[\hat{x},\hat{p}]|x,p\rangle. \tag{3.30}$$

A state $|\psi\rangle$ in the (unspecified) representation space has a coherent state wave function

$$\psi(x,p) = \langle x,p|\psi\rangle, \tag{3.31}$$

and prequantization gives the representation of the HW algebra

$$\Gamma(x) = x + i\hbar\frac{\partial}{\partial p}, \quad \Gamma(p) = -i\hbar\frac{\partial}{\partial x}, \quad \Gamma(x) = 1, \tag{3.32}$$

as linear operators on these wave functions. This representation is reducible. To get an irreducible representation, choose a functional $\langle\varphi|$ on the dense subspace of continuous wave functions in the representation space which is an eigenstate of the operator $\hat{x}$ with zero eigenvalue. The coherent state wave functions

(3.33) $$\psi(x,p) = \langle\varphi|e^{-(i/\hbar)p\hat{x}}e^{(i/\hbar)x\hat{p}}|\psi\rangle$$

then reduce to the subset of x-independent functions and the coherent state representation reduces to

(3.34) $$\Gamma(x) = x, \quad \Gamma(p) = -i\hbar\frac{\partial}{\partial x}, \quad \Gamma(x) = 1.$$

This representation is now irreducible.

For the above prequantization, the isotropy subalgebra $\mathfrak{h}$ is the subalgebra of $\mathfrak{g}$ spanned by the identity element 1. For the full quantization, $\mathfrak{h}$ is extended to the polarization $\mathfrak{p}$ spanned by 1 and x. Since $\mathfrak{p}$ is in fact a subalgebra of $\mathfrak{g}$ it is said to be a *real polarization*. The Bargmann–Segal quantization is the complex polarization spanned by 1 and $x + ip$.

4. A microscopic model of nuclear rotations and vibrations

The above methods can be applied to a non-trivial microscopic model of the rotational-vibrational dynamics of a nucleus proposed by Weaver, Biedenharn, and Cusson [**14**] and known as the CM(3) model, where the acronym stands for Collective Motion in 3 dimensions. The quantization of this model was first given by Rosensteel [**11**].

The set of observables for a model of rotations and vibrations naturally includes six Cartesian quadrupole moments which, for a many-particle nucleus, have a natural microscopic expression

(4.1) $$Q_{ij} = \sum_n x_{ni}x_{nj} = Q_{ji}, \quad i,j = 1,2,3,$$

where (x_{n1}, x_{n2}, x_{n3}) are Cartesian coordinates for the nth particle. These observables define the deformation and orientation of an ellipsoidally shaped nucleus. They span an Abelian real Lie algebra which we denote by $\mathbb{R}^6$. We also need vibrational and angular (rotational) momentum observables of the nucleus which can be interpreted as infinitesimal generators of shape change. A natural choice is the nine moment-of-momentum observables

(4.2) $$X_{ij} = \sum_n x_{ni}p_{nj}, \quad i,j = 1,2,3,$$

which are infinitesimal generators of a general linear group $\mathrm{GL}(3,\mathbb{R})$ and span a $\mathrm{gl}(3,\mathbb{R})$ Lie algebra.

The span of these combined sets of observables $\{Q_{ij}, X_{ij}\}$ is the semi-direct sum Lie algebra $[\mathbb{R}^6]\,\mathrm{gl}(3,\mathbb{R})$ which we take as the spectrum generating algebra for our model. An element of the corresponding dynamical group is conveniently expressed in the form

(4.3) $$(B, g) \equiv e^{-(i/\hbar)\sum_{nij} B_{ij}Q_{ij}}g,$$

where B is a real symmetric matrix and $g \in \mathrm{GL}(3,\mathbb{R})$.

A question that arises is: why are there nine momentum observables $\{X_{ij}\}$ in this model but only six deformation observables $\{Q_{ij}\}$? In fact, this is a common

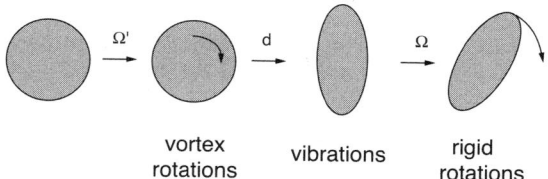

FIGURE 2. Application of the general linear transformation $g = \Omega' d\Omega$ to a spherical nucleus. The first rotation produces an intrinsic rotational motion but no observable effect on the quadrupole moments of the nucleus.

situation and is automatically looked after in the construction when the elements of the isotropy subgroup H are factored out to give a phase space $H\backslash G$. A physical interpretation of how this might be achieved in the present example is given by considering a standard factorization of a $GL(3,\mathbb{R})$ group element as a product of a rotation, a diagonal matrix, and another rotation:

$$(4.4) \qquad g = \Omega' d\Omega.$$

The result of applying such a sequence of transformations to a spherical nuclear density distribution is illustrated in Figure 2. It is seen that the first rotation Ω' produces an intrinsic rotational motion but no observable effect on the quadrupole moments of the nucleus. We refer to such unobservable rotations as *intrinsic vortex rotations*. If H is set equal to SO(3) and identified with the subgroup of vortex rotations, the factor space $H\backslash G$ becomes a 12-dimensional phase space with six local coordinates and six local momentum in the neighbourhood of any point on its surface.

The above microscopic expressions of the model observables, give the dynamical group of the model an immediate unitary representation on the space of many-particle wave functions; we denote this representation by

$$(4.5) \qquad (B,g) \to e^{-(i/\hbar)\sum_{nij} B_{ij}\hat{Q}_{ij}} T(g),$$

where $T(g)$ is the unitary representation of an element $g \in \mathrm{GL}(3,\mathbb{R})$ and $\hat{Q}_{ij}$ is defined by

$$(4.6) \qquad [\hat{Q}_{ij}\psi] = \sum_n x_{ni} x_{nj} \psi(x).$$

However, this representation is highly reducible and cannot be considered a quantization of the model. A quantization is obtained, for example, by choosing a functional $|\varphi\rangle$ that is a zero-eigenstate of the quadrupole moments and has angular momentum zero. The coherent state wave functions

$$(4.7) \qquad \psi(g) \equiv \langle\varphi|e^{-(i/\hbar)\sum_{nij} B_{ij}\hat{Q}_{ij}} T(g)|\psi\rangle = \langle\varphi|T(d\Omega)|\psi\rangle$$

are then seen to be independent of both B and Ω'; i.e., they satisfy

$$(4.8) \qquad \psi(\Omega g) = \psi(g).$$

The coherent state representation of the quadrupole operators on these wave functions is given by

$$(4.9) \qquad [\Gamma(Q_{ij})\psi](g) = (\tilde{g}g)_{ij} \psi(g),$$

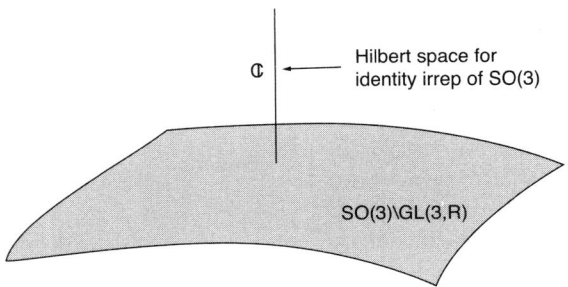

FIGURE 3. Coherent state wave functions for a zero vorticity representation of the rotor-vibrator model are sections of a complex line bundle associated to the principle $GL(3, \mathbb{R}) \to SO(3) \backslash GL(3, \mathbb{R})$ bundle by the identity representation of $SO(3)$.

and the general linear transformations by

(4.10) $$[\Gamma(g)\psi](g') = \psi(g'g).$$

This coherent state representation, is equivalent to a Mackey induced representation and is known to be irreducible. Thus, it is a quantization of the CM(3) model.

5. Intrinsic degrees of freedom in a vector coherent state extension

From a geometrical perspective, one can think of the manifold $SO(3) \backslash GL(3, \mathbb{R})$ as the configuration space for the above model. One can also regard the coherent state wave functions as sections of a complex line bundle over $SO(3) \backslash GL(3, \mathbb{R})$. This bundle is, in fact, a line bundle that is associated to the principle $GL(3, \mathbb{R}) \to SO(3) \backslash GL(3, \mathbb{R})$ bundle by the identity representation of the subgroup $SO(3) \subset GL(3, \mathbb{R})$.

The latter perspective suggests a more general construction with non-zero vorticity in which the wave functions are sections of a vector bundle associated to the principle $GL(3, \mathbb{R}) \to SO(3) \backslash GL(3, \mathbb{R})$ bundle by a non-trivial representation of $SO(3)$. Such a generalization sounds complicated but, in fact, it is a natural extension of a coherent state representation to a *vector coherent state representation*.

Such an irrep is readily constructed as follows. Let $\hat{\rho}$ denote an $SO(3)$ irrep of angular momentum v on a $(2v+1)$-dimensional Hilbert space with orthonormal basis $\{\xi_{v\nu} \mid \nu = -v, \ldots, +v\}$. Thus, we suppose that

(5.1) $$\hat{\rho}(\Omega)\xi_{v\nu} = \sum_{\mu} \xi_{v\mu} \mathcal{D}^v_{\mu\nu}(\Omega), \quad \Omega \in SO(3).$$

Now, instead of a single functional $|\varphi\rangle$, we consider a set of functionals $\{|v\nu\rangle \mid \nu = -v, \ldots, +v\}$ which are all zero eigenstates of the quadrupole moments and transform under rotations as states of angular momentum v and z-component ν, i.e.,

(5.2) $$T(\Omega)|v\nu\rangle = \sum_{\mu} |v\mu\rangle \mathcal{D}^v_{\mu\nu}(\Omega), \quad \Omega \in SO(3) \subset GL(3, \mathbb{R}).$$

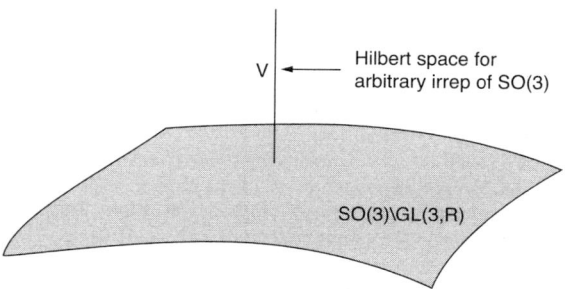

FIGURE 4. Vector coherent state wave functions for a non-zero vorticity representation of the rotor-vibrator model are sections of a vector bundle associated to the principle $GL(3,\mathbb{R}) \to SO(3)\backslash GL(3,\mathbb{R})$ bundle by a multidimensional irrep of $SO(3)$.

Vector-valued coherent state wave functions are then defined by

$$(5.3) \qquad \Psi(g) \equiv \sum_\nu \xi_{v\nu} \langle v\nu | e^{-(i/\hbar) \sum_{nij} B_{ij} \hat{Q}_{ij}} T(g) | \psi \rangle = \sum_\nu \xi_{v\nu} \psi_\nu(g),$$

where

$$(5.4) \qquad \psi_\nu(g) = \langle v\nu | T(g) | \psi \rangle.$$

Thus, VCS wave functions are seen to have intrinsic (vortex spin) degrees of freedom. They continue to be independent of B. However, instead of (4.8), they satisfy the identity

$$(5.5) \qquad \Psi(\Omega g) = \sum_\nu \hat{\rho}(\Omega) \xi_{v\nu} \psi_\nu(g) = \hat{\rho}(\Omega) \Psi(g).$$

VCS wave functions are, in fact, sections of the vector bundle associated to the principle $GL(3,\mathbb{R}) \to SO(3)\backslash GL(3,\mathbb{R})$ bundle by the irrep $\hat{\rho}$ of $SO(3)$. Moreover, as shown in [1], VCS representations can be expressed in the language of geometrical quantization. However, to use them, it is not necessary to understand this geometrical interpretation. As a result of their simplicity, they have been widely used by physicists to construct the explicit matrices of a wide range of irreducible Lie algebra and Lie group representations and in the quantization of algebraic models.

References

1. S. D. Bartlett, D. J. Rowe, and J. Repka, *Vector coherent state representations, induced representations and geometric quantization. II. Vector coherent state representations*, J. Phys. A **35** (2002), 5599–5623.
2. Harish-Chandra, *Representations of semisimple Lie groups. IV*, Amer. J. Math. **77** (1955), 743–777.
3. _____, *Representations of semisimple Lie groups. V*, Amer. J. Math. **78** (1956), **78** 1–41.
4. _____, *Representations of semisimple Lei groups. VI. Integrable and square-integrable representations*, Amer. J. Math. **78** (1956), 564–628.
5. B. Kostant, *On certain unitary representations which arise from a quantization theory*, Group Representations in Mathematics and Physics, Lecture Notes in Phys., vol. 6, Springer, Berlin, 1970, pp. 237–253.
6. _____, *Quantization and unitary representations*, Letures in Modern Analysis and Applications. III, Lecture Notes in Math., vol. 170, Springer, Berlin, 1970, pp. 87–208.
7. G. W. Mackey, *Induced representations of locally compact groups. I*, Ann. of Math. (2) **55** (1952) 101–139.

8. _____, *Induced representation of groups and quantum mechanics*, Benjamin, New York, 1968.
9. _____, *Unitary group representations in physics, probability and number theory*, Math. Lecture Note Ser., vol. 55, Benjamin, Reading, MA, 1978.
10. A. Perelomov, *Generalized coherent states and their applications*, Texts Monogr. Phys., Springer-Verlag, Berlin, 1986.
11. G. Rosensteel and D. J. Rowe, *The algebraic CM(3) model*, Ann. Physics **96** (1976), 1–42.
12. J.-M. Souriau, *Quantification géométrique*, Comm. Math. Phys. **1** (1966), 374–398.
13. _____, *Structure des systèmes dynamiques*, Dunod, Paris, 1970.
14. L. Weaver, L. C. Biedenharn, and R. Y. Cusson, *Rotational bands in nuclei as induced group representations*, Ann. Physics **77** (1973), 250–278.

DEPARTMENT OF PHYSICS, UNIVERSITY OF TORONTO, TORONTO, ON M5S 1A7, CANADA
E-mail address: `rowe@physics.utoronto.ca`

Symmetry Math Video Game Used to Train Profound Spatial-Temporal Reasoning Abilities Equivalent to Dynamical Knot Theory

Mark Bodner and *Gordon L. Shaw*

ABSTRACT. Spatial-temporal (ST) reasoning (making a mental image and thinking ahead in space and time using symmetry operations, as in chess, music and math) is of considerable benefit to children in understanding math and science. Big Seed, an ST computer game requiring thinking ahead up to 10 steps with 27 possible moves at each step, is challenging, demanding and relevant, even for researchers. *The large ST reasoning capabilities on Big Seed demonstrated by groups of young children far exceeded even the most optimistic expectations.* The equivalence of Big Seed to a rich dynamical knot theory is presented here. The dynamical knot theory is shown to represent complex DNA structures and thus is suggested to be useful in understanding DNA function. Other applications of this knot theory to science are presented. Thus while *clearly* 2nd graders do not have the formal background to research in math and science. most have the trainable ability to do relevant underlying ST reasoning.

Spatial-temporal (ST) reasoning—"thinking in pictures using symmetry"—has long been recognized as essential to how we think in math and science, and has been crucial in the intellectual history of technological development [5]. The recognition of its importance has been stressed. Despite this, ST reasoning is essentially neglected in comparison to the complementary quantitative language based (word problems, equations and symbols) reasoning in schools at all grade levels [13, 15]. This puts all children at a disadvantage in our high-tech world where math reasoning is crucial. It has been emphasized [10] that math proficiency is "the new gatekeeper for access to economic participation {in the information age} ...and people who don't have it are like the people who couldn't read and write in the industrial age." Of large educational importance is that children i) learn to understand math concepts *and* ii) be able to perform well on standardized math tests.

The concept that built into our highly structured brain is the innate ability to due sophisticated symmetry operations looking for patterns in space and time, and is at the heart of how we think and reason is now being established. This seminal role of symmetry is why we are pleased to dedicate this paper to the memory of a great mathematician and human being, Robert T. Sharp.

2000 *Mathematics Subject Classification.* 57M60.
This is the final form of the paper.

Based on Mountcastle's columnar organizational principle for cortical function [11, 12, 16], the structured trion model [21] predicted that inherent families of ST firing patterns related by symmetries would occur. (Experimental evidence was found [3] in primate cortical data during memory tasks for families of temporal firing patterns related by symmetries.) These trion model firing patterns can be selectively enhanced [22] through experience and thus are defined as memory patterns. These inherent memory families of firing patterns have the surprising built-in ability to recognize, compare, and find relationships among stimuli using symmetry properties [9]. This ability was proposed to underlie performance on spatial recognition tasks, such as classifying and detecting similarities among objects. The time development of these memory patterns into specific temporal sequences over several cortical areas and tens of seconds was proposed to underlie ST reasoning. *It was predicted that ST reasoning ability exploiting symmetry operations is "innate" or "built-in" to cortex as represented by the Mountcastle columnar principle* [8].

Although brain functions are typically associated with specific, localized regions of cortex, all higher brain functions such as ST reasoning draw upon a wide range of cortical areas [2, 19]. In the trion model, all cortical areas share a common cortical language [18]. It was predicted [8] that music training could enhance the innate cortical ability to do ST reasoning. It was confirmed that piano keyboard training with 3-year-olds enhanced ST reasoning [17]. Piano keyboard training along with training on STAR (ST Animation Reasoning) math video game software enhanced learning of fractions and proportional math by 2nd graders [6]. Adding math integration (bridging the ST approach to language based learning) to music and STAR training completes the Music + Math Program [20]. In general this Program is highly successful in helping young children visualize and master difficult math concepts. At present there are over 8,000 2nd–4th graders (see www.mindinstitue.net). We have presented the remarkable results on one especially challenging STAR game [14].

Big Seed game

One of the STAR math video games, Seed, provides excellent ST training for Music+Math 2nd graders in thinking ahead many steps using symmetry operations. (An interactive version of some STAR games including Seed is on the CD-Rom enclosed in Ref. 20.) An extremely challenging version, Big Seed, has proven to be highly demanding, stimulating and relevant even for research scientists and mathematicians.

Three examples of the 76 puzzles in Big Seed are shown in Figure 1, ranging from a simple introductory puzzle Figure 1(A) to avery difficult middle level puzzle Figure 1(B) to a "master" level puzzle Figure 1(C). In order to appreciate the profound nature of the results from four groups of children, please solve puzzle Figure 1(a) then examine puzzles Figures 1(B)–(C). It is difficult to solve them without having been trained on the earlier puzzles.

Puzzle difficulty is based on increasing complexity of interleaved structures that need to be identified and segmented or unified. These features include an increased number of steps, increase in the number of alterations to the initial Seed at early stages to prepare it for final space filling; increased number of interleaved separate structures, discontinuity of interleaved structures (that is the symmetric structure is not completely connected see, e.g., Figure 1(B)).

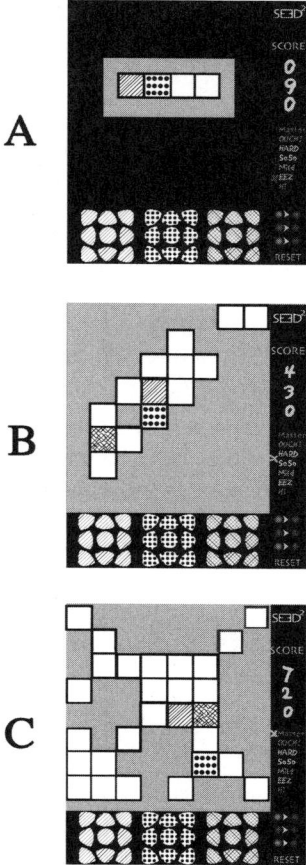

FIGURE 1. Examples of the 76 Big Seed ST puzzles. The Big Seed game requires that all the white squares are covered by the textured "Seed" squares without going outside the given boundaries and without overlaying a textured square. A texture arrow unfolds all the squares of that texture along the direction of the arrow about the axis perpendicular to the arrow and touching the nearest square of that texture. Transformations of the textures (all the squares of a given texture) may be carried out as indicated to the right of the direction arrow buttons. There are 27 possible operations at any given step. Solutions have up to 10 steps. Big Seed will challenge anyone. (A) "Introductory" puzzle (number 10). This puzzle is solved by making a texture change in step 1. For example, first make a texture change from diagonals to dots and then unfold the dots squares to the right. (B) "Very Difficult" puzzle (number 44). In order to appreciate the striking nature of the results from the four studies in Ref. 14, please try to solve puzzle B. It is difficult to solve without having been trained in the previous puzzles. (C) "Master level" puzzle (number 73—difficulty increases with puzzle number).

Four studies of Big Seed with children from 2 nd grade to middle school with remarkable results have been reported [**14**]. The main findings of these studies were:

(a) Rapid speed of improvement
(b) Universality of the Big Seed performance improvements: every child had substantial improvement independent of age, gender or background including language based math performance.
(c) The large magnitude of the children's ST performance after less than 7 hours of training which far exceeded all expectations
(d) Dramatic retention of Big Seed performance after 7 months.
(e) Anecdotal reports from many of the children and from their teachers that the ability to solve a new level of difficulty Big Seed puzzle appeared to just "click in" for a child and then remained (see (d)).

We now demonstrate the equivalence of Big Seed (and ST reasoning) to mathematical reasoning, through showing its equivalence to the specific sub-branch of topology that is knot theory. A specific application of the framework developed to modeling and understanding problems in DNA structure and function will be presented. Subsequent links to other branches of mathematics and applications will be briefly discussed as well.

The planar representation of knots realized in this version of Big Seed allows the exploration of particular classes of knots grouped according to their dynamic evolution. This introduces the idea of applying a dynamical knot theory in the study in exploring physical processes. We concentrate in this paper on the implications and structures generated within the planar representation of knots realized in the present Big Seed game. While this is clearly a subgroup of possible larger and more general classes (which are currently being developed—for example, affine Lie groups represented can be further generalized by allowing initial Seeds other than used to generate the planar graphs in the present game, i.e., those consisting of 2 vertices connected by a single edge), a very rich and broad class of structure and dynamics is already obtained within this restricted set.

Fundamental Big Seed knot representation

We begin by introducing the planar representation of a knot. This representation is simply a graph (which therefore connects with Graph Theory—which in turn implicates the direct connection of dynamic knot theory with such diverse areas of studies as combinatorial mathematics, neural networks, and formal languages). The planar representation of knots has proven to be a powerful tool in identifying unique knots [**1**].

The fundamental "Seed" is the square planar graph shown in Figure 2(A). The graph mathematically defined is simply a set of vertices, and a set of edges that connect those vertices in some fashion. The basic square is a graph with 4 vertices (the corners of the square) and 4 edges (connecting the 4 corners) as shown in Figure 2(A). All of the graphs developed from this basic Seed are by definition regular graphs—that is, graphs having no vertices with edges that connect directly back onto themselves.

The transition from a planar graph representation to a knot representation is realized by associating a knot crossing with each edge in the graph. We then obtain the knot to which the planar graph corresponds by connecting the crossings

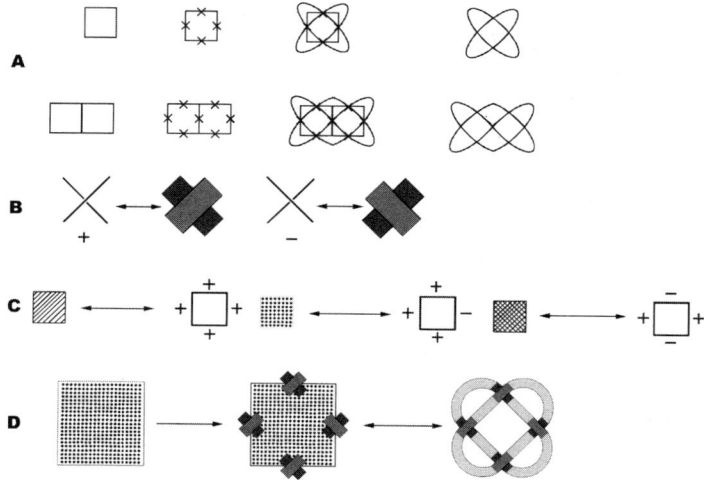

FIGURE 2. The equivalence of the knot projections and the planar Big Seed representation. (A) The basic Seed planar graph. Each connection (edge) between the four corners (vertices) corresponds to a knot crossing. Linking the as indicated crossings gives the knot projection. Top row shows the correspondence (left to right) between an initial Seed square and its knot. Bottom row shows the two square unfolding. (B) Left: The two possible orientations of each crossing of a knot in the standard convention. The "+" and "−" convention is obtained as follows: Choosing a direction to transverse a knot gives direction to the branches of the crossing. The branch on top is rotated into alignment with the branch on the bottom through the angle less than 180 degrees. If the rotation that enables this is clockwise, the crossing is "+", if the rotation is counterclockwise, the crossing is "−". The sign is independent of the choice of direction chosen for transversing the knot. Right: Relative to the grid of a Big Seed Puzzle, "+" orientations indicate that the strand running bottom left to top right passes over top; "−" orientations indicate that the strand running bottom right to top left passes over top. (C) The orientations of each crossing are set by the initial textures of the Seeds at the beginning of the puzzles. The possibilities are: (i) all edges have the same sign , (ii) 3 edges have the same sign, and the 4th the opposite sign of the other 3, (iii) 2 edges are present of each sign, with adjacent edges alternating between + and − , and (iv) 2 edges of each sign with 2 edges of one of the signs both connecting on one end to a single vertex, and the other edges of opposite sign both connecting at one end to the single vertex diagonal to the first (not shown—Big Seed is readily generalized to include this 4 th orientation/texture). (D) For the orientations in C, the specific knot representations for the 2 crossing conventions. Note that all initial Seeds correspond to links including the unlink, and the Hopf link.

as shown in Figure 2(A). The final step in determining the topology of the knot is to determine for each crossing which strand runs over the top, and which strand runs underneath. This can be accomplished by associating with each edge of the graph (or equivalently each crossing) a binary signing of + or −. This binary signing of each edge takes into account that each crossing has one of two possible orientations. The determination of the "over" and "under" strands and the final topology of the knot can then obtained either by choosing a convention relative to the reference frame grid of the puzzle, or by the standard way of choosing an orientation for the knot. That is, selecting any starting point on the knot, a direction is chosen for transversing the knot. The arbitrary direction that one chooses is the orientation. The conventions for determining the over and under strands for crossings relative to a grid or from knot orientation are illustrated in Figure 2(B).

There are 4 different possible unique types of initial signed Seed graphs (Figure 2(C)). Applying the prescription for obtaining the equivalent knot for the possible initial Seeds, we find that they all initially correspond to a link (Figure 2(D))—that is, a pair of tangled knots.

Operations on knots such as cutting and pasting of strands, and Reidemeister moves (a manipulation that changes a knot's projection without changing the knot itself) are realized as symmetry operations on signed planar graphs (the Seeds). We therefore examine the set of possible symmetry operations. These fall into two categories: 1) spatial expansions through mirror reflections, and 2) orientation changes. The first operation consists of doublings along symmetry axes as shown below in Figure 3. The second operation consists of changing the orientation of the Seed's crossings which in the Big Seed framework is represented by changes in texture (Figure 2(D)) and thus knot representation.

The primary concept now is to identify classes of knots based on the evolution of basic Seed projections, and classifying the knots according to these dynamics. That is, identifying sets of knots which can be obtained from a given Seed and a given set of basic symmetry operations, within a particular type of evolution. The final step in classifying the Seed classes is determined by the type of evolution considered.

Seed propagation evolutions

In a first type of evolution, the Seed replicates along the direction determined by the symmetry axis (Figure 3(A)). In this type of evolution the texture of the evolving structure remains relevant. However, ambiguities in the sign of the crossing can occur at junctions where unfolding takes place along an edge. These ambiguities be resolved by the convention that the propagated Seed is rotated in 90 degree increments (either a clockwise or counterclockwise) until the ambiguity is resolved. In contrast to the evolution of knots along shared edges (unfoldings), the second type of spatial operation, unfolding along vertices results in no ambiguities. This evolution is not exploited in this paper.

Geometric evolutions

In the second type of evolution we consider a geometric propagation in which Seeds are considered to "unfold" along symmetry planes running through edges or vertices. (This is the type of evolution that will be considered in all further discussion). That is, we suspend the line topology of the Seed's edges and imagine

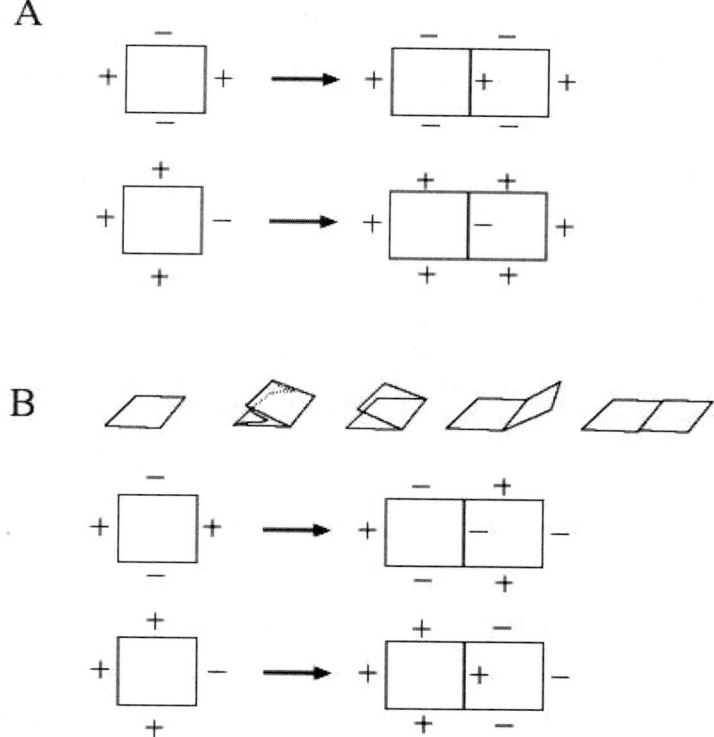

FIGURE 3. Evolution types of Big Seed Planar projection. (A) Seed propagation evolution: In this type of evolution, the Seed is replicated along the symmetry axis (in the example, this is done along the vertical right hand edge). For the upper orientation the sign is maintained. In the lower Seed, the sign of the shared edge results in a mismatch. This is resolved by rotating the new Seed 180 degrees until no mismatch occurs. Subsequent further unfoldings along this axis of the new structure results in no further mismatches. (B) Geometric Seed evolution: In this type of evolution, the Seed is considered to unfold as illustrate in the upper sequence from left to right. The lower sequence illustrates the effect on the sign of the crossings for the Seeds illustrated in A. Note that the sign of the crossings changes as is obvious from flipping the upper strand to become the lower strand (and vice versa) for each crossing. This evolution will be used in this paper.

that the material making up the Seed can be stretched, separated and unfold along an axis as shown in (Figure 3(B)). The process will have the characteristic that when unfolding occurs along a symmetry axis running through an edge, the crossings corresponding to such axes will change sign. Thus the texture of the Seed only fixes the original crossing signs. Depending on the Seed and the symmetry axis along which it is unfolded, the result can be a new knot, or a new link.

Junctions of knots generated from different initial Seed types can be considered as separate links or as a continuous knot. The distinction is whether the solution space to be filled is done so by a specific knot, or by a set of knots. If a single knot is considered, ambiguities in edges of mismatched sign must be resolved. For simplicity therefore, we concentrate on the case were different Seed types define disjoint separate knots and links.

As the initial Seed evolves, subsequent generations of canonical Seeds (knots) are generated. Canonical structures up to 4th generation are illustrated in Figure 4. These subsequent sets can be viewed as a Seeds themselves, which in turn "evolve" until the space defining a Puzzle's solution is generated. Generations of knots are related to each other by symmetry operations in the Seed space and form a symmetry group. The Puzzle solution space is defined by symmetry, or interleaved spaces based on well-defined symmetries. Once the geometry of the solution space is set, a finite number of solution evolutions is fixed as well (Figure 5). Some of these solutions are more "efficient" than others, in the sense that a fewer number of steps in the evolution are required to complete the puzzle than for others. This is of potential great importance in applications.

Dynamic knot theory

When viewed from the standpoint of dynamics, different knots become elements of a symmetry group. The set of knots that are the elements of a given group are those knots that may be reached from one another through an evolution of the planar Seed projections through the application of the symmetry operations. The visualization of solutions however becomes possible in the planar Seed representation, whereas in the knot representation the solution would be (at least initially) far more difficult.

Even with this constrained system, already interesting processes can be realized that mirror those seen in physical applications of considerable interest.

Big Seed, knot theory and DNA

In addition to the well-known double helical structure of DNA, the molecule also possesses a knotted structure [23, 24]. Indeed, DNA in the nucleus is extremely tangled, exhibiting highly knotted and coiled structure. This knotted topology has been found to be crucial to the normal functioning of DNA. The shape and compactness of a DNA molecule will affect how it interacts with enzymes and other DNA molecules. Changing the topology of the knotted structure affects the behavior of DNA. This is of interest in defining particular pathologies. For example viruses can work by changing that knotted structure.

While the complex knotted topology of DNA is vital to its function, in order to perform replication, transcription and recombination, DNA knots must be untangled. Nature accomplishes this through enzymes called topoisomerases that manipulate DNA topologically by cutting at specific locations, untangling, and reattaching strands. Examples of activity which the enzymes can facilitate include those shown in Figure 6.

Topological manipulations on knots correspond to addition of vertices and edges in the planar graph. This is accomplished on the basic Seed by the symmetry operations. In terms of the initial Seed knot, these operation consists of cutting

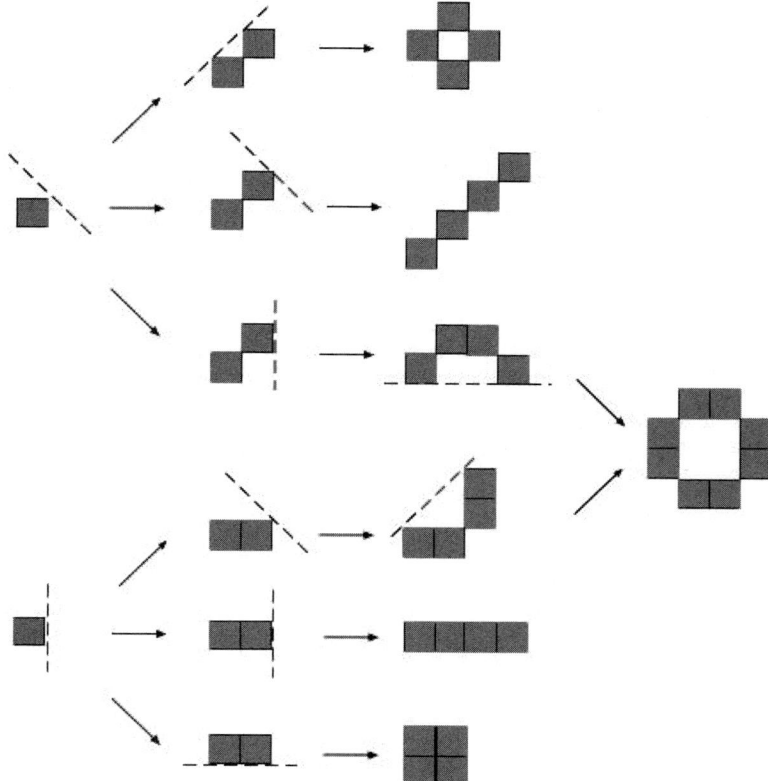

FIGURE 4. Canonical Seed evolutions. Top: Canonical Seeds obtainable after a symmetry operation through a vertex. Bottom: Canonical Seeds obtainable after a symmetry operation through an edge. Note that the bottom progression of the vertex unfolded generation Seed, and the top progression of the edge unfolded generation Seed produce the same canonical Seed, while all other paths diverge.

each of the initial 2 links and reattaching them to their mirror image links as shown in Figure 6.

The sequence of symmetry operations on given initial Seeds resulting in the solution of particular Big Seed puzzles (Figure 7) produces knots (equivalent up to Reidemeister moves) to products of topoisomerases on certain cyclic DNA molecules. Thus the elements of the symmetry group can map to products in the reactions of these enzymes on DNA. While the solution set (the set of knots leading to the completed puzzle) are not necessarily the identical projections of knots that actually occur (as viewed for example in an electron micrograph—see Fig 7), the particular knots are isomorphic to those actual products. This suggests a potential methodology for studying the process and determining the possible specific activities of enzymes on the basis of symmetry principles. The final product(s) are fixed by the symmetry of structure of the final projection (solution space) and the possible evolutions determined by the initial Seed(s). Pathological evolutions (such as the

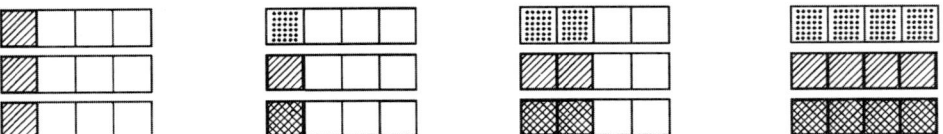

FIGURE 5. Illustration of a repertoire of possible solutions to a particular Big Seed puzzle, requiring different numbers of steps, and evolutions resulting in an incomplete puzzle. These different solutions can result in different knots defined as corresponding to normal or pathological evolutions. (Also see Figure 7.)

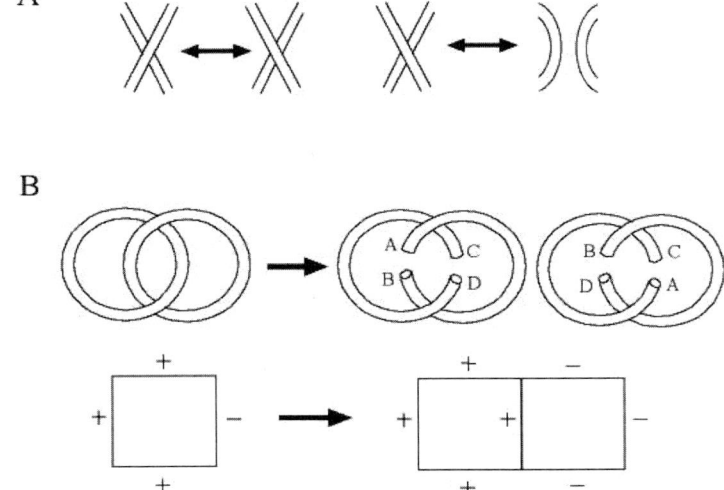

FIGURE 6. Examples of possible actions of topoisomerases on crossings in DNA molecules. These same types of actions on knots as a result of the symmetry operations carried out on the planar projections. (B) Unfolding creates new knots by cutting and pasting. On the right unfolded seed with corresponding knot above: End "A" pastes to "A", etc. (See Figure 2).

case of virus activity) may be defined by the structures resulting in less efficient solutions as defined by the number of steps required, or in evolutions that do not permit complete solution (Figure 5).

Big Seed and other mathematics

Intuitively, it is clear that the ability to visualize relationships through a temporal sequence as exemplified (and "trained") in the Big Seed game are precisely

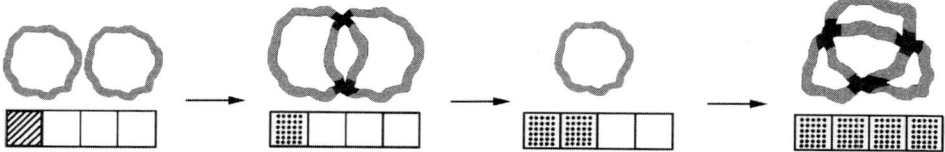

FIGURE 7. Cylcic DNA reactions in knot representation [**23**] equivalent to elementary Big Seed operations. In the topological approach to enzymology, a typical experiment is to observe the changes to the knotting and linking (topology), and supercoiling (Euclidean geometry) of DNA that are caused by specific enzymes (topoisomerases), and to deduce enzyme mechanism from those changes. In this figure, the primary helical structure is not shown, and supercoiling is omitted [**1**]. (A) Starting from an unknotted circular DNA structure, we show the spectrum of possible products that can result in a particular reaction which includes a respectively a linked molecule (known as the Hopf link), and unknotted molecule (the unknot), and a "pretzel shaped" knot (known as the trefoil knot). A schematic example of this knot for DNA is shown. Below, the corresponding spectrum of corresponding planar knot representations as they occur in a possible solution of a particular simple beginning Big Seed puzzle (analogous to Figure 1(A)) are shown. A specific realization of the above is the action ot the enzyme Tn3 resolvase, which acts on a particular duplex cyclic DNA molecule (the substrate). Among the products after the enzyme acts are the unkot and the Hopf link. The figure illustrates explicitly the connection between specific applications of a difficult scientific and mathematical problem and the ability to visual solutions in the Big Seed representation.

those core cognitive operations carried out in the performance of advanced mathematics. Typically these operations consist of grouping together entities (various mathematics quantities) represented by symbols, recognizing the relationships between these groups (specifically for example the symmetry relationships between and within these groups) and then manipulating them through a sequence by applying specific operations in a particular order obtaining a particular form which yields accessible analytic results. It is difficult to see how such a procedure could be carried out with any conceptual understanding from rote analytic (equation based) approach. An initial ST approach is crucial.

The graphic representation of Big Seed has an equivalence with languages and computational machines. Formally we define a quintuplet that consists of a set of elements $\{A\}$ (the alphabet), a set of states $\{S\}$, a subset of those states which are "accepting" states $\{T\}$, a starting state $S0$, and a set of transition rules such that given an input from the set $\{A\}$ a transition is made to a new state from $\{S\}$. A language is the complete set of all phrases that end in an accepting state [**7**]. In this sense, Big Seed has an equivalence to automata, and thus languages and grammar [**4**]. We make a connection then that the languages have a knot group representation as well. In the Big Seed representation, the accepting state is the correctly completed puzzle, the starting state is the initial puzzle and Seeds, the set of elements are the canonical shapes that can be formed, and the production rules are the symmetry operations acting upon the Seeds. The language consists of all sequences of canonical shapes that lead to the completed puzzle. A sequence ending in a completed puzzle is a legal sentence accepted. In this sense we introduce geometry and symmetry into languages which can be grouped according to those properties. Consequently, these languages and grammars have a knot theoretical representation.

Discussion

ST reasoning is crucial to children in *understanding* math and science. It was predicted from the trion columnar model of cortex that ST reasoning *ability is "innate"* to the structured brain. Big Seed, a ST video game exploiting symmetry operations, is challenging, demanding and *relevant*, even for research mathematicians and scientists. We have demonstrated this relevance by showing the equivalence of Big Seed to the mathematics of knot theory and have illustrated its application in solving and understanding problems in DNA structure and function. Further connections with other branches of mathematics were indicated.

In demonstrating these equivalences, we suggested a potentially powerful approach to the disciplines themselves. More importantly, however, is the proof that ST reasoning ability is an essential aspect of mathematical reasoning. By demonstrating the high level of ST skill already present in young children to carry out the fundamentals of such mathematics, we believe represents a paradigm shift of high societal importance. The results of behavioral studies with Big Seed with middle school and elementary school children [14] provides strong evidence that ST reasoning ability is indeed *innate to our structured brain*. The implication of this is that mathematical reasoning is an innate cognitive ability of our brains. While clearly a large formal base of learning and techniques need to be acquired far beyond what is acquired by young children. *The large ST reasoning capabilities demonstrated by the different groups of children far exceed even the most optimistic expectations or understanding from a neurophysiological basis.* This is in regard to the young age of the children, the high level of their ST performance, and the high percentage of children achieving exceptional ST performance. In the context of identifying 3rd grade children that are able to think 8 to 10 moves in advance in playing chess game, or to carry out such operations in the context of advanced mathematics or physics calculations (as in the equivalences outlined in this paper) involving that number of steps, it would not be controversial to label such children as prodigies. Of course as indicated earlier, the performance of such endeavors also requires a considerable amount of formal learning of the "language" involved, particularly to communicate the information. Big Seed bypasses these sophisticated knowledge based training issues.

We suggest that Big Seed be used for training and assessing "creativity" (functionally defined [14]) and ST reasoning as well as discovering genius (functionally defined). Big Seed is totally language free, independent of culture and of education of the player. Thus we suggest that it could be used to readily identify (a non-trivial fraction of) young children with exceptionally large or genius-level potential ST reasoning abilities and creativity who might otherwise be overlooked. There is a profound responsibility of society to nurture the abilities of these truly exceptional young children. We suggest that these genius-level children are perhaps the world's greatest untapped resource.

Acknowledgements. We thank Edward G. Jones, Jiri Patera and Matthew Peterson for helpful discussions. Further, we are indeed indebted to Matthew for his creating Big Seed. This research was supported by Paul and Daranne Folino, Ralph and Leona Gerard Family Trust, Herbert and Ann Lucas, Marjorie Rawlins, Samueli Foundation, Ted and Janice Smith.

References

1. C. C. Adams, *The knot book*, W. H. Freeman, New York, 1994.
2. M. Bodner, L. T. Muftuler, O. Nalcioglu, and G. L. Shaw, *fMRI study relevant to the Mozart effect: brain areas involved in spatial-temporal reasoning*, Neurological Research **23** (2001), 683–690.
3. M. Bodner, Y.-D. Zhou, G. L. Shaw, and J. M. Fuster, *Symmetric temporal patterns in cortical spike trains during performance of a short-term memory task*, Neurological Research **19** (1997), 509–514.
4. N. Chomsky and G. A. Miller, *Finite state language*, Inform. and Control **1** (1958), 91–112.
5. E. S. Ferguson, *The mind's eye: Nonverbal thought in technology*, Science **197** (1977), 827–835.
6. A. B. Graziano, M. Peterson, and G. L. Shaw, *Enhanced learning of proportional math through music training and spatial-temporal training*, Neurological Research **21** (1999), 139–152.
7. J. E. Hopcroft and J. D. Ullman, *Introduction to automata theory, languages, and computation*, Addison-Wesley Ser. Comput. Sci., Addison-Wesley, Reading, MA, 1979.
8. X. Leng X. and G. L. Shaw, *Toward a neural theory of higher brain function using music as a window*, Concepts in Neuroscience **2** (1991), 229–258.
9. J. V. McGrann, G. L. Shaw, K. V. Shenoy, X. Leng, X. and R. B. Mathews, *Computation by symmetry operations in a structured model of the brain: Recognition of rotational invariance and time reversal.* Phys. Rev. E **49** (1994), 5830–5839.
10. R. Moses and C. E. Cobb, *Radical equations: Math literacy and civil rights*, Beacon, Boston, 2001.
11. V. B. Mountcastle, *An organizing principle forfor cerebral function: The unit module and the distributed system*, The Mindful Brain (G .M. Edelman, and V. B. Mountcastle, eds.), MIT Press, Cambridge, 1977, pp. 1–50.
12. _____ *The columnar organization of the neocortex*, Brain **120** (1997), 701–722.
13. S. Papert, *Mindstorms: Children, computers and powerful ideas*, Basic, New York, 1980.
14. M. Peterson, D. Balzarini, M. Bodner, E. G. Jones, T. Phillips, D. Richardson, and G. L. Shaw, *Innate spatial-temporal reasoning and the identification of genius* (submitted).
15. J. Piaget, *The child's concept of number*, Norton, New York, 1965.
16. P. Rakic, *Specifications of cerebral cortical areas*, Science **241** (1988), 170–176.
17. F. H. Rauscher, G. L. Shaw, L. J. Levine, E. L. Wright, W. R. Dennis, and R. L. Newcomb, *Music training causes long-term enhancement of preschool children's reasoning*, Neurologial Research **19** (1997), 2–8.
18. M. Sardesai, C. Figge, M. Bodner, M. Crosby, J. Hansen, J. A. Quillfeldt, S, Landau, A. Ostling, S. Vuong, and G. L. Shaw, *Reliable short-term memory in the trion model: toward a cortical language and grammar*, Biological Cybernetics **84** (2001), 173–182.
19. J. Sarnthein, A. von Stein, A., P. Rappelsberger, H. Petsche, F. H. Rauscher, and G. L. Shaw, *Persistent patterns of brain activity: An EEG coherence study of the positive effect of music on spatial-temporal reasoning*, Neurological Research **19** (1997), 107–116.
20. G. L. Shaw, *Keeping Mozart in mind*, Academic Press, San Diego, CA, 2000.
21. G. L. Shaw, D. J. Silverman, and J. C. Pearson, *Model of cortical organization embodying a basis for a theory of information processing and memory recall*, Proc. Natl. Acad. Sci. USA **82** (1985), 2364–2368.
22. K. V. Shenoy, J. Kaufman, J. V. McGrann, and G. L. Shaw, *Learning by selection in the trion model of cortex*, Cerebral Cortex **3** (1993), 239–248.
23. D. W. Sumners (ed.), *New scientific applications of geometry and topology*, Proc. Sympos. Appl. Math., vol. 45, Amer. Math. Soc., Providence, RI, 1992.

24. S. A. Wasserman, J. M. Dungan, and N. R. Cozzarelli, *Discovery of a predicted DNA knot substantiates a model for site-specific recombination*, Science **229** (1985), 171–174.

M.I.N.D. INSTITUTE, COSTA MESA, CA 92626, USA
URL: www.mindinstitute.net
E-mail address, M. Bodner: mbodner@mindinstitute.net
E-mail address, G. L. Shaw: gshaw@mindinstitute.net

Polytope Sums and Lie Characters

Mark A. Walton

ABSTRACT. A new application of polytope theory to Lie theory is presented. Exponential sums of convex lattice polytopes are applied to the characters of irreducible representations of simple Lie algebras. The Brion formula is used to write a polytope expansion of a Lie character, that makes more transparent certain degeneracies of weight-multiplicities beyond those explained by Weyl symmetry.

1. Introduction

A polytope is the convex hull of finitely many points in $\mathbb{R}^d$ (see [16], for example). A lattice polytope (also known as an integral polytope) is a polytope with all its vertices in an integral lattice $\Lambda \in \mathbb{R}^d$ (see [3], for example). The exponential sum of a lattice polytope Pt,

$$\sum_{x \in \mathrm{Pt} \cap \Lambda} \exp\{\langle c, x \rangle\}, \tag{1.1}$$

is a useful tool. Here c is a vector in $\mathbb{R}^d$, and $\langle \cdot, \cdot \rangle$ is the usual inner product. For simplicity of notation, we'll consider *formal* exponential sums

$$E[\mathrm{Pt}; \Lambda] := \sum_{x \in \mathrm{Pt} \cap \Lambda} e^x. \tag{1.2}$$

Here the formal exponential e^x satisfies

$$e^x e^y = e^{x+y}, \tag{1.3}$$

for all $x, y \in \Lambda$, and simply stands for the function $e^x(c) := e^{\langle c, x \rangle}$. These exponential sums are the generating functions for the integral points in a lattice polytope.

EXAMPLE. As an extremely simple example, consider the one-dimensional lattice polytope with vertices (2) and (7). Its exponential sum is

$$e^{(7)} + e^{(6)} + e^{(5)} + e^{(4)} + e^{(3)} + e^{(2)}. \tag{1.4}$$

2000 *Mathematics Subject Classification.* Primary 17B10, 52B20; Secondary 22E46, 13F25.
This work was supported in part by NSERC.
This is the final form of the paper.

©2004 American Mathematical Society

As a very simple example, consider the lattice polytope in $\mathbb{Z}^2$ with vertices $(0,0)$, $(1,0)$ and $(1,1)$. Its exponential sum is just

$$e^{(1,1)} + e^{(1,0)} + e^{(0,0)}, \tag{1.5}$$

since its vertices are the only lattice points it contains.

We will apply knowledge of these polytope sums to the calculation of the characters of representations of simple Lie algebras (Lie characters, for short). Polytope theory and Lie theory may have much to teach each other, and we hope this contribution will prompt others.

This work is preliminary. Proofs are not given, and we concentrate on low-rank examples. A fuller treatment will be given in [8].

We are grateful to Chris Cummins, Terry Gannon and Pierre Mathieu for comments on the manuscript.

2. Brion formula

Brion [5] has proved a formula for the exponential sum of a convex lattice polytope, that expresses it as a sum of simpler terms, associated with each of the vertices of the polytope. It reads

$$E[\mathrm{Pt}; \Lambda] = \sum_{x \in \mathrm{Pt} \cap \Lambda} e^x = \sum_{v \in \mathrm{Vert}\,\mathrm{Pt}} e^v \sigma_v. \tag{2.1}$$

Here $\mathrm{Vert}\,\mathrm{Pt}$ is the set of vertices of Pt.

The vertex term σ_v is determined by the cone K_v, defined by extending the vertex $v \in \mathrm{Vert}\,\mathrm{Pt}$,

$$K_v = \{x \mid v + \epsilon x \in \mathrm{Pt}, \text{ for all sufficiently small } \epsilon > 0\}. \tag{2.2}$$

In other words, K_v is generated by the vectors $u_i = v_i - v$, where $[v_i, v]$ is an edge of Pt, that indicate the feasible directions at v; it is the cone of feasible directions at v.[1]

σ_v is associated with the exponential sum of K_v. Precisely, if K_v is generated by vectors $u_1, \ldots, u_k \in \Lambda$, then the series $\sum_{x \in K_v \cap \Lambda} \exp\{\langle c, x \rangle\}$ converges for any c such that $\langle c, u_i \rangle < 0$, for all $i = 1, \ldots, k$. The series defines a meromorphic function of c, $\sigma_v(c)$ [3].

A cone K_v is unimodular if the fundamental parallelopiped bounded by its feasible directions u_i contains no lattice points in its interior. If K_v is unimodular, its exponential sum takes the very simple form of a multiple geometric series

$$\sigma_v = \prod_{i=1}^{d} (1 - e^{u_i})^{-1}. \tag{2.3}$$

Pt is totally unimodular if K_v is unimodular for all $v \in \mathrm{Vert}\,\mathrm{Pt}$. If a cone is not unimodular, its exponential sum is not so simple, but is still easy to write; the fundamental parallelopiped of a non-unimodular cone is always a finite union of unimodular parallelopipeds.

---

[1] We will assume that the vectors u_i are linearly independent for all $v \in \mathrm{Vert}\,\mathrm{Pt}$, i.e., that all the K_v are simple cones, and so Pt is a simple polytope.

EXAMPLE. For the extremely simple one-dimensional example considered above, the Brion formula gives

$$\frac{e^{(7)}}{1-e^{(-1)}} + \frac{e^{(2)}}{1-e^{(1)}}. \tag{2.4}$$

If we use

$$\frac{1}{1-e^x} = \frac{-e^{-x}}{1-e^{-x}}, \tag{2.5}$$

this becomes

$$\frac{e^{(7)} - e^{(1)}}{1 - e^{-(1)}}, \tag{2.6}$$

i.e., the simple geometric series, so that (1.4) is recovered.

For the $d = 2$ example above, the Brion formula (2.1), (2.3) gives

$$\frac{e^{(1,1)}}{(1-e^{(-1,-1)})(1-e^{(0,-1)})} + \frac{e^{(1,0)}}{(1-e^{(-1,0)})(1-e^{(0,1)})} + \frac{e^{(0,0)}}{(1-e^{(1,0)})(1-e^{(1,1)})}. \tag{2.7}$$

This simplifies to (1.5).

3. Weyl character formula

Let $P = \mathbb{Z}\{\Lambda^i \mid i = 1, \ldots, r\}$ denote the weight lattice of a simple Lie algebra X_r, of rank r. Here Λ^i stands for the ith fundamental weight. $R_>$ ($R_<$) will denote the set of positive (negative) roots of X_r, and $S = \{\alpha_i \mid i = 1, \ldots, r\}$ its simple roots.

The highest weights of integrable irreducible representations of X_r belong to the set $P_\geq = \{\lambda = \sum_{j=1}^r \lambda_j \Lambda^j \mid \lambda_j \in \mathbb{Z}_{\geq 0}\}$. The formal character of the irreducible representation $L(\lambda)$ of highest weight λ is

$$\mathrm{ch}_\lambda = \sum_{\mu \in P} \mathrm{mult}_\lambda(\mu) e^\mu, \tag{3.1}$$

where $\mathrm{mult}_\lambda(\mu)$ denotes the multiplicity of weight μ in $L(\lambda)$. The famous Weyl character formula is

$$\mathrm{ch}_\lambda = \frac{\sum_{w \in W} (\det w) e^{w.\lambda}}{\prod_{\alpha \in R_>} (1 - e^{-\alpha})}. \tag{3.2}$$

Here $W = \langle r_i \mid i = 1, \ldots, r \rangle$ is the Weyl group of X_r, and r_i is the i-th primitive Weyl reflection, with action $r_i \mu = \mu - \mu_i \alpha_i$ on $\mu \in P$. $\det w$ is the sign of $w \in W$, and $w.\lambda = w(\lambda + \rho) - \rho$ is the shifted action of w on λ, with the Weyl vector $\rho = \sum_{i=1}^r \Lambda^i = \sum_{\alpha \in R_>} \alpha/2$.

After using $\mathrm{ch}_0 = 1$ to derive the Weyl denominator formula, we can rewrite the character formula as

$$\mathrm{ch}_\lambda = \frac{\sum_{w \in W} (\det w) e^{w(\lambda+\rho)}}{\sum_{w \in W} (\det w) e^{w\rho}}. \tag{3.3}$$

Comparing to (3.1) reveals the Weyl symmetry

$$\mathrm{mult}_\lambda(\mu) = \mathrm{mult}_\lambda(w\mu) \quad (\forall w \in W) \tag{3.4}$$

of the weight multiplicities.

Alternatively, we can rewrite (3.2) in a different form that is also manifestly W symmetric:

$$\text{ch}_\lambda = \sum_{w \in W} e^{w\lambda} \prod_{\alpha \in R_>} (1 - e^{-w\alpha})^{-1}. \tag{3.5}$$

The usual formula (3.2) is recovered from this using (2.5). Each Weyl element $w \in W$ separates the positive roots into two disjoint sets:

$$\begin{aligned} R^w_> &:= \{\alpha \in R_> \mid w\alpha \in R_>\}, \quad R^w_< := \{\alpha \in R_> \mid w\alpha \in R_<\}, \\ R^w_> \cup R^w_< &= R_>, \quad R^w_> \cap R^w_< = \varnothing, \\ wR^w_> &= R^w_>, \quad wR^w_< = -R^w_<. \end{aligned} \tag{3.6}$$

It can be shown that $\det w = (-1)^{\|R^w_<\|}$, and

$$-w\rho + \rho = \sum_{\beta \in R^w_<} \beta. \tag{3.7}$$

Using these results, (3.5) becomes

$$\begin{aligned} \text{ch}_\lambda &= \sum_{w \in W} e^{w\lambda} \prod_{\beta \in R^w_<} (-e^{w\beta})(1 - e^{w\beta})^{-1} \prod_{\alpha \in R^w_>} (1 - e^{-w\alpha})^{-1} \\ &= \sum_{w \in W} (\det w) e^{w\lambda - w\sum_{\gamma \in R^w_<} \gamma} \prod_{\beta \in R^w_<} (1 - e^{w\beta})^{-1} \prod_{\alpha \in R^w_>} (1 - e^{-w\alpha})^{-1}, \end{aligned} \tag{3.8}$$

so that (3.2) follows.

4. Character polytope-expansion

The form (3.5) of the Weyl character formula is similar to the Brion formula. To make this more precise, we'll write the Brion formula for the exponential sum of the lattice polytope Pt_λ with vertices in the Weyl orbit of a highest weight $\lambda \in P_\geq$, i.e., $\text{Vert}\,\text{Pt}_\lambda = W\lambda$. The appropriate lattice here is the root lattice Q of X_r, shifted by λ: $\lambda + Q \subset P$. If we define

$$P(\lambda) := \{\mu \mid \text{mult}_\lambda(\mu) \neq 0\} \subset P, \tag{4.1}$$

then

$$\text{Pt}_\lambda \cap (\lambda + Q) = P(\lambda), \tag{4.2}$$

and the polytope sum will be

$$E[\text{Pt}_\lambda; \lambda + Q] = \sum_{\mu \in P(\lambda)} e^\mu. \tag{4.3}$$

Let us call the so-defined polytope Pt_λ the *weight polytope*.

If the weight λ is regular, and the weight polytope is totally unimodular, then it is easy to see that the Brion formula gives

$$B_\lambda = \sum_{w \in W} e^{w\lambda} \prod_{\alpha \in S} (1 - e^{-w\alpha})^{-1}, \tag{4.4}$$

since the feasible directions at the vertex $w\lambda$ are just the Weyl-transformed simple roots $\{w\alpha \mid \alpha \in S\}$.

The last formula is remarkably similar to (3.5). We propose here to exploit this similarity, by expanding

$$\text{ch}_\lambda = \sum_{\mu \leq \lambda} A_{\lambda,\mu} B_\mu. \tag{4.5}$$

Here $\mu \leq \lambda$ means $\lambda - \mu \in \mathbb{Z}_{\geq 0} R_>$, as usual. The constraint on the sum implies that the $A_{\lambda,\mu}$ are the entries of a triangular matrix. The character polytope-expansion (4.5) will manifest weight-multiplicity degeneracy beyond Weyl symmetry.

Before considering examples, let us mention two possible complications. First, the weight λ may not be regular, so that some subgroup of the Weyl group W stabilizes it. We still find that the exponential sum for Pt_λ equals B_λ, as written in (4.4). Also, when the algebra is not simply-laced, and λ is not regular, then the corresponding weight polytope Pt_λ may not be totally unimodular. Remarkably, even in that case, (4.4) seems to be the appropriate formula.

EXAMPLE. Consider a simple example, the adjoint representation of G_2. If α_2 is the short simple root, then the highest weight λ of the adjoint representation is $\Lambda^1 = \theta$. (Here θ will be used to denote the highest root.) The feasible directions at the vertex $\lambda = \Lambda^1$ are $-\alpha_1$ and $-\alpha_1 - 3\alpha_2$, *not* $-\alpha_1$ and $-\alpha_2$. The cone is not unimodular. Using (2.3), the cone function would be

$$(1 - e^{-\alpha_1})^{-1}(1 - e^{-\alpha_1 - 3\alpha_2})^{-1}. \tag{4.6}$$

This cone function would miss points—in this example, the consequent formula for the polytope Pt_λ would not have contributions from the short roots.

As pointed out in [3], a cone can be decomposed into a set of unimodular ones, and (2.3) can then be applied. In general, however, it is more efficient to use a signed decomposition into unimodular cones. In this example, drawing a G_2 root diagram shows that

$$e^\lambda \sigma_\lambda = e^{\Lambda^1}(1 - e^{-\alpha_1})^{-1}(1 - e^{-\alpha_2})^{-1} \tag{4.7}$$
$$- e^{\Lambda^1 - \alpha_2}(1 - e^{-\alpha_1 - 3\alpha_2})^{-1}(1 - e^{-\alpha_2})^{-1}.$$

But this is just

$$\begin{aligned}
e^\lambda \sigma_\lambda &= e^{\Lambda^1}(1 - e^{-\alpha_1})^{-1}(1 - e^{-\alpha_2})^{-1} \\
&\quad + e^{r_2 \Lambda^1}(1 - e^{-r_2 \alpha_1})^{-1}(-e^{-\alpha_2})(1 - e^{-\alpha_2})^{-1} \\
&= e^{\Lambda^1}(1 - e^{-\alpha_1})^{-1}(1 - e^{-\alpha_2})^{-1} \\
&\quad + e^{r_2 \Lambda^1}(1 - e^{-r_2 \alpha_1})^{-1}(1 - e^{-r_2 \alpha_2})^{-1}.
\end{aligned} \tag{4.8}$$

By Weyl invariance, similar formulas work at the other vertices, and so (4.4) agrees with $E[\text{Pt}_\lambda; Q]$.

We will proceed with our study of the expansion (4.5), postponing to [8] a proof that for all $\lambda \in P_\geq$, $E[\text{Pt}_\lambda; \lambda + Q] = B_\lambda$. Let us emphasize, however, that we are only relying on (4.4) and (4.5). Even if the sums of (4.3) and (4.4) are not identical, the expansion we are studying seems a natural one, and so should still be of value.

EXAMPLE. Consider the simplest nontrivial case: $X_r = A_2$. Then $S = \{\alpha_1, \alpha_2\}$, and most importantly, $R_> \setminus S = \{\alpha_1 + \alpha_2\} = \{\theta\}$. For A_2 then, we can write (4.4) as

(4.9)
$$\text{ch}_\lambda = \sum_{w \in W} e^{w\lambda} \left[\prod_{\alpha \in S}(1 - e^{-w\alpha})^{-1}\right](1 - e^{-w\theta})^{-1}$$
$$= B_\lambda + \sum_{w \in W} e^{w(\lambda-\theta)} \left[\prod_{\alpha \in S}(1 - e^{-w\alpha})^{-1}\right](1 - e^{-w\theta})^{-1},$$

so that

(4.10)
$$\text{ch}_\lambda = B_\lambda + \text{ch}_{\lambda-\theta}.$$

It is also simple to show that if either $\nu_1 = 0$ or $\nu_2 = 0$, then $\text{ch}_\nu = B_\nu$. Since $\theta = \Lambda^1 + \Lambda^2$ for A_2, we find that if $\lambda_{\min} := \min\{\lambda_1, \lambda_2\}$, then

(4.11)
$$\text{ch}_\lambda = B_\lambda + B_{\lambda-\theta} + B_{\lambda-2\theta} + \cdots + B_{\lambda-\lambda_{\min}\theta}.$$

Equation (4.11) manifests the weight multiplicity pattern of A_2 representations.[2]

It is difficult to generalize the derivation just given of (4.11) to other algebras. Equation (4.10) is easier, however. Consider $X_r = C_2$, with the short simple root labelled as α_1. Then

(4.12)
$$R_> \setminus S = \{2\alpha_1 + \alpha_2 = 2\Lambda^1, \alpha_1 + \alpha_2 = \Lambda^2\}.$$

Therefore

(4.13)
$$\text{ch}_\lambda = \sum_{w \in W} e^{w\lambda} \left[\prod_{\alpha \in S}(1 - e^{-w\alpha})^{-1}\right](1 - e^{-w(2\Lambda^1)})^{-1}(1 - e^{-w\Lambda^2})^{-1}$$
$$= B_\lambda + \sum_{w \in W} e^{w\lambda}\left(e^{w(-2\Lambda^1)} + e^{w(-\Lambda^2)} - e^{w(-2\Lambda^1+\Lambda^2)}\right)$$
$$\times \left[\prod_{\alpha \in R_>}(1 - e^{-w\alpha})^{-1}\right],$$

so that

(4.14)
$$\text{ch}_\lambda = B_\lambda + \text{ch}_{\lambda-2\Lambda^1} + \text{ch}_{\lambda-\Lambda^2} - \text{ch}_{\lambda-2\Lambda^1-\Lambda^2}.$$

This recurrence relation is a bit more complicated than that of A_2, but can be analysed easily. We need the relation

(4.15)
$$\text{ch}_\lambda = (\det w)\,\text{ch}_{w.\lambda},$$

derived from either (3.2) or (3.3). Using it with (4.14), we can establish

(4.16)
$$\text{ch}_{\lambda_j \Lambda^j} = B_{\lambda_j \Lambda^j} + \text{ch}_{(\lambda_j - 2)\Lambda^j} \quad (j = 1, 2).$$

This immediately shows that $\text{ch}_\lambda = B_\lambda$ for $\lambda \in \{0, \Lambda^1, \Lambda^2\}$, and that

(4.17)
$$\text{ch}_{\lambda_j \Lambda^j} = B_{\lambda_j \Lambda^j} + B_{(\lambda_j - 2)\Lambda^j} + \cdots + B_{[\lambda_j]_2 \Lambda^j} \quad (j = 1, 2).$$

Here $[\lambda_j]_2 := 0\ (1)$ if λ_j is even (odd). Now, if we define

(4.18)
$$v_\lambda := \text{ch}_\lambda - \text{ch}_{\lambda - 2\Lambda^1},$$

---

[2]This formula was derived in [2], where weight multiplicity patterns were studied for low-rank algebras, using methods close in spirit to the Kostant multiplicity formula [9]. The A_2 multiplicity pattern was known long before, however, by Wigner, for example; I thank Professor R. King for so informing me.

then the recursion relation (4.14) becomes
$$v_\lambda = B_\lambda + v_{\lambda - \Lambda^2} \tag{4.19}$$
$$= B_\lambda + B_{\lambda - \Lambda^2} + \cdots + B_{\lambda_1 \Lambda^1},$$
since (4.17) implies $v_{\lambda_1 \Lambda^1} = B_{\lambda_1 \Lambda^1}$. Solving (4.18) yields
$$\text{ch}_\lambda = v_\lambda + v_{\lambda - 2\Lambda^1} + \cdots + v_{\Lambda^1 + \lambda_2 \Lambda^2} \tag{4.20}$$
for λ_1 odd; and
$$\text{ch}_\lambda = v_\lambda + v_{\lambda - 2\Lambda^1} + \cdots + v_{2\Lambda^1 + \lambda_2 \Lambda^2} + \text{ch}_{\lambda_2 \Lambda^2} \tag{4.21}$$
for λ_1 even. Using first (4.20), then (4.19), we find
$$\text{ch}_\lambda = B_\lambda + B_{\lambda - \Lambda^2} + \cdots + B_{\lambda_1 \Lambda^1} \tag{4.22}$$
$$+ B_{\lambda - 2\Lambda^1} + B_{\lambda - 2\Lambda^1 - \Lambda^2} + \cdots + B_{(\lambda_1 - 2)\Lambda^1}$$
$$+ \cdots + B_{\Lambda^1 + \lambda_2 \Lambda^2} + B_{\Lambda^1 + (\lambda_2 - 1)\Lambda} + \cdots + B_{\Lambda^1}$$
for λ_1 odd. Replacing (4.20) by (4.21) and (4.17), we get instead
$$\text{ch}_\lambda = B_\lambda + B_{\lambda - \Lambda^2} + \cdots + B_{\lambda_1 \Lambda^1} \tag{4.23}$$
$$+ B_{\lambda - 2\Lambda^1} + B_{\lambda - 2\Lambda^1 - \Lambda^2} + \cdots + B_{(\lambda_1 - 2)\Lambda^1}$$
$$+ \cdots + B_{\lambda_2 \Lambda^2} + B_{(\lambda_2 - 2)\Lambda^2} + \cdots + B_{[\lambda_2]_2 \Lambda^2}$$
for λ_1 even. These C_2 results confirm those of [2].

Re-writing (4.10) as $B_\lambda = \text{ch}_\lambda - \text{ch}_{\lambda - \theta}$ gives us a hint as to how to generalize the method of computation of the $A_{\lambda, \mu}$ to all algebras. Expanding
$$B_\lambda = \sum_{\mu \leq \lambda} A^{-1}_{\lambda, \mu} \text{ch}_\mu, \tag{4.24}$$
is straightforward. Then finding $A_{\lambda, \mu}$ by diagonalizing the triangular matrix $(A^{-1}_{\lambda, \mu})$ is relatively easy.[3]

First, re-write (4.4) as
$$B_\lambda = \sum_{w \in W} e^{w\lambda} \left[\prod_{\gamma \in R_> \setminus S} (1 - e^{-w\gamma}) \right] \left[\prod_{\alpha \in R_>} (1 - e^{-w\alpha})^{-1} \right]. \tag{4.25}$$
Comparing to (3.5), we can therefore write
$$B_\lambda = \widehat{\text{ch}} e^\lambda \prod_{\gamma \in R_> \setminus S} (1 - e^{-\gamma}), \tag{4.26}$$
where we have defined
$$\widehat{\text{ch}} e^\lambda := \text{ch}_\lambda. \tag{4.27}$$
Of course, some of the terms of the expansion just written may be of the form ch_λ, but with $\lambda \notin P_\geq$. To find the coefficients $A^{-1}_{\lambda, \mu}$ with $\lambda, \mu \in P_\geq$, therefore, it is necessary to use the relation (4.15).

To write an explicit formula for $A^{-1}_{\lambda, \mu}$, we first define a partition function F by
$$\prod_{\gamma \in R_> \setminus S} (1 - e^{-\gamma}) =: \sum_{\beta \in \mathbb{Z}_{\geq 0} R_>} F(\beta) e^\beta. \tag{4.28}$$

---

[3]The author learned this trick from a work of Professors J. Patera and R. T. Sharp [12], and has also used it elsewhere [7].

Then

(4.29) $$B_\lambda = \sum_{\beta \in \mathbb{Z}_{\geq 0} R_>} F(\beta) \operatorname{ch}_{\lambda - \beta}.$$

Using (4.24) and (4.15), we find

(4.30) $$A_{\lambda,\mu}^{-1} = \sum_{w \in W} (\det w) F(\lambda - w.\mu).$$

EXAMPLE. For $X_r = G_2$, we have

(4.31) $$R_> \setminus S = \{\alpha_1 + \alpha_2, 2\alpha_1 + 3\alpha_2, \alpha_1 + 2\alpha_2, \alpha_1 + 3\alpha_2\},$$
$$= \{\Lambda^1 - \Lambda^2, \Lambda^1, \Lambda^2, -\Lambda^1 + 3\Lambda^2\}.$$

Therefore (4.26) yields

(4.32) $$B_\lambda = \operatorname{ch}_\lambda - \operatorname{ch}_{\lambda+\Lambda^1-3\Lambda^2} - \operatorname{ch}_{\lambda-\Lambda^2} + \operatorname{ch}_{\lambda+\Lambda^1-4\Lambda^2} + \operatorname{ch}_{\lambda-\Lambda^1-\Lambda^2}$$
$$- \operatorname{ch}_{\lambda-4\Lambda^2} - \operatorname{ch}_{\lambda-\Lambda^1+\Lambda^2} + \operatorname{ch}_{\lambda-2\Lambda^2} + \operatorname{ch}_{\lambda-2\Lambda^1+\Lambda^2}$$
$$- \operatorname{ch}_{\lambda-\Lambda^1-2\Lambda^2} - \operatorname{ch}_{\lambda-2\Lambda^1} + \operatorname{ch}_{\lambda-\Lambda^1-3\Lambda^2}.$$

A simple calculation then yields the matrix $(A_{\lambda,\mu})$:

(4.33) $$\begin{bmatrix}
1 & 0 & 0 & 0 & 0 & 0 & 0 & 0 & 0 & 0 & 0 & 0 & 0 & 0 & 0 \\
0 & 1 & 0 & 0 & 0 & 0 & 0 & 0 & 0 & 0 & 0 & 0 & 0 & 0 & 0 \\
1 & 0 & 1 & 0 & 0 & 0 & 0 & 0 & 0 & 0 & 0 & 0 & 0 & 0 & 0 \\
1 & 1 & 0 & 1 & 0 & 0 & 0 & 0 & 0 & 0 & 0 & 0 & 0 & 0 & 0 \\
0 & 2 & 0 & 1 & 1 & 0 & 0 & 0 & 0 & 0 & 0 & 0 & 0 & 0 & 0 \\
2 & 1 & 1 & 1 & 0 & 1 & 0 & 0 & 0 & 0 & 0 & 0 & 0 & 0 & 0 \\
1 & 2 & 0 & 2 & 1 & 0 & 1 & 0 & 0 & 0 & 0 & 0 & 0 & 0 & 0 \\
2 & 2 & 1 & 2 & 1 & 1 & 0 & 1 & 0 & 0 & 0 & 0 & 0 & 0 & 0 \\
2 & 2 & 0 & 2 & 1 & 1 & 0 & 0 & 1 & 0 & 0 & 0 & 0 & 0 & 0 \\
1 & 3 & 0 & 2 & 2 & 1 & 0 & 1 & 0 & 1 & 0 & 0 & 0 & 0 & 0 \\
3 & 4 & 1 & 4 & 2 & 2 & 1 & 1 & 1 & 0 & 1 & 0 & 0 & 0 & 0 \\
3 & 2 & 1 & 2 & 1 & 2 & 0 & 1 & 1 & 0 & 0 & 1 & 0 & 0 & 0 \\
2 & 2 & 1 & 2 & 1 & 1 & 1 & 1 & 0 & 0 & 0 & 0 & 1 & 0 & 0 \\
3 & 5 & 0 & 5 & 3 & 2 & 1 & 2 & 1 & 1 & 1 & 0 & 0 & 1 & 0 \\
4 & 4 & 1 & 4 & 3 & 3 & 0 & 2 & 2 & 1 & 1 & 1 & 0 & 0 & 1 & 0 \\
3 & 4 & 0 & 4 & 2 & 2 & 1 & 1 & 2 & 0 & 1 & 1 & 0 & 0 & 0 & 1
\end{bmatrix}$$

Here all weights $\lambda =: (\lambda_1, \lambda_2)$ with $\lambda \cdot \theta = 2\lambda_1 + \lambda_2 \leq 6$ are included, in the order

(4.34) $(0,0), (0,1), (1,0), (0,2), (1,1), (0,3), (2,0), (1,2),$
$(0,4), (2,1), (1,3), (0,5), (3,0), (2,2), (1,4), (0,6).$

The character polytope expansion (4.5) can be combined with the Weyl dimension formula for the dimension d_λ of the representation of highest weight λ:

(4.35) $$d_\lambda = \prod_{\alpha \in R_>} \frac{(\lambda + \rho) \cdot \alpha}{\rho \cdot \alpha}.$$

The result is a formula for the number b_λ of lattice points counted by B_λ, that provides helpful checks on any expansions derived, since (4.24) and (4.5) imply

$b_\lambda = \sum_\mu A^{-1}_{\lambda,\mu} d_\mu$ and $d_\mu = \sum_\sigma A_{\mu,\sigma} b_\sigma$, respectively. The formulas relevant to the simple rank-two algebras are

(4.36)
$$\begin{aligned} A_2 : & \quad b_\lambda = (\lambda_1^2 + 4\lambda_1\lambda_2 + \lambda_2^2 + 3\lambda_1 + 3\lambda_2 + 2)/2, \\ C_2 : & \quad b_\lambda = \lambda_1^2 + 4\lambda_1\lambda_2 + 2\lambda_2^2 + 2\lambda_1 + 2\lambda_2 + 1, \\ G_2 : & \quad b_\lambda = 9\lambda_1^2 + 12\lambda_1\lambda_2 + 3\lambda_2^2 + 3\lambda_1 + 3\lambda_2 + 1. \end{aligned}$$

EXAMPLE. As our final example, consider $X_r = A_3$. The important subset of positive roots is

(4.37)
$$\begin{aligned} R_> \setminus S &= \{\alpha_{12}, \alpha_{123}, \alpha_{23}\} \\ &= \{\Lambda^1 + \Lambda^2 - \Lambda^3, \Lambda^1 + \Lambda^3, -\Lambda^1 + \Lambda^2 + \Lambda^3\}, \end{aligned}$$

where $\alpha_{12} := \alpha_1 + \alpha_2$, etc. We therefore find

(4.38) $B_\lambda = \mathrm{ch}_\lambda - \mathrm{ch}_{\lambda+\Lambda^1-\Lambda^2-\Lambda^3} - \mathrm{ch}_{\lambda-\Lambda^1-\Lambda^3} + \mathrm{ch}_{\lambda-\Lambda^2-2\Lambda^3}$
$$- \mathrm{ch}_{\lambda-\Lambda^1-\Lambda^2+\Lambda^3} + \mathrm{ch}_{\lambda-2\Lambda^2} + \mathrm{ch}_{\lambda-2\Lambda^1-\Lambda^2} - \mathrm{ch}_{\lambda-\Lambda^1-2\Lambda^2-\Lambda^3}.$$

Using the Weyl dimension formula,

(4.39) $b_\lambda = 1 + \frac{1}{6}(11\lambda_1 + 14\lambda_2 + 11\lambda_3)$
$$+ 4\lambda_1\lambda_2 + 3\lambda_1\lambda_3 + 4\lambda_2\lambda_3 + \lambda_1^2 + 2\lambda_2^2 + \lambda_3^2$$
$$+ \frac{1}{6}(36\lambda_1\lambda_2\lambda_3 + 12\lambda_1\lambda_2^2 + 12\lambda_2^2\lambda_3 + 6\lambda_1^2\lambda_2 + 6\lambda_2\lambda_3^2$$
$$+ 9\lambda_1^2\lambda_3 + 9\lambda_1\lambda_3^2 + \lambda_1^3 + 4\lambda_2^3 + \lambda_3^3)$$

follows.

We present our results for all weights $\lambda =: (\lambda_1, \lambda_2, \lambda_3) \in P_\geq$ of A_3, with $\lambda \cdot \theta = \lambda_1 + \lambda_2 + \lambda_3 \leq 5$. The weights separate into four congruence classes, corresponding to the four shifted root lattices $\{0, \Lambda^1, \Lambda^2, \Lambda^3\} + Q$ that combine to form the weight lattice P.

For weights $\lambda, \mu \in Q$, we find the matrix $(A_{\lambda,\mu})$ is

(4.40)
$$\begin{bmatrix} 1 & 0 & 0 & 0 & 0 & 0 & 0 & 0 & 0 & 0 & 0 & 0 & 0 & 0 \\ 2 & 1 & 0 & 0 & 0 & 0 & 0 & 0 & 0 & 0 & 0 & 0 & 0 & 0 \\ 1 & 0 & 1 & 0 & 0 & 0 & 0 & 0 & 0 & 0 & 0 & 0 & 0 & 0 \\ 1 & 1 & 0 & 1 & 0 & 0 & 0 & 0 & 0 & 0 & 0 & 0 & 0 & 0 \\ 1 & 1 & 0 & 0 & 1 & 0 & 0 & 0 & 0 & 0 & 0 & 0 & 0 & 0 \\ 0 & 0 & 0 & 0 & 0 & 1 & 0 & 0 & 0 & 0 & 0 & 0 & 0 & 0 \\ 1 & 0 & 1 & 0 & 0 & 0 & 1 & 0 & 0 & 0 & 0 & 0 & 0 & 0 \\ 0 & 0 & 0 & 0 & 0 & 0 & 0 & 1 & 0 & 0 & 0 & 0 & 0 & 0 \\ 2 & 1 & 1 & 1 & 1 & 0 & 0 & 0 & 1 & 0 & 0 & 0 & 0 & 0 \\ 3 & 2 & 0 & 0 & 0 & 0 & 0 & 0 & 1 & 0 & 0 & 0 & 0 & 0 \\ 2 & 2 & 0 & 1 & 0 & 1 & 0 & 0 & 0 & 1 & 1 & 0 & 0 & 0 \\ 2 & 2 & 0 & 0 & 1 & 0 & 0 & 1 & 0 & 1 & 0 & 1 & 0 & 0 \\ 1 & 1 & 0 & 1 & 1 & 0 & 0 & 0 & 1 & 0 & 0 & 0 & 1 & 0 \\ 1 & 1 & 0 & 1 & 1 & 0 & 0 & 0 & 1 & 0 & 0 & 0 & 0 & 1 \end{bmatrix}.$$

In the order used, the weights are

(4.41)
$$(0,0,0), (1,0,1), (0,2,0), (2,1,0), (0,1,2), (4,0,0), (0,4,0),$$
$$(0,0,4), (1,2,1), (2,0,2), (3,1,1), (1,1,3), (2,3,0), (0,3,2).$$

For weights $\lambda, \mu \in \Lambda^1 + Q$, we find $(A_{\lambda,\mu})$ to be

(4.42)
$$\begin{bmatrix}
1 & 0 & 0 & 0 & 0 & 0 & 0 & 0 & 0 & 0 & 0 & 0 & 0 & 0 \\
1 & 1 & 0 & 0 & 0 & 0 & 0 & 0 & 0 & 0 & 0 & 0 & 0 & 0 \\
0 & 0 & 1 & 0 & 0 & 0 & 0 & 0 & 0 & 0 & 0 & 0 & 0 & 0 \\
1 & 1 & 0 & 1 & 0 & 0 & 0 & 0 & 0 & 0 & 0 & 0 & 0 & 0 \\
2 & 0 & 0 & 0 & 1 & 0 & 0 & 0 & 0 & 0 & 0 & 0 & 0 & 0 \\
1 & 1 & 0 & 1 & 0 & 1 & 0 & 0 & 0 & 0 & 0 & 0 & 0 & 0 \\
2 & 1 & 1 & 0 & 1 & 0 & 1 & 0 & 0 & 0 & 0 & 0 & 0 & 0 \\
1 & 0 & 0 & 0 & 1 & 0 & 0 & 1 & 0 & 0 & 0 & 0 & 0 & 0 \\
1 & 0 & 1 & 0 & 1 & 0 & 1 & 0 & 1 & 0 & 0 & 0 & 0 & 0 \\
0 & 0 & 2 & 0 & 0 & 0 & 0 & 0 & 0 & 1 & 0 & 0 & 0 & 0 \\
1 & 1 & 0 & 1 & 0 & 1 & 0 & 0 & 0 & 0 & 1 & 0 & 0 & 0 \\
2 & 1 & 1 & 1 & 1 & 0 & 1 & 1 & 0 & 0 & 0 & 1 & 0 & 0 \\
3 & 0 & 0 & 0 & 2 & 0 & 0 & 0 & 0 & 0 & 0 & 0 & 1 & 0 \\
0 & 0 & 0 & 0 & 0 & 0 & 0 & 0 & 0 & 0 & 0 & 0 & 0 & 1
\end{bmatrix}$$

for weights $(\lambda_1, \lambda_2, \lambda_3)$ in the order

(4.43)
$$\begin{aligned}
&(1,0,0),\ (0,1,1),\ (0,0,3),\ (1,2,0),\ (2,0,1),\ (0,3,1),\ (1,1,2),\\
&(3,1,0),\ (0,2,3),\ (1,0,4),\ (1,4,0),\ (2,2,1),\ (3,0,2),\ (5,0,0).
\end{aligned}$$

The results for weights in $\Lambda^3 + Q$ can be obtained from this, by replacing all weights $(\lambda_1, \lambda_2, \lambda_3)$ by their charge conjugates $(\lambda_3, \lambda_2, \lambda_1)$.

The matrix $(A_{\lambda,\mu})$ is

(4.44)
$$\begin{bmatrix}
1 & 0 & 0 & 0 & 0 & 0 & 0 & 0 & 0 & 0 & 0 & 0 & 0 & 0 \\
0 & 1 & 0 & 0 & 0 & 0 & 0 & 0 & 0 & 0 & 0 & 0 & 0 & 0 \\
0 & 0 & 1 & 0 & 0 & 0 & 0 & 0 & 0 & 0 & 0 & 0 & 0 & 0 \\
1 & 0 & 0 & 1 & 0 & 0 & 0 & 0 & 0 & 0 & 0 & 0 & 0 & 0 \\
1 & 1 & 1 & 0 & 1 & 0 & 0 & 0 & 0 & 0 & 0 & 0 & 0 & 0 \\
0 & 2 & 0 & 0 & 0 & 1 & 0 & 0 & 0 & 0 & 0 & 0 & 0 & 0 \\
0 & 0 & 2 & 0 & 0 & 0 & 1 & 0 & 0 & 0 & 0 & 0 & 0 & 0 \\
0 & 1 & 1 & 0 & 1 & 0 & 0 & 1 & 0 & 0 & 0 & 0 & 0 & 0 \\
0 & 1 & 1 & 0 & 1 & 0 & 0 & 0 & 1 & 0 & 0 & 0 & 0 & 0 \\
0 & 1 & 0 & 0 & 0 & 1 & 0 & 0 & 0 & 1 & 0 & 0 & 0 & 0 \\
1 & 0 & 0 & 1 & 0 & 0 & 0 & 0 & 0 & 1 & 0 & 0 & 0 & 0 \\
1 & 1 & 1 & 1 & 1 & 0 & 0 & 1 & 1 & 0 & 0 & 1 & 0 & 0 \\
1 & 2 & 2 & 0 & 1 & 1 & 1 & 0 & 0 & 0 & 0 & 0 & 1 & 0 \\
0 & 0 & 1 & 0 & 0 & 0 & 1 & 0 & 0 & 0 & 0 & 0 & 0 & 1
\end{bmatrix}$$

for weights $\lambda, \mu \in \Lambda^2 + Q$. The order of weights is

(4.45)
$$\begin{aligned}
&(0,1,0),\ (0,0,2),\ (2,0,0),\ (0,3,0),\ (1,1,1),\ (1,0,3),\ (3,0,1),\\
&(0,2,2),\ (2,2,0),\ (0,1,4),\ (0,5,0),\ (1,3,1),\ (2,1,2),\ (4,1,0).
\end{aligned}$$

The preliminary calculations we have done for G_2 and A_3 confirm that the polytope-expansion multiplicities $A_{\lambda,\mu}$ can be computed easily by computer. Some patterns can already be seen from their results. To make some general statements, however, we plan to attempt an analysis of the recursion relations modeled on the one done above for C_2 [8]. Perhaps a relatively simple algorithm, of a combinatorial type, can be found for the calculation of the $A_{\lambda,\mu}$.

5. Conclusion

First, we'll summarize the main points. Then we'll discuss what still needs to be done, and what might be done.

We point out that the Brion formula applied to the weight polytopes Pt_λ produces a formula (4.4) that is remarkably similar to the Weyl character formula (3.5). The polytope expansion (4.5) of the Lie characters was therefore advocated. It makes manifest certain degeneracies in weight multiplicities beyond those explained by Weyl group symmetry, for example. The expansion multiplicities $A_{\lambda,\mu}$ were studied. A closed formula (4.30) was given for them, and methods for their computation were discussed and illustrated with the examples of A_2, C_2, G_2 and A_3.

Two conjectures need to be established. A general proof that the polytope-expansion coefficients are non-negative integers,

$$(5.1) \qquad A_{\lambda,\mu} \in \mathbb{Z}_{\geq 0},$$

would be helpful. It should also be determined if the exponential sum $E[\mathrm{Pt}_\lambda; \lambda + Q]$ of the weight polytope Pt_λ is given by the Brion expression B_λ in (4.4), i.e., if

$$(5.2) \qquad \sum_{\mu \in P(\lambda)} e^\mu = \sum_{w \in W} e^{w\lambda} \prod_{\alpha \in S} (1 - e^{-w\alpha})^{-1},$$

for all integrable highest weights $\lambda \in P_\geq$. As mentioned above, however, we use only (4.4) and (4.5). Our results therefore have value even if this last equality is not always obeyed.

We'll now mention a few other possible applications of polytope theory to Lie theory.

This author was introduced to polytope theory in different contexts—in the study of tensor product multiplicities and the related affine fusion multiplicities (see [14, 15] and references therein, and [4]), the fusion of Wess–Zumino–Witten conformal field theories. It would be interesting to consider applications of the Brion formula in those subjects.

For example, the Brion formula also allows the derivation of a character generating function relevant to affine fusion. Its derivation is a simple adaptation of that of the Patera–Sharp formula for the character generator of a simple Lie algebra [13]. We hope to report on it and its applications elsewhere.

As another example, consider the tensor-product multiplicity patterns for A_2 that were studied thoroughly long ago, in [11] and [10]. Not surprisingly, the pattern is similar to the weight-multiplicity pattern of a single irreducible A_2 representation. Perhaps a polytope expansion in the spirit of (4.5) could manifest properties of the tensor-product patterns for all simple Lie algebras.

Recent work on tensor products has provided results that are valid for general classes of algebras, but on the simpler question of which weights appear in a given tensor product, ignoring their multiplicities. See [6] for a review. The Brion formula might be useful for this question, and perhaps also for the corresponding problem in affine fusion [1].

Finally, it is our hope that applications in the other direction will also be found, i.e., that certain techniques from Lie theory can also help in the study of polytopes. For example, the Brion formula has similarities with the Weyl character formula, as we have discussed. Are there polytope formulas that correspond to other formulas for Lie characters?

References

1. S. Agnihotri and C. Woodward, *Eigenvalues of products of unitary matrices and quantum Schubert calculus*, Math. Res. Lett. **5** (1998), no. 6, 817–836.
2. J.-P. Antoine and D. Speiser, *Characters of irreducible representations of the simple groups.* I. *General theory*, J. Mathematical Phys. **5** (1964), 1226–1234; II. *Application to classical groups*, 1560–1572.
3. A. Barvinok, *Lattice points and lattice polytopes.* Handbook of Discrete and Computational Geometry, CRC Press Ser. Discrete Math. Appl., CRC, Boca Raton, FL, 1997, pp. 133–152,.
4. L. Bégin, C. Cummins, P. Mathieu, and M. A. Walton, *Fusion rules and the Patera–Sharp generating-function method*, in this book.
5. M. Brion, *Polyèdres et réseaux*, Enseign. Math. (2) **38** (1992), no. 1-2, 71–88.
6. W. Fulton, *Eigenvalues, invariant factors, highest weights, and Schubert calculus*, Bull. Amer. Math. Soc. (N.S.) **37** (2000), no. 3, 209–249 (electronic).
7. T. Gannon, C. Jakovljevic, and M. A. Walton, *Lie group weight multiplicities from conformal field theory*, J. Phys. A **28** (1995), no. 9, 2617–2625.
8. T. Gannon and M. A. Walton (in preparation).
9. B. Kostant, *A formula for the multiplicity of a weight*, Trans. Amer. Math. Soc. **93** (1959), 53–73.
10. A. M. Perelomov and V. S. Popov, *Expansion of the direct product of irreducible representations of* $SU(3)$ *in irreducible representations*, Jadernaja Fiz. **2** (1965), 294–306 (Russian); English transl., Soviet J. Nuclear Phys. **2** (1965), 210–218.
11. B. Preziosi, A. Simoni, and B. Vitale, *A general analysis of the reduction of the direct product of two irreducible representations of* SU_3 *and of its multiplicity structure*, Nuovo Cimento (10) **34** (1964), no. 10, 1101–1113.
12. J. Patera and R. T. Sharp, *Branching rules for representations of simple Lie algebras through Weyl group orbit reduction*, J. Phys. A **22** (1989), no. 13, 2329–2340.
13. _____, *Generating function techniques pertinent to spectroscopy and crystal physics*, Recent Advances in Group Theory and Their Application to Spectroscopy, NATO Adv. Study Inst. Ser., Ser. B: Physics, vol. 43, Plenum, New York–London, 1979, pp. 219–248.
14. J. Rasmussen and M. A. Walton, $su(N)$ *tensor product multiplicities and virtual Berenstein–Zelevinsky triangles*, J. Phys. A **34** (2001), no. 49, 11095–11105.
15. _____, *Affine* $su(3)$ *and* $su(4)$ *fusion multiplicities as polytope volumes*, J. Phys. A **35** (2002), no. 32, 6939–6952.
16. G. Ziegler, *Lectures on polytopes*, Grad. Texts in Math., vol. 152, Springer-Verlag, New York, 1995.

DEPARTMENT OF PHYSICS, UNIVERSITY OF LETHBRIDGE, LETHBRIDGE, ALBERTA T1K 3M4, CANADA

E-mail address: `walton@uleth.ca`

Subalgebras of Lie Algebras. Example of sl(3, ℝ)

Pavel Winternitz

ABSTRACT. Methods of classifying subalgebras of finite-dimensional Lie algebras are briefly reviewed and applied to the Lie algebra sl(3, ℝ). The results for sl(3, ℝ) are summed up in a table. Sublalgebras of sl(3, ℂ) are classified as a byproduct.

1. Introduction

One of the scientific programs that Bob Sharp was involved in for several years was the classification of the continuous subgroups of the "fundamental groups of physics." These were the Poincaré and similitude group, the de Sitter groups and the conformal group of Minkowski space-time [1, 12–18]. The methods developed in this series of papers amount to an algorithm that can be used to classify the subgroups of any finite-dimensional Lie group. The methods are algebraic and the actual procedure is to classify the subalgebras of the Lie algebra $\mathfrak{g}$ of the considered Lie group G. They are classified into conjugacy classes under the action of the group of inner automorphisms of the Lie algebras $\mathfrak{g}$, i.e., the group G itself.

The purpose of this article is to review the subgroup classification algorithm and then to apply it to a case not treated previously, namely the Lie algebra sl(3, ℝ). This example is sufficiently simple to be treated in a short article. On the other hand, it is rich enough to illustrate all aspects of the algorithm.

The Lie groups SL(3, ℂ) and SL(3, ℝ) play an important role in the theory of ordinary differential equations. Indeed, S. Lie himself showed that the largest symmetry group that a second order ordinary differential equation can allow is SL(3, F) with $F = \mathbb{R}$, or $F = \mathbb{C}$ (depending on whether real or complex variables are being considered [9]). Moreover, he showed that the equation

$$(1.1) \qquad y'' = F(x, y, y')$$

allows an SL(3, F) symmetry group if and only if it is linearizable by a point transformation. A classification of subgroups of SL(3, F) provides a classification of possible ways that the symmetry of a linear, or linearizable ODE can be broken by imposing additional conditions on solutions.

2000 *Mathematics Subject Classification.* Primary 81R05; Secondary 22E70, 22E60.

The author's research is partly supported by research grants from NSERC of Canada and FQRNT of Québec.

This is the final form of the paper.

2. Subalgebra classification algorithms

2.1. General strategy. Let us consider a Lie algebra $\mathfrak{g}$ and the corresponding Lie group $G = \langle \exp \mathfrak{g} \rangle$. Three different strategies should be used depending on the type of algebra being considered. The different cases that occur are

(i) The algebra $\mathfrak{g}$ is simple.
(ii) The algebra $\mathfrak{g}$ is a direct sum of two Lie algebras

(2.1) $$\mathfrak{g} \approx \mathfrak{g}_1 \oplus \mathfrak{g}_2, \quad [\mathfrak{g}_1, \mathfrak{g}_2] = 0, \quad [\mathfrak{g}_a, \mathfrak{g}_a] \subseteq \mathfrak{g}_a, \quad a = 1, 2$$

(iii) The algebra $\mathfrak{g}$ is a semidirect sum of two Lie algebras

(2.2) $$\mathfrak{g} \approx \mathfrak{g}_1 \supsetplus \mathfrak{g}_2, \quad [\mathfrak{g}_1, \mathfrak{g}_2] \subseteq \mathfrak{g}_2, \quad [\mathfrak{g}_a, \mathfrak{g}_a] \subseteq \mathfrak{g}_a, \quad a = 1, 2.$$

These three cases cover all possibilities. For instance, in case (ii) the algebras $\mathfrak{g}_1$ and $\mathfrak{g}_2$ may themselves be direct sums. Then the algorithm is applied recursively, until we obtain a direct sum of indecomposable components. Case (iii) may correspond to a Levi decomposition, in which case $\mathfrak{g}_1$ would be semisimple and $\mathfrak{g}_2$ would be the radical of $\mathfrak{g}$ (the maximal solvable ideal). On the other hand $\mathfrak{g}$ may itself be solvable, then $\mathfrak{g}_2$ is some ideal in $\mathfrak{g}$ (for instance the nilradical).

Let us consider the three cases separately.

2.2. Maximal subalgebras of simple Lie algebras. Let $\mathfrak{g}$ be a finite dimensional simple Lie algebra over $\mathbb{R}$, or $\mathbb{C}$. We choose a finite-dimensional faithful irreducible representation of $\mathfrak{g}$, i.e., a representation by real, or complex matrices, $E(\mathfrak{g})$. Usually it is convenient to use the lowest dimensional representation available. In particular, for one of the classical Lie algebras we would choose the defining matrix representation.

The chosen representation will act on a linear space V of dimension $\dim V = n$. A subalgebra $\mathfrak{g}_0 \subset \mathfrak{g}$ can be imbedded into the representation $E(\mathfrak{g})$ reducibly, or irreducibly. Let us consider the two cases separately

(a) *Irreducibly imbedded subalgebras.* The set of matrices $E(\mathfrak{g}_0) \subset E(\mathfrak{g})$ leaves no proper nontrivial subspace of V invariant. Then $\mathfrak{g}_0$ must be a reductive algebra, i.e., simple, semisimple, or the direct sum of a semisimple Lie algebra with an Abelian one. The question of finding all such subalgebras $\mathfrak{g}_0 \subset \mathfrak{g}$ thus becomes a question of Lie algebra representation theory: $\mathfrak{g}_0$ must have a faithful irreducible representation of dimension equal to $\dim E(\mathfrak{g})$. The representations of the simple and semisimple Lie algebras are well studied and tabled [2, 6, 11]. Those of the reductive algebras can be obtained by constructing the centralizers of the semisimple algebras in $E(\mathfrak{g})$. This is a problem of linear algebra.

(b) *Reducibly imbedded subalgebras.* The set of matrices $E(\mathfrak{g}_0)$ leaves a nontrivial linear subspace $V_0 \subset V$ invariant. In this case it suffices to classify the spaces V_0, construct a representative of each class and then to find the algebra $E(\mathfrak{g}_0)$ leaving V_0 invariant. This again is a problem of linear algebra. If $\mathfrak{g}$ is $\mathrm{sl}(n, F)$ then V_0 is completely characterized by its dimension. If $\mathfrak{g}$ is an orthogonal, or a symplectic Lie algebra, then V_0 is also characterized by its signature (number of positive, negative or zero length vectors in any orthogonal basis).

Once the maximal subalgebras of a simple Lie algebra are found, we proceed in the same manner if $\mathfrak{g}_0$ is simple. Otherwise, we apply one of the other two algorithms described below.

2.3. Subalgebras of direct sums. The Goursat method. E. Goursat proposed a method for classifying discrete subgroups of direct products of continuous and discrete groups [5, 7]. It has been adapted to classify subalgebras of direct sums of Lie algebras [15].

The algorithm consists of several steps. Consider a Lie algebra $\mathfrak{g}$ that is a direct sum of two Lie algebras, as in eq. (2.1). Changing notation, we put

(2.3) $\qquad \mathfrak{g} \sim A \oplus B, \quad [A, A] \subseteq A, \quad [B, B] \subseteq B, \quad [A, B] = 0,$

(A and B are not necessarily indecomposable). We wish to classify subalgebras of $\mathfrak{g}$ into conjugacy classes under the action of the group G

(2.4) $\qquad G \sim G_A \otimes G_B \quad G_A = \langle \exp A \rangle, \quad G_B = \langle \exp B \rangle.$

We distinguish two types of subalgebras of the direct sums (2.3).

(a) *Nontwisted subalgebras.* Such subalgebras are G conjugate to direct sums of subalgebras $A_0 \subseteq A$, $B_0 \subseteq B$. They can be represented by direct sums themselves:

(2.5) $\qquad L_0 \sim A_0 \oplus B_0, \quad A_0 \subseteq A, \quad B_0 \subseteq B$

(b) *Twisted subalgebras.* These are subalgebras that are not G-conjugate to any subalgebra of the form (2.5).

To obtain a representative list of all G-conjugacy classes of subalgebras of L, we proceed in 4 steps.

Step 1. Construct representatives of all G_A and G_B conjugacy classes of subalgebras of A and B, respectively. Denote them $A_{j,a} \subseteq A$, $B_{k,b} \subseteq B$, where the first label is equal to the dimension of the subalgebra, the second label enumerates all mutually nonconjugate subalgebras of the same dimension. The trivial subalgebras $\{\emptyset\}$ and A (and B) must be included. For each representative subalgebra in the list, find its normalizer group in the corresponding group G_A (or G_B):

(2.6) $\qquad \text{Nor}\{A_{j,a}, G_A\} \sim \{g \in G_A \mid g A_{ja} g^{-1} \subseteq A_{ja}\}$

Step 2. Form a representative list S_1 of all all nontwisted subalgebras of $\mathfrak{g}$:

(2.7) $\qquad S_1 : \{A_{j,a} \oplus B_{k,b}\}, \quad \forall j, k, a, b$

Step 3. Construct a representative list S_2 of all twisted subalgebras of $\mathfrak{g}$. We start from the list S_1. Two subalgebras $A_{j,a} \subseteq A$, $B_{k,b} \subseteq B$ can be twisted together if a homomorphism (that may be an isomorphism) exists from one to the other, e.g.:

(2.8) $\qquad \tau(A_{j,a}) \sim B_{k,b}$

if a homomorphism of $A_{j,a}$ into $B_{k,b}$ exists, then choose a basis for $A_{j,a} \sim \{a_1, \ldots a_j\}$ and construct the most general homomorphic mapping

(2.9) $\qquad \tau : a_i \to \tau(a_i) \in B_{k,b}.$

Taking

(2.10) $\qquad \tilde{\mathfrak{g}}_{j,a} = \{a_i + \tau(a_i)\}, \quad i = 1, \ldots j$

we obtain a twisted subalgebra of $\mathfrak{g}$. To obtain a G-representative list S_2 of all twisted subalgebras of $\mathfrak{g}$ we must now classify the subalgebras (2.10) into conjugacy classes under the normalizer group

(2.11) $\qquad \text{Nor}(A_{j,a}, G_A) \otimes \text{Nor}(B_{k,b}, G_B).$

This group is used to annul as many of the parameters introduced by the mapping τ as possible and to normalize as many of the remaining ones as possible.

Step 4. Form a final representative list S of all subalgebras of $\mathfrak{g}$ by merging the lists S_1 and S_2. It is useful to organize the final list by dimension and isomorphism class. It is also convenient to provide a "normalized list." That means that the list S should contain the normalizer of each algebra in the list: for each $\mathfrak{g}_{j,a}$ in the list we should also have

$$(2.12) \qquad \operatorname{nor}(\mathfrak{g}_{j,a}, \mathfrak{g}) \sim \{x \in \mathfrak{g} \mid [X, \mathfrak{g}_{j,a}] \subseteq \mathfrak{g}_{j,a}\}$$

in the list S (rather than an algebra that is conjugate, but not equal to the normalizer).

2.4. Subalgebras of semidirect sums. The method for classifying subalgebras of semidirect sums was elaborated and applied in Refs. [1, 12–18]. Let L have the form of eq. (2.2), i.e., it contains an ideal $\mathfrak{g}_2 = N$ and a factor algebra $\mathfrak{g}_1 = F$ that is itself a Lie algebra. In particular, this may be a Levi decomposition; then F is semisimple and N is solvable.

We again proceed in a series of steps.

Step 1. Form a representative list $S(F)$ of subalgebras of the factor algebra F, classified under the action of the group $G_F = \exp F$:

$$(2.13) \qquad S(F) = \{F_1 = \{\varnothing\}, F_2, \ldots, F_N \equiv F\}.$$

The list should be normalized, i.e. for each subalgebra F_i its normalizer algebra in F should also be in the list $S(F)$. For each subalgebra F_i construct its normalizer group in G_F.

Step 2. Classify all *splitting* subalgebras of L. A subalgebra of a semidirect sum is called splitting if it is itself a semidirect sum of subalgebras of the two components:

$$(2.14) \qquad L_0 = F_0 \uplus N_0, \quad F_0 \subseteq F, \quad N_0 \subseteq N.$$

Splitting subalgebras of semidirect sums are thus analogs of the untwisted subalgebras of direct sums.

The procedure for finding all splitting subalgebras is as follows.

(1) For each subalgebra F_i in the list $S(F)$ find all invariant subspaces $\widetilde{N}_{i,\alpha}$ that are also subalgebras:

$$(2.15) \qquad [F_i, \widetilde{N}_{i,\alpha}] \subseteq \widetilde{N}_{i,\alpha}, \quad \widetilde{N}_{i,\alpha} \subseteq N, \quad [\widetilde{N}_{i,\alpha}, \widetilde{N}_{i,\alpha}] \subseteq \widetilde{N}_{i,\alpha},$$

(if N is Abelian then any subspace is a subalgebra). The trivial invariant subspaces $\widetilde{N}_{i,0} = \{\varnothing\}$ and $\widetilde{N}_{i,n} = N$ must be included for each F_i.

(2) For each F_i classify all invariant subalgebras $\widetilde{N}_{i,\alpha}$ into conjugacy classes under the action of the normalizer group $\operatorname{Nor}(F_i, G_F)$. Choose a representative $N_{i,\alpha}$ of each conjugacy class.

(3) Form a representative list of all splitting subalgebras of L

$$(2.16) \qquad S_1(L) = \{L_{i,\alpha} \subseteq L \mid L_{i,\alpha} = F_i \uplus N_{i,\alpha}\}.$$

Note that the list $S(F)$ is contained in $S_1(L)$.

(4) For each subalgebra $L_{i,\alpha}$ in the list $S_1(L)$ find its normalizer group $\operatorname{Nor}(L_{i,\alpha}, G)$ in the group G. The list $S_1(L)$ should be normalized.

We have obtained the list $S_1(L)$ in such a manner that each splitting subalgebra $L_{i,\alpha} \subseteq L$ has a basis satisfying.

(2.17) $\quad L_{i,\alpha} = \{B_a, X_j\}, \quad B_a \in F, \ X_j \in N, \ 1 \le a \le f_i = \dim F_i,$
$$1 \le j \le r = \dim N_{i,\alpha}.$$

Step 3. Classify all *nonsplitting* subalgebras of L. By definition, a nonsplitting subalgebra is not conjugate to any splitting one. Nonsplitting subalgebras of a semidirect sum are generalizations of Goursat twisted subalgebras of direct sums.

The procedure for finding all nonsplitting subalgebras is as follows.

(1) Run through the list $S_1(L)$ of all splitting subalgebras of L. For each member $L_{i,\alpha}$ of the list take a basis as in eq. (2.17). Complement the basis $\{X_j\}$ of $N_{i,\alpha}$ to a basis of N

(2.18) $\quad\quad\quad\quad N = \{X_1, \ldots, X_r, Y_1, \ldots, Y_s\}, \quad r + s = \dim N.$

(2) For each splitting subalgebra $L_{i,\alpha}$ form the vector space

(2.19) $\quad\quad V = \left\{B_a + \sum_{\mu=1}^{s} c_{a\mu} X_\mu, X_j\right\}, \quad a = 1, \ldots, f_i, \ j = 1, \ldots, r,$

where the constants $c_{a\mu}$ are such that V is a Lie algebra

(2.20) $\quad\quad\quad\quad\quad\quad [V, V] \subseteq V.$

In order to obtain equations for the constants $c_{a\mu}$ following from eq. (2.20) we must now specify the commutation relations for L in the chosen basis:

(2.21) $\quad\begin{aligned} & [B_a, B_b] = f_{ab}^c B_c, & & [B_a, X_k] = \alpha_{ak}^l X_l, \\ & [B_a, X_\mu] = \rho_{a\mu}^\nu Y_\nu + \sigma_{a\mu}^m X_m, & & [X_i, X_j] = \omega_{ij}^m X_m, \\ & [Y_\mu, Y_\nu] = \beta_{\mu\nu}^\sigma Y_\sigma + \gamma_{\mu\nu}^m X_m, & & [X_i, Y_\mu] = \lambda_{i\mu}^\nu Y_\nu + \tau_{i\mu}^m X_m. \end{aligned}$

The condition (2.20) implies that the constants $c_{a\mu}$ in eq. (2.19) must satisfy a set of algebraic equations

(2.22) $\quad\quad\quad\quad c_{b\nu}\rho_{a\nu}^\alpha - c_{a\mu}\rho_{b\mu}^\alpha - c_{c\alpha}f_{ab}^c = -c_{a\mu}c_{b\nu}\beta_{\mu\nu}^\alpha,$
(2.23) $\quad\quad\quad\quad c_{a\mu}\lambda_{j\mu}^\nu = 0,$

for all a, b, α, j and ν.

In general, equations (2.22) are bilinear, rather than linear and solving them is a non-trivial task in algebraic geometry. It reduces to linear algebra if the ideal N is such that

(2.24) $\quad\quad\quad\quad\quad \beta_{\mu\nu}^\alpha = 0, \quad \forall \mu, \nu, \alpha.$

In the simplest case, when the ideal N is Abelian, we have

(2.25) $\quad\quad\quad\quad \omega_{ij}^m = \beta_{\mu\nu}^\sigma = \gamma_{\nu\mu}^m = \lambda_{i\mu}^\nu = \tau_{i\mu}^\nu = 0,$

and the equations for $c_{a\mu}$ reduce to a system of homogeneous linear equations

(2.26) $\quad\quad\quad\quad c_{b\nu}\rho_{a\nu}^\alpha - c_{a\mu}\rho_{b\mu}^\alpha - c_{c\alpha}f_{ab}^c = 0.$

In mathematical terms eq. (2.26) means that the quantities $c_{a\mu}$ form 1-cocycles.

(3) Once the constants $c_{a\mu}$ are obtained as general solutions of the system (2.22), (2.23), or of (2.26) if N is Abelian, the vector space V of eq. (2.19) becomes a Lie algebra. The Lie algebras V must now be classified into conjugacy classes under the group

$$\widetilde{G} = \mathrm{Nor}(L_{i,\alpha}, G) \oslash \mathrm{Nor}(N_{i,\alpha}, G_N), \tag{2.27}$$

where $G_N = \exp N$.

This classification is again simplified if N is Abelian. Then we have $\mathrm{Nor}(N_{i,\alpha}, G_N) = G_N$.

In the Abelian case we can generate trivial nonzero cocycles by conjugating the splitting subalgebras by elements of G_N:

$$\begin{aligned} \exp(\lambda_\mu Y_\mu) B_a \exp(-\lambda_\mu Y_\mu) &= B_a + \lambda_\mu [Y_\mu, B_a], \\ \exp(\lambda_\mu Y_\mu) X_j \exp(-\lambda_\mu Y_\mu) &= X_j. \end{aligned} \tag{2.28}$$

We obtain an algebra conjugate to the splitting one, namely:

$$L_{i,\alpha} \sim \{B_a + \lambda_\mu \rho_{a\mu}^\nu X_\nu, X_j\}. \tag{2.29}$$

The constants λ_μ can be freely chosen and the trivial cocycles

$$\delta_{a\mu} = \sum_{\alpha=1}^{s} \lambda_\alpha \rho_{a\alpha}^\mu \tag{2.30}$$

are called coboundaries. Any cocycle $c_{a\mu}$ can be replaced by $c_{a\mu} + \delta_{a\mu}$. We choose the constants λ_α to annul as many as possible of the cocycles. If all cocycles can be annuled, the subalgebra is splitting. The remaining cocycles $c_{a\mu}$ (mod $\delta_{a\mu}$) are to be further classified under the action of the normalizer $\mathrm{Nor}(L_{i,\alpha}, G)$ of the corresponding splitting subalgebra $L_{i,\alpha}$.

Step 4. Form the final representative list $S(L) = S_1(L) \cup S_2(L)$ of all G-conjugacy classes of splitting and nonsplitting subalgebras.

We note that most of the work in the classification procedure concerns nonsplitting algebras. Which algebras are splitting depends on the decomposition of L and this is usually not unique. Whenever possible, N should be chosen to be Abelian. The algorithm starts with the assumption that the subalgebras of F are known. If this is not the case, then one of the methods described above should be applied to F: this is a lower dimensional problem, so the procedure is iterative.

3. Subalgebras of sl(3, ℝ)

3.1. General comments. All three algorithms described in Section 2 can be examplified on the case of the algebra $\mathfrak{g} \sim \mathrm{sl}(3, \mathbb{R})$, the Lie algebra of 3×3 real traceless matrices. We shall use a basis given by the following elements

$$K_1 = \frac{1}{2}\begin{pmatrix} 1 & & \\ & -1 & \\ & & 0 \end{pmatrix}, \quad K_2 = \frac{1}{2}\begin{pmatrix} 0 & 1 & \\ 1 & 0 & \\ & & 0 \end{pmatrix}, \quad K_3 = \frac{1}{2}\begin{pmatrix} 0 & -1 & \\ 1 & 0 & \\ & & 0 \end{pmatrix}$$

$$D = \begin{pmatrix} 1 & & \\ & 1 & \\ & & -2 \end{pmatrix}, \quad P_1 = \begin{pmatrix} 0 & 0 & 1 \\ 0 & 0 & 0 \\ 0 & 0 & 0 \end{pmatrix} \quad P_2 = \begin{pmatrix} 0 & 0 & 0 \\ 0 & 0 & 1 \\ 0 & 0 & 0 \end{pmatrix} \tag{3.1}$$

$$R_1 = \begin{pmatrix} 0 & 0 & 0 \\ 0 & 0 & 0 \\ 1 & 0 & 0 \end{pmatrix}, \quad R_2 = \begin{pmatrix} 0 & 0 & 0 \\ 0 & 0 & 0 \\ 0 & 1 & 0 \end{pmatrix}$$

Let us first look at one- and two-dimensional subalgebras directly. An element of sl$(3, \mathbb{R})$ can be taken in its Jordan canonical form. As a basis element of a one-dimensional Lie algebra, the corresponding matrix can be multiplied by an arbitrary nonzero real constant. We obtain the following representatives of one-dimensional subalgebras of sl$(3, \mathbb{R})$:

(3.2)
$$L_{1,1}(a): \begin{pmatrix} 1 & & \\ & a & \\ & & -1-a \end{pmatrix}, \quad -1 \leq a \leq -\tfrac{1}{2}, \qquad L_{1,2}: \begin{pmatrix} 1 & 1 & \\ & 1 & \\ & & -2 \end{pmatrix}$$

$$L_{1,3}: \begin{pmatrix} 0 & 1 & \\ 0 & 0 & \\ & & 0 \end{pmatrix}, \qquad L_{1,4}(a): \begin{pmatrix} a & 1 & \\ -1 & a & \\ & & -2a \end{pmatrix}, \quad 0 \leq a.$$

For sl$(3, \mathbb{C})$ the algebra $L_{1,4}(a)$ is conjugate to $L_{1,1}(\tilde{a})$ and should be eliminated from the list. Moreover, the condition on a in (3.2) should be replaced by $|a| \leq 1$, $\operatorname{Re} a \leq -\tfrac{1}{2}$. In $L_{1,1}(a)$ and $L_{1,4}(a)$ the letter a denotes a fixed parameter.

A two-dimensional Lie algebra can either be Abelian, or solvable with a basis satisfying $[A, B] = A$.

Much is known about maximal Abelian subalgebras of the classical Lie algebras [**4, 8, 19, 21, 23**]. Let us list those of sl$(3, \mathbb{R})$; they are all two-dimensional

$$L_{2,1}: \left\{\begin{pmatrix} a+b & & \\ & -a+b & \\ & & -2b \end{pmatrix}\right\}, \quad L_{2,2}: \left\{\begin{pmatrix} a & b & \\ -b & a & \\ & & -2a \end{pmatrix}\right\}, \quad L_{2,3}: \left\{\begin{pmatrix} a & b & \\ & a & \\ & & -2a \end{pmatrix}\right\},$$

$$L_{2,4}: \left\{\begin{pmatrix} 0 & 0 & a \\ 0 & 0 & b \\ 0 & 0 & 0 \end{pmatrix}\right\}, \quad L_{2,5}: \left\{\begin{pmatrix} 0 & a & b \\ 0 & 0 & 0 \\ 0 & 0 & 0 \end{pmatrix}\right\}, \quad L_{2,6}: \left\{\begin{pmatrix} 0 & a & b \\ 0 & 0 & a \\ 0 & 0 & 0 \end{pmatrix}\right\}$$

(in all cases $a \in \mathbb{R}$, $b \in \mathbb{R}$ are group parameters). Algebras $L_{2,1}$ and $L_{2,2}$ are both Cartan subalgebras, mutually conjugate over $\mathbb{C}$, but not over $\mathbb{R}$. The algebra $L_{2,1}$ generates a noncompact Abelian group, $L_{2,2}$ an Abelian group with one compact basis element. The algebra $L_{2,3}$ is not a Cartan subalgebra because the element corresponding to $a = 0$, $b \neq 0$ is not diagonalizable. The algebras $L_{2,4}$, $L_{2,5}$ and $L_{2,6}$ are maximal Abelian nilpotent subalgebras (MANS). In any finite-dimensional representation they are represented by nilpotent matrices. A MANS of sl$(n, \mathbb{R})$ (or sl$(n, \mathbb{C})$) is partly characterized by its Kravchuk signature [**21, 23**], a triplet of integers

(3.3) $\qquad (\lambda, \mu, \nu) \quad \lambda \geq 1, \nu \geq 1, \mu \geq 0, \lambda + \mu + \nu = n.$

The invariant definition of the signature is that λ is the dimension of the kernel of X and ν is the codimension of the image of X, where X is the set of matrices representing the MANS, viewed as a mapping. In our case $L_{2,4}$, $L_{2,5}$ and $L_{2,6}$ have Kravchuk signature $(2, 0, 1)$, $(1, 0, 2)$, and $(1, 1, 1)$, respectively. Hence they are not conjugate under SL$(3, \mathbb{R})$ (nor SL$(3, \mathbb{C})$).

For non-Abelian algebras satisfying $[A, B] = A$ the matrix A must be nilpotent. Four SL$(3, \mathbb{R})$ classes of such subalgebras exist in sl$(3, \mathbb{R})$, represented by

$$L_{2,7}(a): \left\{\begin{pmatrix} \cos\alpha & & \\ & \sin\alpha & \\ & & -\cos\alpha - \sin\alpha \end{pmatrix}, \begin{pmatrix} 0 & 1 & 0 \\ 0 & 0 & 0 \\ 0 & 0 & 0 \end{pmatrix}\right\}, \quad 0 \leq \alpha < \pi$$

where α is a fixed parameter,

$$L_{2,8}: \left\{\begin{pmatrix} -2 & & \\ & 1 & 0 \\ & 1 & 1 \end{pmatrix}, \begin{pmatrix} 0 & 1 & 0 \\ 0 & 0 & 0 \\ 0 & 0 & 0 \end{pmatrix}\right\}, \quad L_{2,9}: \left\{\begin{pmatrix} 1 & 0 & 1 \\ 0 & -2 & 0 \\ 0 & 0 & 1 \end{pmatrix}, \begin{pmatrix} 0 & 1 & 0 \\ 0 & 0 & 0 \\ 0 & 0 & 0 \end{pmatrix}\right\},$$

$$L_{2,10}: \left\{\begin{pmatrix} 1 & 0 & 0 \\ 0 & 0 & 0 \\ 0 & 0 & -1 \end{pmatrix}; \begin{pmatrix} 0 & 1 & 0 \\ 0 & 0 & 1 \\ 0 & 0 & 0 \end{pmatrix}\right\}.$$

3.2. Maximal subalgebras of $\mathrm{sl}(3,\mathbb{R})$. Following the method outlined in Section 2.2 we choose the defining representation of $\mathrm{sl}(3,\mathbb{R})$ by real 3×3 traceless matrices as $E(\mathfrak{g})$.

(a) *Irreducibly imbedded subalgebras.* The only semisimple Lie algebras that have real irreducible faithful representations of dimension 3 are the simple Lie algebras $\mathrm{o}(3)$ and $\mathrm{o}(2,1)$. They can be represented as subalgebras of $\mathrm{sl}(3,\mathbb{R})$ as

$$(3.4) \quad \mathrm{o}(3) \sim \begin{pmatrix} 0 & a & b \\ -a & 0 & c \\ -b & -c & 0 \end{pmatrix}, \quad \mathrm{o}(2,1) \sim \begin{pmatrix} 0 & a & b \\ -a & 0 & c \\ b & c & 0 \end{pmatrix},$$

respectively. Neither of them has a nontrivial centralizer in $\mathrm{SL}(3,\mathbb{R})$. They are conjugate to each other in $\mathrm{sl}(3,\mathbb{C})$.

(b) *Reducibly imbedded maximal subalgebras.* An invariant subspace $V_0 \subset V$ can be either two-dimensional, or one-dimensional. We consider the two separately, and construct the subalgebras $\mathfrak{g}_0$ satisfying $\mathfrak{g}_0 V_0 \subseteq V_0$.

$$(3.5) \quad V_2 \sim \begin{pmatrix} x \\ y \\ 0 \end{pmatrix} \quad \mathfrak{g}_{7,1} \sim \begin{pmatrix} a_{11} & a_{12} & a_{13} \\ a_{21} & a_{22} & a_{23} \\ 0 & 0 & -a_{11}-a_{22} \end{pmatrix}$$

$$(3.6) \quad V_1 \sim \begin{pmatrix} x \\ 0 \\ 0 \end{pmatrix} \quad \mathfrak{g}_{7,2} \sim \begin{pmatrix} a_{11} & a_{12} & a_{13} \\ 0 & a_{22} & a_{23} \\ 0 & a_{32} & -a_{11}-a_{22} \end{pmatrix}.$$

The two algebras $\mathfrak{g}_{7,1}$ and $\mathfrak{g}_{7,2}$ are both isomorphic to the affine Lie algebra $\mathrm{aff}(2,\mathbb{R})$. They are however not conjugate to each other under $\mathrm{SL}(3,\mathbb{R})$ since the spaces V_2 and V_1 have different dimensions and hence cannot be mutually conjugate. The same is true in the case of $\mathrm{sl}(3,\mathbb{C})$.

We shall classify all subalgebras of $\mathfrak{g}_{7,1}$ under the action of the corresponding affine group, realized by the matrices

$$(3.7) \quad g \sim \begin{pmatrix} g_{11} & g_{12} & g_{13} \\ g_{21} & g_{22} & g_{23} \\ 0 & 0 & g_{33} \end{pmatrix}, \quad \det g = 1.$$

Most subalgebras of $\mathfrak{g}_{7,2}$ will be $\mathrm{SL}(3,\mathbb{R})$ conjugate to some subalgebra of $\mathfrak{g}_{7,1}$. Indeed, any subalgebra of $\mathfrak{g}_{7,2}$ that leaves a two-dimensional vector space invariant will be conjugate to a subalgebra of $\mathfrak{g}_{7,1}$. It follows that a necessary condition for $\mathfrak{g}_0 \in \mathfrak{g}_{7,2}$ not to be conjugate to some $\tilde{\mathfrak{g}}_0 \in \mathfrak{g}_{7,1}$ is that it must contain an element with $a_{32}a_{23} \neq 0$ in eq. (3.6). We also have

$$(3.8) \quad \begin{pmatrix} & & 1 \\ & -1 & \\ 1 & & \end{pmatrix} \begin{pmatrix} a & & \\ & b & d \\ & e & c \end{pmatrix} \begin{pmatrix} & & 1 \\ & -1 & \\ 1 & & \end{pmatrix} = \begin{pmatrix} c & -e & \\ -d & b & \\ & & a \end{pmatrix},$$

so every subalgebra of gl(2, ℝ) in $\mathfrak{g}_{7,2}$ is conjugate to a subalgebra of $\mathfrak{g}_{7,1}$. Similarly, any proper subalgebra of the algebras o(3), or o(2, 1) of eq. (3.4) will also be a subalgebra of $\mathfrak{g}_{7,1}$ (and also $\mathfrak{g}_{7,2}$).

3.3. Subalgebras of the affine Lie algebra $\mathfrak{g}_{7,1}$. We apply the algorithm of Section 2.4 and write

(3.9) $$\text{aff}(2, \mathbb{R}) \sim \text{gl}(2, \mathbb{R}) \rhd T_2$$

with

(3.10) $$\text{gl}(2, \mathbb{R}) \sim \{K_1, K_2, L_3\} \oplus D, \quad T_2 = \{P_1, P_2\}.$$

Step 1. The subalgebras of gl(2, ℝ) $\sim$ sl(2, ℝ) $\oplus \{D\}$ can be constructed using the Goursat method of Section 2.3. We represent them by

(3.11)
$$\begin{aligned}
&F_{4,1} \sim \{L_3, K_1, K_2, D\}, &&F_{3,1} \sim \{L_3, K_1, K_2\},\\
&F_{3,2} \sim \{K_1, K_2 - L_3, D\}, &&F_{2,1} = \{K_1, K_2 - L_3\},\\
&F_{2,2} = \{K_1 + aD, K_2 - L_3, a \neq 0\} &&F_{2,3} = \{L_3, D\},\\
&F_{2,4} = \{K_1, D\}, &&F_{2,5} = \{K_2 - L_3, D\},\\
&F_{1,1} = \{L_3\}, &&F_{1,2} = \{K_1\},\\
&F_{1,3} = \{K_2 - L_3\}, &&F_{1,5} = \{L_3 + aD, a \neq 0\},\\
&F_{1,1} = \{K_1 + aD, a > 0\}, &&F_{1,2} = \{D + \varepsilon(K_2 - L_3), \varepsilon = \pm 1\},\\
&F_{0,1} = \{\varnothing\}.
\end{aligned}$$

Each algebra in (3.11) represents a GL(2, ℝ) class of subalgebras of gl(2, ℝ).

Step 2. We omit all details and just give a representative list of all Aff(2, ℝ) conjugacy classes of splitting subalgebras of aff(2, ℝ). The notation is the following:

(3.12)
$$\begin{aligned}
&S_{a,b;1} \sim \{F_{a,b}, P_1, P_2\}, &&S_{a,b;2} \sim \{F_{a,b}\}\\
&S_{a,b;k} \sim \{F_{a,b}, X, X \in T_2\}, &&k = 3, 4, \ldots;
\end{aligned}$$

with $F_{a,b}$ running through the list (3.11). We only give the cases with $k \geq 3$. In addition to (3.12) (for all a and b), we have:

(3.13)
$$\begin{aligned}
&S_{3,2;3} \sim \{K_1, K_2 - L_3, D, P_1\}, &&S_{2,1;3} \sim \{K_1, K_2 - L_3, P_1\},\\
&S_{2,2;3} \sim \{K_1 + aD, K_2 - L_3, P_1, a \neq 0\}, &&S_{2,4;3} \sim \{K_1, D, P_1\},\\
&S_{2,5;3} \sim \{K_2 - L_3, D, P_1\}, &&S_{1,2;3} \sim \{K_1, P_1\},\\
&S_{1,3;3} \sim \{K_2 - L_3, P_1\}, &&S_{1,4;3} \sim \{D, P_1\},\\
&S_{1,6;3} \sim \{K_1 + aD, P_1, a \neq 0, \pm\tfrac{1}{6}\}, &&S_{1,6;4} \sim \{K_1 + \varepsilon\tfrac{1}{6}D, P_1, \varepsilon = \pm 1\},\\
&S_{1,7;3} \sim \{D + \varepsilon(K_2 - L_3), P_1, \varepsilon = \pm 1\}, &&S_{0,1;3} \sim \{P_1\}.
\end{aligned}$$

Step 3. Starting from the splitting algebras (3.12) and (3.13), we form a representative list of all nonsplitting ones. We only present the final Aff(2, ℝ) representative list. The notation again indicates the subalgebra $F_{a,b} \subset$ gl(2, ℝ) that gives rise to the nonsplitting subalgebra $S_{a,b,k} \in$ aff(2, ℝ).

(3.14)
$$\begin{aligned}
&S_{2,2;4} \sim \{K_1 + \tfrac{1}{2}D, K_2 - L_3 + P_2\}, &&S_{2,2;5} \sim \{K_1 + \tfrac{1}{2}D, K_2 - L_3 + P_2, P_1\},\\
&S_{2,2;6} \sim \{K_1 + \tfrac{1}{6}D + P_2, K_2 - L_3, P_1\}, &&S_{2,2;7} \sim \{K_1 - \tfrac{1}{6}D + P_1, K_2 - L_3\},\\
&S_{1,6;5} \sim \{K_1 + \tfrac{1}{2}D + P_2\}, &&S_{1,6;6} \sim \{K_1 - \tfrac{1}{6}D + P_1, P_2\}.
\end{aligned}$$

Step 4. This step consists of merging the subalgebras (3.12), (3.13) and (3.14) into one table, ordered by dimension and isomorphism class. We will not display the Table of representatives of Aff$(2,\mathbb{R})$ classes of subalgebras of aff$(2,\mathbb{R})$ here. Instead we proceed to the final list of SL$(3,\mathbb{R})$ classes of subalgebras of sl$(3,\mathbb{R})$.

3.4. All subalgebras of sl(3, $\mathbb{R}$). To obtain representatives of all SL$(3,\mathbb{R})$ classes of subalgebras of sl$(3,\mathbb{R})$ we must take the four maximal subalgebras and then combine together their lists of subalgebras. We proceed as follows:

(1) Take the list of subalgebras of $\mathfrak{g}_{7,1} \sim$ aff$(2,\mathbb{R})$, classified under Aff$(2,\mathbb{R})$ and establish which subalgebras in the list are mutually SL$(3,\mathbb{R})$ conjugate. Among those choose one and drop all others from the list.

(2) Add representatives of subalgebras of $\mathfrak{g}_{7,2}$, not SL$(3,\mathbb{R})$ conjugate to subalgebras of $\mathfrak{g}_{7,1}$. As established in Section 3.2 such subalgebras must contain an element of the type $P_2 - R_2 + aD$.

(3) Add the irreducibly imbedded subalgebras $o(3)$ and $o(2,1)$. Their proper subalgebras should not be included since they are conjugate to subalgebras of $\mathfrak{g}_{7,1}$.

The final list of subalgebras of sl$(3,\mathbb{R})$ is given in Table 1, ordered by dimension and isomorphism class. The notations in the last column are that $A_{2,1}$ is a solvable non-Abelian two-dimensional algebra, $A_{3,1}$ is the nilpotent three-dimensional algebra (the Heisenberg algebra) $A_{3,2}$, $A_{3,3}$ and $A_{3,4}$ are all solvable algebras with a two-dimensional Abelian ideal, the nilradical (denoted NR(L) in the Table). The action of the non-nilpotent element on the nilradical is given by a matrix M with

$$M_2 = \begin{pmatrix} 1 & \\ & \alpha \end{pmatrix}, \ 0 < |\alpha| \leq 1, \quad M_3 = \begin{pmatrix} \alpha & 1 \\ -1 & \alpha \end{pmatrix}, \ 0 \leq \alpha, \quad M_4 = \begin{pmatrix} 1 & 1 \\ 0 & 1 \end{pmatrix},$$

for $A_{3,2}$, $A_{3,3}$ and $A_{3,4}$, respectively.

The Borel subalgebra $W_{5,3}$ is a maximal solvable subalgebra of sl$(3,\mathbb{R})$, isomorphic to the algebra of traceless upper triangular 3×3 matrices. Interestingly, sl$(3,\mathbb{R})$ has two further classes of maximal solvable subalgebras, namely $W_{4,4}$ and $W_{4,5}$ (over $\mathbb{C}$ they are no longer maximal and are both contained in the complexification of $W_{5,3}$).

The only subalgebras of sl$(3,\mathbb{R})$, not contained in $W_{6,1}$ are $W_{6,2}$, $W_{5,2}$, $W_{4,5}$, $W_{3,9}(a)$, $W_{3,13}$ and $W_{3,14}$.

4. Conclusion

The subgroup classification algorithms described in Section 2 are very general and powerful. The application to sl$(3,\mathbb{R})$ is rather simple, though the results are, to my knowledge, new. For other applications, see the original articles [**1, 12–18**] and also the lecture series [**22**]. The method has also been generalized to the case of infinite dimensional Lie algebras [**3, 10**].

The subalgebras of sl$(3,\mathbb{C})$ can also be read off from Table 1. Certain subalgebras should be dropped from the list. Thus, over $\mathbb{C}$ the subalgebras $W_{4,4}$ and $W_{4,5}$ are equivalent to $W_{4,3}$ and $W_{4,2}$, respectively. Similarly $W_{3,8}$ and $W_{3,9}$ are conjugate to $W_{3,6}$ and $W_{3,5}$, respectively. The $o(3)$ and $o(2,1)$ algebras are conjugate under SL$(3,\mathbb{C})$, so $W_{3,14}$ can be dropped. Other mutually conjugate subalgebras (over $\mathbb{C}$) are $W_{2,1}$ and $W_{2,2}$ and also $W_{1,1}$ and $W_{1,2}$. Moreover, the ranges of parameters have to be adapted to the complex case.

TABLE 1. Representative list of all subalgebras of $sl(3,\mathbb{R})$ classified under the group $SL(3,\mathbb{R})$

Dim	Name	Basis	Isomorphism class and comments
6	$W_{6,1}$	$\{K_1, K_2, L_3, D, P_1, P_2\}$	$\mathrm{aff}(2,\mathbb{R})$
	$W_{6,2}$	$\{K_1 - D/2, P_2, R_2, K_1 + D/6, K_2 - L_3, P_2\}$	$\mathrm{aff}(2,\mathbb{R})$
5	$W_{5,1}$	$\{K_1, K_2, L_3, P_1, P_2\}$	$s\,\mathrm{aff}(2,\mathbb{R})$
	$W_{5,2}$	$\{K_1 - D/2, P_2, R_2, K_2 - L_3, P_2\}$	$s\,\mathrm{aff}(2,\mathbb{R})$
	$W_{5,3}$	$\{K_1, D, K_2 - L_3, P_1, P_2\}$	Borel subalgebra
4	$W_{4,1}$	$\{L_3, K_1, K_2\} \oplus \{D\}$	$\mathrm{gl}(2,\mathbb{R})$
	$W_{4,2}$	$\{D, P_1\} \oplus \{K_1 - D/6, K_2 - L_3\}$	$A_{2,1} \oplus A_{2,1}$
	$W_{4,3}$	$\{K_1 + D/6, P_1\} \oplus \{K_1 - D/6, P_2\}$	$A_{2,1} \oplus A_{2,1}$
	$W_{4,4}$	$\{L_3, D, P_1, P_2\}$	solvable, $NR(L) \sim 2A_1$
	$W_{4,5}$	$\{P_2 - R_2, K_1 + D/6, K_2 - L_3, P_1\}$	solvable, $NR(L) \sim 2A_1$
	$W_{4,6}$	$\{\cos\alpha(2K_1) + \sin\alpha(D/3), K_2 - L_3, P_1, P_2\}, 0 \leq \alpha \leq \pi$	solvable, $NR(L) \sim A_{3,1}$
3	$W_{3,1}$	$\{K_1, L_2 - L_3\} \oplus \{D\}$	$A_{2,1} \oplus A_1$
	$W_{3,2}$	$\{D, P_1\} \oplus \{K_2 - L_3\}$	$A_{2,1} \oplus A_1$
	$W_{3,3}$	$\{K_1 + D/6, P_1\} \oplus \{P_2\}$	$A_{2,1} \oplus A_1$
	$W_{3,4}$	$\{K_2 - L_3, P_1, P_2\}$	$A_{3,1}$, nilpotent
	$W_{3,5}^{(a)}$	$\{K_1 + aD, K_2 - L_3, P_1\}, \frac{1}{2} \leq a \leq \frac{1}{6}, a \neq -\frac{1}{6}$	$A_{3,2}(\frac{1}{2} + 3a)$
	$W_{3,6}^{(\alpha)}$	$\{\cos\alpha(2K_1) + \sin\alpha(D/3), P_1, P_2\}, 0 \leq \alpha \leq \pi/2, \alpha \neq \pi/4$	$A_{3,2}([\sin\alpha - \cos\alpha]/[\sin\alpha + \cos\alpha])$
	$W_{3,7}$	$\{K_1 + D/2, K_2 - L_3 + P_2, P_1\}$	$A_{3,2}(\frac{1}{2})$
	$W_{3,8}^{(a)}$	$\{L_3 + aD, P_1, P_2\}, 0 \leq a$	$A_{3,3}(a)$
	$W_{3,9}^{(a)}$	$\{P_2 - R_2 + aD, K_2 - L_3, P_1\}, 0 \leq a$	$A_{3,3}(a)$
	$W_{3,10}$	$\{D + K_2 - L_3, P_1, P_2\}$	$A_{3,4}$
	$W_{3,11}$	$\{K_1 + D/3 + P_2, K_2 - L_3, P_1\}$	$A_{3,4}$
	$W_{3,12}$	$\{K_1, K_2, L_3\}$	$\mathrm{sl}(2,\mathbb{R})$
	$W_{3,13}$	$\{L_3, P_1 + R_1, P_2 + R_2\}$	$\mathrm{o}(2,1)$,

TABLE 1. (continued)

Dim	Name	Basis	Isomorphism class and comments
	$W_{3,14}$	$\{L_3, P_1 - R_1, P_2 - R_2\}$	$o(3)$,
2	$W_{2,1}$	$\{L_3, D\}$	$2A_1$, Cartan, semicompact
	$W_{2,2}$	$\{K_1, D\}$	$2A_1$, Cartan, noncompact
	$W_{2,3}$	$\{K_2 - L_3, D\}$	not Cartan, decomposable
	$W_{2,4}$	$\{K_2 - L_3, P_1\}$	MANS, (102)
	$W_{2,5}$	$\{P_1, P_2\}$	MANS, (201)
	$W_{2,6}$	$\{K_2 - L_3 + P_2, P_1\}$	MANS, (111)
	$W_{2,7}^{(a)}$	$\{K_1 + aD, K_2 - L_3\}, a \in \mathbb{R}$	$A_{2,1}$
	$W_{2,8}$	$\{K_1 - D/6 + P_1, K_2 - L_3\}$	$A_{2,1}$
	$W_{2,9}$	$\{K_1 + D/6 + P_2, P_1\}$	$A_{2,1}$
	$W_{2,10}$	$\{K_1 + D/2, K_2 - L_3 + P_2\}$	$A_{2,1}$
1	$W_{1,1}$	$\cos\alpha(2K_1) + \sin\alpha(3D), 0 \leq \alpha < \pi$	
	$W_{1,2}$	$L_3 + aD, a \geq 0$	
	$W_{1,3}$	$D + K_2 - L_3$	
	$W_{1,4}$	$K_2 - L_3$	
	$W_{1,5}$	$K_2 - L_3 + P_2$	

References

1. G. Burdet, J. Patera, M. Perrin, and P. Winternitz, *The optical group and its subgroups*, J. Math. Phys. **19** (1978), 1758–1780.
2. J. F. Cornwell, *Group theory in physics*, Vol. II, Techniques Phys., vol. 7, Academic Press, London, 1984.
3. D. David, N. Kamran, D. Levi, and P. Winternitz, *Symmetry reduction for the Kadomtsev–Petviashvili equation using a loop algebra*, J. Math. Phys. **27** (1986), 1225–1237.
4. M. A. del Olmo, M. A. Rodriguez, P. Winternitz, and H. Zassenhaus, *Maximal Abelian subalgebras of pseudounitary Lie algebras*, Linear Algebra Appl. **135**, 79–151 (1990).
5. P. DuVal, *Homographies, quaternions and rotations*, Clarendon Press, Oxford, 1964.
6. E. B. Dynkin, *Semisimple subalgebras of semisimple Lie algebras*, Five Papers on Algebra and Group Theory, Amer. Math. Soc. Transl. Ser. 2, vol. 6, Amer. Math. Soc. Providence, RI, 1957, pp. 111–244.
7. E. Goursat, *Sur les substitutions orthogonales*, Ann. Sci. École Norm. Sup. (3) **6** (1889), 9–102.
8. V. Hussin, P. Winternitz, and H. Zassenhaus, *Maximal Abelian subalgebras of complex orthogonal Lie algebras*, Linear Algebra Appl. **141** (1990), 183–220.
9. S. Lie, *Klassifikation und Integration von Gewohnlichen Differentialgleichungen zwischen x, y die eine Gruppe von Transformationen gestatten*, Gessamelte Abhandlungen, Vol. 5, Teubner, Leipzig, 1924.
10. L. Martina and P. Winternitz, *Analysis and applications of the symmetry group of the multidimensional three wave resonant interactions problem*, Ann. Physics **196** (1989), 231–277.
11. W. G. McKay and J. Patera, *Tables of dimensions, indices and branching rules for representations of simple Lie algebras*, M. Dekker, New York, 1981.
12. J. Patera, R. T. Sharp, P. Winternitz, and H. Zassenhaus, *Subgroups of the similitude group of three-dimensional Minkowski space*, Canad. J. Phys. **54** (1976), 950–561.
13. _____, *Subgroups of the Poincaré group and their invariants*, J. Math. Phys. **17** (1976), 977–985.
14. _____, *Invariants of real low dimension Lie algebras*, J. Math. Phys. **17** (1976), 986–994.
15. _____, *Continuous subgroups of the fundamental groups of physics. III. The de Sitter groups*, J. Math. Phys. **18** (1977), 2259–2288.
16. J. Patera, P. Winternitz, and H. Zassenhaus, *Continuous subgroups of the fundamental groups of physics. I. General method and the Poincaré group*, J. Math. Phys. **16** (1975), 1597–1614.
17. _____, *Continuous subgroups of the fundamental groups of physics. II. The similitude group*, J. Math. Phys. **16** (1975), 1615–1624.
18. _____, *Quantum numbers for particles in de Sitter space*, J. Math. Phys **17** (1976), 717–728.
19. _____, *Maximal Abelian subalgebras of real and complex symplectic Lie algebras*, J. Math. Phys. **24** (1983), 1973–1985.
20. D. Rand, P. Winternitz, and H. Zassenhaus, *On the indentification of a Lie algebra given by its structure constants I. Direct decompositions, Levi decompositions and nilradicals*, Linear Algebra Appl. **19** (1988), 197–246.
21. D. A. Suprunenko and R. I. Tyshkevich, *Commutative matrices*, Academic Press, New York, 1968.
22. P. Winternitz, *Lie groups and solutions of nonlinear partial differential equations*, Integrable Systems, Quantum Groups and Quantum Field Theories (A. Ibort and M. A. Rodriguez, ed.), NATO Adv. Sci. Inst. Ser. C Math. Phys. Sci., Kluwer, Dordrecht, 1993, pp. 429–495.
23. P. Winternitz and H. Zassenhaus, *Decomposition theorems for maximal Abelian subalgebras of the classical Lie algebras*, Report no. CRMA-1199, Centre de recherche en mathématiques appliquées, Montréal, 1984.

CENTRE DE RECHERCHES MATHÉMATIQUES AND DÉPARTEMENT DE MATHÉMATIQUES ET DE STATISTIQUE, UNIVERSITÉ DE MONTRÉAL, C.P. 6128, SUCCURSALE CENTRE-VILLE, MONTRÉAL, QC, CANADA H3C 3J7

E-mail address: `wintern@crm.umontreal.ca`

Titles in This Series

34 P. Winternitz, J. Harnad, C. S. Lam, and J. Patera, Editors, Symmetry in physics, 2004

33 André D. Bandrauk, Michel C. Delfour, and Claude Le Bris, Editors, Quantum control: Mathematical and numerical challenges, 2003

32 Vadim B. Kuznetsov, Editor, The Kowalevski property, 2002

31 John Harnad and Alexander Its, Editors, Isomonodromic deformations and applications in physics, 2002

30 John McKay and Abdellah Sebbar, Editors, Proceedings on moonshine and related topics, 2001

29 Alan Coley, Decio Levi, Robert Milson, Colin Rogers, and Pavel Winternitz, Editors, Bäcklund and Darboux transformations. The geometry of solitons, 2001

28 J. C. Taylor, Editor, Topics in probability and Lie groups: Boundary theory, 2001

27 I. M. Sigal and C. Sulem, Editors, Nonlinear dynamics and renormalization group, 2001

26 J. Harnad, G. Sabidussi, and P. Winternitz, Editors, Integrable systems: From classical to quantum, 2000

25 Decio Levi and Orlando Ragnisco, Editors, SIDE III—Symmetries and integrability of difference equations, 2000

24 B. Brent Gordon, James D. Lewis, Stefan Müller-Stach, Shuji Saito, and Noriko Yui, Editors, The arithmetic and geometry of algebraic cycles, 2000

23 Pierre Hansen and Odile Marcotte, Editors, Graph colouring and applications, 1999

22 Jan Felipe van Diejen and Luc Vinet, Editors, Algebraic methods and q-special functions, 1999

21 Michel Fortin, Editor, Plates and shells, 1999

20 Katie Coughlin, Editor, Semi-analytic methods for the Navier-Stokes equations, 1999

19 Rajiv Gupta and Kenneth S. Williams, Editors, Number theory, 1999

18 Serge Dubuc and Gilles Deslauriers, Editors, Spline functions and the theory of wavelets, 1999

17 Olga Kharlampovich, Editor, Summer school in group theory in Banff, 1996, 1998

16 Alain Vincent, Editor, Numerical methods in fluid mechanics, 1998

15 François Lalonde, Editor, Geometry, topology, and dynamics, 1998

14 John Harnad and Alex Kasman, Editors, The bispectral problem, 1998

13 Michel C. Delfour, Editor, Boundaries, interfaces, and transitions, 1998

12 Peter C. Greiner, Victor Ivrii, Luis A. Seco, and Catherine Sulem, Editors, Partial differential equations and their applications, 1997

11 Luc Vinet, Editor, Advances in mathematical sciences: CRM's 25 years, 1997

10 Donald E. Knuth, Stable marriage and its relation to other combinatorial problems: An introduction to the mathematical analysis of algorithms, 1997

9 D. Levi, L. Vinet, and P. Winternitz, Editors, Symmetries and integrability of difference equations, 1996

8 J. Feldman, R. Froese, and L. M. Rosen, Editors, Mathematical quantum theory II: Schrödinger operators, 1995

7 J. Feldman, R. Froese, and L. M. Rosen, Editors, Mathematical quantum theory I: Field theory and many-body theory, 1994

6 Guido Mislin, Editor, The Hilton Symposium 1993: Topics in topology and group theory, 1994

5 D. A. Dawson, Editor, Measure-valued processes, stochastic partial differential equations, and interacting systems, 1994

TITLES IN THIS SERIES

4 **Hershy Kisilevsky and M. Ram Murty, Editors,** Elliptic curves and related topics, 1994
3 **Rémi Vaillancourt and Andrei L. Smirnov, Editors,** Asymptotic methods in mechanics, 1993
2 **Philip D. Loewen,** Optimal control via nonsmooth analysis, 1993
1 **M. Ram Murty, Editor,** Theta functions: From the classical to the modern, 1993